CW01262457

ECOLOGY OR CATASTROPHE

Ecology or Catastrophe

THE LIFE OF MURRAY BOOKCHIN

Janet Biehl

OXFORD
UNIVERSITY PRESS

OXFORD
UNIVERSITY PRESS

Oxford University Press is a department of the University of
Oxford. It furthers the University's objective of excellence in research,
scholarship, and education by publishing worldwide.

Oxford   New York
Auckland   Cape Town   Dar es Salaam   Hong Kong   Karachi
Kuala Lumpur   Madrid   Melbourne   Mexico City   Nairobi
New Delhi   Shanghai   Taipei   Toronto

With offices in
Argentina   Austria   Brazil   Chile   Czech Republic   France   Greece
Guatemala   Hungary   Italy   Japan   Poland   Portugal   Singapore
South Korea   Switzerland   Thailand   Turkey   Ukraine   Vietnam

Oxford is a registered trademark of Oxford University Press
in the UK and certain other countries.

Published in the United States of America by
Oxford University Press
198 Madison Avenue, New York, NY 10016

© Janet Biehl 2015

All rights reserved. No part of this publication may be reproduced, stored in
a retrieval system, or transmitted, in any form or by any means, without the prior
permission in writing of Oxford University Press, or as expressly permitted by law,
by license, or under terms agreed with the appropriate reproduction rights organization.
Inquiries concerning reproduction outside the scope of the above should be sent to the
Rights Department, Oxford University Press, at the address above.

You must not circulate this work in any other form
and you must impose this same condition on any acquirer.

Library of Congress Cataloging-in-Publication Data
Biehl, Janet, 1953–
Ecology or catastrophe : the life of Murray Bookchin / Janet Biehl.
   pages cm.
Includes bibliographical references and index.
ISBN 978–0–19–934248–8 (alk. paper)
1. Bookchin, Murray, 1921–2006.   2. Ecologists—United States—
Biography.   3. Environmentalists—United States—Biography.   4. Social ecology—United
States.   5. Environmentalism—United States.   I. Title.
GF16.B66B54 2014
363.7092—dc23
[B]
2013029714

9 8 7 6 5 4 3 2 1
Printed in the United States of America
on acid-free paper

Contents

Acknowledgments vii
Prologue ix

1. *Young Bolshevik* 1

2. *Labor Organizer* 28

3. *Rethinker* 52

4. *Eco-decentralist* 69

5. *Eco-anarchist* 82

6. *Counterculture Elder* 105

7. *Man of the Moment* 131

8. *Social Ecologist* 156

9. *Antinuclear Activist* 178

10. *Municipalist* 205

11. *Green Politico* 234

12. *Assembly Democrat* 256

13. *Historian* 284

Epilogue 315
Index 319

Acknowledgments

I MET MURRAY late in his life, in 1987, and as I did the research for this book, I was fortunate that many who knew him well before I did gave generously of their time and their thoughts. I could not have written it without extensive interviews from Dan Chodorkoff, Barry Costa-Pierce, Bob D'Attilio, the late Dave Eisen, Joy Gardner, David Goodway, Jack Grossman, Wayne Hayes, Howie Hawkins, Joseph Kiefer, Ben Morea, Jim Morley, Calley O'Neill, Charles Radcliffe, Dimitri Roussopoulos, Karl-Ludwig Schibel, and Brian Tokar. In some cases they shared papers, books, and tapes as well as their impressions and memories. I'm deeply grateful to all of them.

Heartfelt thanks to others who talked to me about Murray as well and sometimes also shared documents: Flavia Alaya, Steve Baer, Harriet Barlow, the late Peter Berg, David Block, Reni Bob, Horst Brand, Stewart Brand, Frank Bryan, Caitlin Casey, Juan Diego Pérez Cebada, Stuart Christie, Linda Cohen, Jutta Ditfurth, David Dobereiner, Crescent Dragonwagon, Mike Edelstein, Fanis Efthymiadis, Bob Erler, Paolo Finzi, Gil Friend, Carlos "Chino" Garcia, Vincent Gerber, Rafa Grinfeld, Richard Grossman, Susan Harding, Anne Harper, Wolfgang Haug, James Herod, Annette Jacobson, Robert Kadar, Jerry Kaplan, Stavros Karageorgakis, Ken Knabb, Bill Koehnlein, Makis Korakianitis, Lucia Kowaluk, Burton Lasky, Ursula K. Le Guin, Sveinung Legard, John Lepper, Mat Little, Arthur Lothstein, Sam Love, Svante Malmström, Vivien Marx, Lisa Max, Lester Mazor, John McHale, Paul McIsaac, John McMillian, Richard Merrill, Stephanie Mills, Roy Morison, Brian Morris, David Morris, Pat Murtagh, Osha (formerly Tom) Neumann, Roz Payne, Ivar Petterson, Paul Prensky, Peter Prontzos, Michalis Protopsaltis, Michael Riordan, Mark Roseland, Meg Seaker, Rick Sharp,

Josh Shortlidge, Chuck Stead, Suzanne Stritzler, James Swan, Jane Thiebaud, and Bruce Wilson.

In the early 2000s Bookchin donated his papers to the Tamiment Library at New York University. Before he did so, I made photocopies of many of them; hence the allusions to "MBPTL and author's collection" in my source notes. The originals are at Tamiment; I am extremely grateful to Peter Filardo and Erika Gottfried for help in accessing those materials. After this manuscript was completed, I turned my own collection over to the International Institute for Social History in Amsterdam, for the convenience of scholars in Europe. I'm grateful to Huub Sanders of the IISH for arranging for that deposit.

I'm also most grateful to the Anarchist Archives Project in Cambridge, Massachusetts; the Institute for Social Ecology archive in Marshfield, Vermont; the Fletcher Free Library in Burlington, Vermont; the Bailey-Howe Library's Special Collections, at the University of Vermont in Burlington; the Goddard College Archives in Plainfield; the Vermont Historical Society in Montpelier; the Vermont State Archives and Records Administration in Waterbury; Milne Special Collections at the University of New Hampshire, in Durham; and the Centro Studi Libertari in Milan, Italy.

For their hospitality on my research trips, I thank Inara De Leon and Todd Norbitz, Tim and Berle Driscoll, Dave and Shirley Eisen, David Goodway and Che Mah, and Stephen Kurman. Inara De Leon, Laura Ramirez, and Andy Price gave me early encouragement. K. K. Wilder and Ted Tedford's tips on interviewing proved helpful, and my writing benefited greatly from conversations with Paula Harrington and Eve Thorsen. Judith Jones, in retirement after her remarkable decades-long editing career at Knopf, read and commented helpfully on the manuscript, as did Stavros Karageorgakis and Karl-Ludwig Schibel. Any errors in the book are my own responsibility.

For their emotional support during my work, I'm indebted to Bronwyn Dunne, Eirik Eiglad, and Eve Thorsen—you were indispensable to this project, my dear friends. Thanks as well to Sezgin Ata, Emet Degermenci, Metin Guven, Marcus Melder, Peter Munsterman, Michael Speidel, and Thodoris Velissaris for their encouragement; to Mary Biehl and Bea Larsen for cheering me on; and to Lynn Brelsford for keeping my spirits up in the home stretch.

Heartfelt thanks as well to Drs. Claudia Berger, Zail Berry, and Matthew Watkins, as well as the Visiting Nurses Association of Chittenden County, for the superb end-of-life care that they provided for Murray. Special thanks to William E. Drislane, Esq.

For her persistence in finding my book an excellent home, hearty thanks to my agent, Anne G. Devlin. It is a great privilege to be published by Oxford University Press. My editor, Jeremy Lewis, stood by my book through thick and thin, for which I'm immensely grateful.

I'm grateful above all to Murray for all he gave me, and to the improbable twist of fate, indeed the astounding good fortune, that brought his life and mine together.

Prologue

THE KEYNOTE SPEAKER, said to be a leading figure in the ecology movement, waited beside the stage. I'd expected to see an aging hippie, that day in 1986. Yes, aging he was, with a shaggy gray mustache and a certain weariness in his sixty-something frame. But his clothes were hardly hippie style—they were industrial dark green, like a polyester uniform, and his shirt pocket was stuffed with mechanical pencils that skewed the suspenders and his houndstooth wool vest.

"He was a Communist as a boy, you know," a woman sitting next to me remarks. I eavesdrop: "Yes, but he's been writing on ecology since before I was born," says her friend.

As the audience settles, the student organizer steps to the podium. Tonight's speaker, he explains, has been writing about environmental issues since the 1950s. His book *Our Synthetic Environment* raised the alarm about pesticides and industrial agriculture, soil depletion, air and water pollution, deforestation, and nuclear power—and it was published in the spring of 1962, a few months before Rachel Carson's *Silent Spring*. In his next book, *Crisis in Our Cities*, he actually warned about global warming and said that to avoid ecological catastrophe, we'd have to wean ourselves from fossil fuels and learn to use renewable energy, as well as eat locally and farm organically.

He was the founder of radical ecology—not of environmentalism, and surely not of conservationism, but ecology as a radical social and political issue.

The student organizer makes the introduction: "Please welcome ... Murray Bookchin!"

He moves with arthritic deliberation to the podium. Grasping it by the edges, he surveys the crowded with large smoldering eyes.

"My friends," he begins in a rumbling New York accent, "the power of capitalism to destroy is unprecedented in human history. It is on a collision course with the environment, threatening our air and water, flora and fauna, the natural cycles on which all life depends. It is destroying diversity, simplifying the natural world, turning forests into deserts, soil into sand, and water to sewage. It's pushing back the clock, undoing countless millennia of biotic evolution."[1]

His craggy countenance comes to life as he continues, his words falling into the rhythm of a street orator. His formidable eyebrows quiver like those of the old 1930s labor leader John L. Lewis, who, I would later learn, he much admired.

"It's not only threatening the integrity of life of earth, it's turning us into commodities. It invades us with advertising, making us think we need things that are actually useless. It's simplifying our social relationships, defining us as buyers and sellers. It's turning our neighborhoods and communities over to the cash nexus, bringing them into the ever-expanding market."

If you listen closely to his baritone cadences, you can hear traces of Russian folksongs and Old Testament prophecy, of Old Left agitprop and the snarky defiance of a Dead End Kid.

"Either we're going to let the grow-or-die market economy massively destroy the planet"—his hands slice the air—"or we will have to make a sweeping reconstruction of society. If we're going live in harmony with the natural world, we will have to change the social world.

"In earlier times society was more communal and small scale—people took responsibility for each other, and for themselves, and had the confidence to stand up for themselves. I'm not saying we should go back in time. But we can learn from some of the old ways. About interdependence, about cooperation and mutual aid."

He is bubbling and rattling like a samovar on the boil.

"We have to organize a movement to create an ecological society, one that's decentralized, democratic, humane. We have to start revitalizing our communities and our neighborhoods, creating a new politics at the local level, bringing them back to life, strengthening them."

It isn't just a speech—it's a performance, inspiring his listeners to take action.

"My critics will tell you that I'm a wild-eyed utopian. But I assure you everything I'm suggesting is immensely realistic. The more we try, on the basis of so-called pragmatic considerations, to change society in a small piecemeal way, the more we lose sight of the larger picture. The real pragmatic solution is the long-range one—the one that gets down to the root causes of the ecological crisis."

By now the audience is leaning forward in their seats. "My friends, either we will create an ecotopia based on ecological principles, or we will simply go under as a species. We have to be realistic and do the impossible—because otherwise we will have the unthinkable!"

* * *

Prologue

The man I met in 1986 believed that ideas can move history forward, that if you come up with a good social-political idea, then people will recognize its rightness and help you try to put it into effect.

That belief came from his childhood in the Communist children's movement, and it stayed with him for the rest of his lifelong journey through the American Left. It stayed with him after he gave up on Marx and became the most important anarchist thinker in the second half of the twentieth century. It stayed with him as he realized that the growing environmental crisis had profound implications for human social organization itself. It stayed with him as he explored the terrain between utopia and reality and found there a vision of a rational, ecological society. It stayed with him as he became a mentor to the counterculture in the 1960s. It stayed with him in the early 1970s, when he founded a school in Vermont that taught students how to farm organically, make solar and wind installations, and create urban gardens. It stayed with him as he helped build the antinuclear movement in the 1970s and the Green parties of the 1980s, calling for face-to-face democracy based on citizens' assemblies, like the town meetings of his adopted state.

He was a genuine political and intellectual independent, living outside the usual spectrum of life choices. Fired with a sense of urgency to spread the message that the ecological crisis required a profound rethinking, he subordinated his personal aims to the larger cause until they merged. He refused to yield to despair, holding firm to his belief that struggling to create the new society would bring to the fore people's potentialities for ethical behavior, a rational outlook, and social cooperation.

Yet he was also ebullient and charming. By lucky happenstance, I met him at a good moment and joined his cause for the last nineteen years of his life, as we collaborated, writing and traveling together. The Communists might have taught him to be combative, but I found that he had an open, guileless, and generous heart.

We agreed that I would one day write his biography. I interviewed him formally a few times, but more often he told me stories about his life, over the kitchen table or in the wee hours. The line between interview and conversation blurred. The stories I absorbed became second nature to me and now form the architecture of this biography.[2]

After he died, a pauper at eighty-five, I tried to keep him in my life for at least another few years by writing his biography, making up for his absence by researching his life, filling in the gaps between those stories, interviewing people who had known him, and studying the movements in which he'd worked. Where people's memories were contradicted by a document, I chose to follow the written record. Similarly, rather than rely entirely on my own memory of his stories, where possible I've cited a written or published source.

Finally I was able to reconstruct his trajectory moving forward in time, and in so doing I discovered its logic and integrity. I make no claim to have written a full flesh-and-bones biography; it is rather a political biography, of a thoroughgoing *zoon politikon*, a man formed by the political actors he knew, by the close-knit political groups to which he belonged, by the broader movements to which he adhered, and by the times in which he lived.

Over the decades Murray himself influenced many people, but to trace that influence, to identify even a fraction of those who felt their lives changed by him, would be beyond the scope of this biography. Rather, I focus on those who influenced him: who altered his way of thinking in some substantive way, or made a concerted effort to put his theoretical ideas into practice. Although an energetic public speaker, he preferred working intimately with small groups of dedicated, educated comrades; above all, he needed them to be writers, able and willing to enter the public sphere with him, at the very least in the periodicals that his various political groups issued. The secondary figures whom I spotlight, then, are those whose work with him is evident in their paper trails.[3]

It's easy to dismiss him as a utopian, but he made a compelling case that utopia has actually become necessary for the continuation of life on earth. The crisis of climate change that we face today is unprecedented, and his framework may yet prove to be the one that we need not only to sustain but also to advance life on earth.

NOTES

1. I've created here a composite from several speeches he made in the 1980s: the keynote address to the Waterloo Public Interest Group, University of Waterloo, Ontario, Mar. 14–16, 1985; "Forms of Freedom," Mar. 23, 1985, at The Farm in San Francisco (parts 1–5 are online on YouTube, http://bit.ly/pRISDj). See also Jeff Mortimer, "No Environmentalist," *Ann Arbor News*, Oct. 10, 1982; John Gushue, "Environmentalist Probes Organic Future," *McGill Daily*, Feb. 10, 1986, 5; and Stephen Hall, "An Interview with Murray Bookchin," *Probe Post*, Winter 1987, 16–18.

2. For a bibliography of published works by Bookchin, see the one I compiled for his seventieth birthday in 1991, at http://bit.ly/11woGBf.

3. After Murray's death in 2006, I became estranged from his first wife, Beatrice (they were married 1951–63), and from their two children. Hence these family members appear only minimally in these pages. It is to be hoped they will someday write about Murray's life from their own point of view.

1

Young Bolshevik

IN OCTOBER 1913 Murray Bookchin's maternal grandmother Zeitel stepped off the SS *Rotterdam* in New York and, with her two children, passed through Ellis Island. Up to now, the tall, stern, steely-eyed woman had spent her life trying to overthrow Russian tsardom. Now she was coming to America, to battle tyranny in a new land.

Born in the Russian Pale in the early 1860s, Zeitel Carlat had received an unusual secular, liberal education as part of the Haskalah, the Jewish Enlightenment. Studying science and mathematics and literature, she had learned to reject traditional religion and to dress in modern clothing. She even disdained the Yiddish language in favor of Russian, the ecumenical language, the tongue of Pushkin and Nekrasov. Her cousin Moishe Kalusky was another child of the Haskalah—he cut his hair, modern style, after which his parents said Kaddish for him.[1]

At the time Zeitel and Moishe were born, Tsar Alexander II ("the kindliest prince who ever ruled Russia," Disraeli called him) had emancipated the serfs and instituted a liberalizing reign—he even eased some of the restrictions on Jews, opening the doors of universities to a small quota. Growing up in those forward-looking times, Zeitel and Moishe felt that they were witnessing the dawn of a new era, in which the Russian people would finally achieve emancipation. Young intellectuals, they read Alexander Herzen and Nikolai Chernyshevsky and became *narodovoltsy*, or revolutionary populists. They dreamed that Russian peasantry would mount an uprising that would spark a revolution to sweep away Russia's oppressive ruling structure and undertake a *chernyi peredel*, or "black redistribution," of land. Then Russia could become a socialist society based on the traditional peasant *obshchina*, or village commune, in which land would be held by all cooperatively.

Zeitel particularly admired Chernyshevsky's 1863 novel *What Is to Be Done?*, a handbook for her generation of radicals. It depicts young Russians—especially women—participating in the communal lifestyles of an emancipated society. To the end of her life, Zeitel would keep a portrait of its author on her wall.

By 1881, some *narodovoltsy* felt that the peasant uprising was too slow in coming. In March a group of them decided to spark it themselves: they assassinated the tsar-liberator, Alexander II. But instead of bringing on the long-awaited revolutionary upheaval, this terrorist act brought Russia's liberalizing process to a screeching halt. The empire's wrath thundered down on the Jews, blaming them as a people for the murder. Incensed Russians carried out pogroms in the shtetls. The successor tsar, Alexander III, slammed shut the doors to Jewish education and entry into professions and issued repressive decrees that made life intolerable. Jews began emigrating to the West en masse.

The most ardent Jewish revolutionary populists, however, were unfazed by the pogroms, regarding themselves as human beings and Russians more than as Jews and determined to continue the fight for a humane, socialist Russia, even against the new wave of repression. When the tsar's secret police, the *okhrana*, crushed their groups in Moscow and St. Petersburg, they simply moved to the provinces. In the 1880s, southwestern Russia became a revolutionary hotbed, churning out reams of radical literature. By the 1890s, Zeitel and Moishe—now married—were living there, at the edge of the Russian empire, in the small city of Yednitz, just across the Prut River from Romania.[2] Revolutionaries who were being pursued by the *okhrana* could make their way to Yednitz, where Zeitel and Moishe would help them cross the river, silently, on moonless nights. Sometimes Zeitel would cross the river herself and bring back literature and exiled agitators.[3]

Their home became a hub of socialist intellectual and political activity. Young revolutionaries came there for advice and education. They would meet at night, by candlelight or kerosene lamps, in forested areas, even cemeteries, to discuss political developments and develop strategies. In 1902 Zeitel and Moishe joined the Socialist Revolutionary Party, the latest incarnation of Russian populism, this one infused with Marxist ideas to attract the growing population of urban workers as well as the peasantry. In 1905 the urban workforce mounted a series of strikes and mutinies that constituted an incipient revolution against tsarist absolutism. To aid the movement, Zeitel ran guns across the border, stashing them in hiding places where comrades could stealthily retrieve them for good use.

But the 1905 revolution was crushed, and soon afterward Moishe died of bladder cancer. Zeitel was left with their two children, Rose or Rachel (who had been born in 1894) and Dan (born a year later).[4] She hoped to mold her offspring into disciplined, self-sacrificing revolutionaries like herself, but alas, neither of them had inherited her temperament, her ability to defer gratification indefinitely for the sake of a great purpose. In fact, her high-spirited daughter Rose tried to run off to the nearby Gypsy camp—Zeitel had to drag her back home. Chubby, impulsive Rose had somehow inherited none of the mother's sharp edges; and where Zeitel's eyes were gray and austere, Rose's were brown and expressive, even dreamy.

Sometime in 1912 or 1913, the tsarist police raided their home. In the summer of 1913, Zeitel packed up her family and crossed the Prut River for the last time. Making their way northward to the Netherlands, they boarded the *Rotterdam* and sailed for New York.[5]

A decade earlier, the US Congress had barred immigration by anarchists. Zeitel might well have been denied entry. But she managed to get through with her teenage children. Border crossings, after all, were her specialty.

Zeitel rented a room in a squalid tenement on the Lower East Side, already packed with immigrant Jews, and found work for herself and her children in the needle trades. The neighborhood and its surroundings seethed with labor unrest. Public meetings abounded—the "greenhorns" likely heard the anarchist Emma Goldman stand and inveigh against the ruling class and the state, and the socialist Eugene V. Debs demand the emancipation of the working class.

Even in this radical milieu, Zeitel was contemptuous of the new country, with its frantic pace and crass materialism. Russian culture and lifeways were far superior to these boorish American ones, she believed, and its radical tradition immensely more advanced. The astonishing events of 1917 only proved her case. In February, strikes and demonstrations in Petrograd (formerly St. Petersburg) led to the abdication of the hated tsar and the end of the Romanov dynasty. A provisional government took power, while the city of Petrograd was governed by a workers' council or soviet. Then in November, a group of disciplined Marxist revolutionaries—Bolsheviks—stormed the Winter Palace, toppled the provisional government, and proclaimed a working-class state. In the old country, socialism was finally at hand.

Zeitel and her family rejoiced, as did radicals all over the world, be they socialist or anarchist. In solidarity, Communist parties were formed in many countries, including the United States. The Kaluskys did not join the American Communist Party—rather, following their populist roots, they joined the anarchistic Union of Russian Workers.[6] Rose, now a young milliner, joined the Wobblies, or the Industrial Workers of the World (IWW).

While attending a summer camp for Communist youth, she met Nathan Bookchin, a fellow Russian Jewish immigrant, two years younger than she. With Nathan, she could speak Russian; together they could wax nostalgic about their homeland and euphoric about its world-shaking revolution. Perhaps she was wearing one of her Russian blouses the day he proposed marriage and she accepted. Her domineering mother didn't like the young man, but despite Zeitel's disapproval—or perhaps because of it—Rose and Nathan married.

Around 1920, the family moved north to the fresher air of the Bronx, whose new IRT subway line, the Third Avenue El, would allow them to commute to work in Manhattan's garment district each day. Newlywed Rose and Nathan moved into a railroad flat on the ground floor at 1843 Crotona Avenue, in East Tremont, while

Zeitel and Dan rented an apartment nearby. East Tremont felt comfortably Russian to them, an ethnic Jewish village laced with Old World traditions.[7] But it also had a radical ambience—Workmen's Circle clubs, union locals, and socialist meeting halls were more numerous and influential in that neighborhood than synagogues. Shops were family owned: the kosher butcher, the greengrocer, the cigar maker, and more.

Rose gave birth to Murray, her first and only child, on January 14, 1921.[8] In keeping with the family's secular, Haskalah tradition, they would give him no Jewish education. He would not be bar mitzvahed. They would observe no religious holidays or rituals. Nor would they make any effort to Americanize him.

Instead, they raised him as a little Russian boy. For his first two years, Rose—who resisted learning English and would never even take US citizenship—spoke to him only in Russian, sang Russian songs to him, and much to his embarrassment, clothed him in Russian dress. His earliest memories were of her playing Glinka on the piano, wearing a Russian blouse. In the evenings, the family would go to Crotona Park to listen to Goldman's brass band play Rachmaninov.

But the Bookchin-Kalusky marriage was troubled. Zeitel's dislike of Nathan evolved into outright hostility. Her animus was well founded—her son-in-law bullied Rose (as Murray would later tell me), beating her when she displeased him and striking his son as well. Finally, when Murray was five or six, Nathan abandoned the family altogether.[9]

Unruly Rose, inexperienced with motherhood, had limited childrearing abilities. Even a child with a placid temperament might have taxed her, but this high-spirited son was more than she could handle alone. Zeitel probably didn't have to be asked twice for help—she moved into the Crotona Avenue apartment, bringing Rose's brother Dan with her.

Zeitel hung her old pictures of Chernyshevsky and Herzen and Tolstoy on the walls and arranged her leather-bound Russian books on the shelves. She suffered, it was understood, from a "weak heart," but she scrubbed and tidied the disheveled apartment energetically nonetheless. She still stood dignified and straight-backed behind her pince-nez. She still possessed that stern revolutionary will.

And she perceived Murray's bright, inquiring mind: here, finally, was the child she could imbue with her own political passion. She taught her grandson, first, what she had learned in school: that all religion is mere superstition. Then she taught him about the Russian revolutionary tradition, going back to Stenka Razin and Emilian Pugachev, those old Cossack firebrands, dazzling his mind with tales of their seventeenth- and eighteenth-century insurrections against despotism—he knew and loved their names and exploits before he'd even heard of Robin Hood. She taught him about populism and the *chernyi peredel*, the black redistribution, and about her own life as a Socialist Revolutionary, and about gun-running and secret strategy sessions in the dark of night.

"Basically," he would tell me, "my family educated me in revolution," especially the greatest revolution in history. Together Zeitel and Murray gazed at picture books of

Bolsheviks marching and conferring and orating. Before young Murray knew who Washington and Lincoln were, he was familiar with Lenin, as well as the German revolutionary leaders Rosa Luxemburg and Karl Liebknecht. He especially admired the dashing Leon Trotsky, who had engineered the Bolshevik takeover, then organized the Red Army, personally commanded it on horseback from the front lines, and led it to stunning victories over the White forces of reaction.

Murray revered his grandmother, who was strict and strong but also warm and loving to him. He shared a bedroom with her, sleeping on a cot next to her bed. On an August night in 1927, they heard newspaper boys running along Crotona Avenue shouting out the terrible news that Sacco and Vanzetti had been electrocuted. The state of Massachusetts had tried and convicted the two Italian immigrant anarchists for murdering a bank official. Their trial had been blatantly unfair—the judge who sentenced them to death had boasted of getting rid of "those anarchist bastards."[10] Now they had been executed. Zeitel rushed outside to get a newspaper, finding their neighbors on the street weeping. When she returned to the flat, she held up the front page—it had a drawing of the two men in the electric chair—for her grandson as if to brand it in his mind. *This is what capitalism does to working people*, she told him. *Don't ever forget.*[11]

By now, other Kalusky family members had immigrated to New York, and on weekends the railroad flat would often be filled with relatives, drinking tea together from a samovar, sipping it from saucers, the Russian way, through sugar cubes held between their teeth. They played Russian music on an accordion or piano and sang songs like the stirring "Whirlwinds of Danger" ("March, march ye toilers, and the world shall be free") and the melancholy strains of "Stenka Razin."

They talked incessantly of politics: about the death of Lenin in 1924 and the rise of Stalin to the revolutionary helm. Doubtless they were stunned to learn in 1927 that Stalin had sent Trotsky into internal exile, condemned for betraying the revolution. They must have shaken their heads in disbelief: could the Red Army commander really be guilty of such a thing?

As much as they were transplanted Russians, however, they could not prevent Murray from becoming American. When he stepped out the door of the railroad flat, he secretly changed out of the Russian clothes that his mother made him wear and put on knickers: he tossed aside the hated Russian hat and put on a snap cap, turning it sideways for panache. Then he could join his friends to play stickball in the shadow of the Third Avenue El or play-act Al Capone stories in alleys and fire escapes and roofs, with water pistols and BB guns.

As he got older, school was the last place he wanted to be. "My truancy was scandalous," he told me. He and a friend much preferred to head west to the Hudson, cross the spanking-new George Washington Bridge, and explore the Palisades. Or they'd go north to the Italian neighborhood, above 182nd Street, where people had gardens and goats. Beyond lay forests and grasslands and farms. Sometimes the farmers brought their produce into his neighborhood, rather like today's farmers' markets.[12]

* * *

One night in May 1930, nine-year-old Murray stretched out on his cot while Zeitel lay reading a volume of her beloved Gorky. Suddenly her book dropped to the floor. When the boy peered over, he found her dead from a heart attack.

Losing her was devastating. She had been his only competent, attentive parent. Childlike Rose, too, had been psychologically dependent on her dominating mother—in fact, without her, she would not do well. Too often she would try to escape from her responsibilities into the fantasy world of the local movie theater, much as she had once tried to run off to the Gypsies. "Instead of raising me," Murray told me, "she took me to movies."

The personal crisis of losing his grandmother coincided, it so happened, with a social crisis: the onset of the Great Depression. The stock market crashed in October 1929, leading to bank closings and factory shutdowns. Rose lost her millinery job. She and Murray now had only Nathan's paltry alimony payments to live on—and she spent them mostly on clothes at Klein's in Union Square or on Orchard Street.

She bullied her son, slapping him for trivial offenses. Once when she raised her hand to hit him yet again, he grabbed his BB gun and fired at her posterior, grazing it—and warned her never to strike him again. Thereafter the two merely coexisted. At the age of nine, he was all but orphaned.

A few months after Zeitel's death, the doorbell rang, and Murray opened the door to find a boy about two years older than he, hawking a children's magazine called *New Pioneer*. He held up a copy for Murray. "It tells young people the truth about what's going on in America . . . that Washington was a drunkard, and Jefferson owned slaves."[13] Murray gave him a few coins and then curled up with the magazine, reading it from cover to cover. It explained that American democracy was a hoax, and that the rich called the shots. A handful of millionaires of "tremendous wealth—Rockefeller, Vanderbilt, the Goulds—owned the American economy," it explained, "while workers were paid only ten to twenty dollars a week." But in the Soviet Union, the story was very different: there workers were "pioneering in a new world," creating a socialist paradise. Wouldn't it be fine, *New Pioneer* urged its young readers, if American workers would "follow in the footsteps of the workers of the Soviet Union!" Fortunately, the Communist Party was ready to lead them.[14]

The magazine-selling boy had told Murray about a group called the Young Pioneers of America that talked about these ideas at regular weekly meetings, and had invited him to the next one. The following night Murray went over to the International Workers Order building on East 180th Street, dashed up the creaky staircase, opened a door—and entered the international Communist movement. The Young Pioneers were the Communist children's section, for those nine to fourteen years old. Perhaps a dozen boys were arrayed behind desks. A few teenagers were standing by, but the children ran the meeting as much as possible.[15]

Comrades and fellow workers! shouted a boy at the front. *This meeting of the Young Pioneers will come to order!* The first part of the meeting, Murray came to understand, would always be educational. The Pioneers would talk about the glorious Soviet Union, or the economic crisis underway in the United States, or cases of injustice like the Scottsboro Boys; or they would analyze popular culture, pointing out the racism in movies like *Tarzan the Ape Man*.

In the latter half of the meeting, the Pioneers would do something practical, like make signs and banners for a demonstration. And at the meeting's close, they would sing "The Internationale." When they came to the last, rousing chorus ("'Tis the final conflict"), they would raise their right hand and hold it with the palm to their temple in a five-finger salute, to symbolize the five-sixths of the world's landmass that socialism had not yet conquered. Not yet.

The Communists rescued Murray from his personal crisis by becoming his surrogate parents. "It was the Communist movement that truly raised me," he would recall, "and frankly they were amazingly thorough." And they continued Zeitel's training regimen: where she had groomed him to become a Russian revolutionary, they would mold him into a leader of the future proletarian revolution in America—a young commissar. And even as they educated him, they provided him with stability and validation. They taught him to subsume his personal distress into an intense devotion to the Communist Party, the Soviet Union, and the coming revolution. They gave him brothers and sisters—his branch comrades—as well as an extended family in the movement's many other branches. The Communist movement became, in effect, his home.[16]

They taught him about Lenin, who had shown the Russian workers "how to organize against their oppressors, how to fight for their rights," then had led them in revolution. They taught him about Rosa Luxemburg, that "flaming symbol of revolutionary courage and devotion to the working class," who had tried to spark an insurrection to create a Soviet Republic in Germany.[17]

But to the fatherless, grandmotherless boy, the persona of Karl Marx, as presented by the Pioneers, must have been irresistible. "There was no one jollier or merrier" than Marx, *New Pioneer* explained. "His eyes twinkled at a good joke or a quick answer," and "when he laughed, it was with a hearty roar which shook him all over." He had been a loving father, and his home in London "was always the meeting place for revolutionists of many countries"—just as Zeitel's had been, back in Russia.[18]

At one meeting, the Young Pioneers were shown a Soviet propaganda movie, *The Road to Life*. Set in 1923, it portrayed the lives of children left orphaned and homeless by the Russian civil war. A group of these children (there were actually thousands) survive by roaming the countryside and robbing people. Soviet police round up the young bandits and bring them to a containment center, from which they will be sent to a reformatory. But just in time a kindly social worker realizes that they can have a better fate: he sends them to an abandoned monastery instead, where they learn trades, set up a factory, and run it collectively. The lost boys become upstanding Soviet citizens.

In one scene, midway through the film, a counterrevolutionary leaves a wounded boy to die on railroad tracks. A locomotive pulls near, but suddenly red flags appear, and a band strikes up "The Internationale"—the boy is saved. That scene of rescue so enthralled Murray that he leaped to his feet and raised his palm in a fervent Pioneer salute. Like the Soviet orphan, the orphan of Crotona Avenue found deliverance with the Communists.

He threw himself into Pioneer activities, especially helping evicted people. Many adults in East Tremont were losing not only their jobs but their homes. Landlords would simply put their possessions out on the street. When that happened, Young Pioneers would swing into action and haul the refrigerators, furniture, pots, and pans back inside.[19] The first time Murray got hit by a cop with a billy club, it was while carrying an evicted family's furniture back up a stairwell.

Like their adult counterparts, the Pioneers marched in demonstrations and parades, most memorably those on May Day, the workers' holiday. They would assemble at the Battery and march uptown, led by the party's central committee, carrying a huge red banner. Then came the red trade unions and the unemployed. Bands played "The Internationale" and European songs of struggle and even folk songs. Then came the benefit societies and various front organizations. Then came the teenagers of the Young Communist League (YCL).[20]

At the very end came the Young Pioneers, wearing blue uniforms, red bandanas, and garrison caps with a red star. Murray's East Tremont branch carried its red and gold banner, with a hammer and sickle. Others carried homemade placards reading, "Toward a Soviet America." As they tried to march in military formation, they sang, "With ordered step, red flag unfurled, / We'll make a new and better world! / We are the youthful guardsmen of the proletariat!"[21] The onlookers would give them a hearty round of applause—after all, they were the next generation.

The Communist movement of the early 1930s was very radical. In fact, it was the most ultraleft political movement in American history. And within that most ultraleft of movements, the two youth organizations, the Young Pioneers and the YCL, were the most ultraleft sectors. The Pioneers, Bookchin later recalled, were "super-revolutionaries."[22]

In 1928 Stalin had announced to Communists worldwide that capitalism was entering its death throes and that the final revolutionary upheaval was at hand. Communist parties, he said, must initiate armed uprisings. They must not work with Socialists, who by refusing to acknowledge the vanguard role of the Communists were overtly counterrevolutionary. Stalin was enraged by the Socialists. In fact, their stubborn wrongheadedness, he proclaimed, made them villains, twins, on a par with capitalists—with fascists even. Stalin advised the international Communist movement that Socialists were "social fascists"—the Communists' outright enemies.

Murray and his fellow Pioneers must have wondered: Were the Socialist kids at school really so evil? Was Norman Thomas, the Socialist Party leader, truly as wicked as a capitalist? Was Franklin Roosevelt actually a political twin of the Nazis? Pioneer meetings taught that they were. According to the *Daily Worker*, the Communist newspaper, FDR was "the leading organizer and inspirer of Fascism in this country," fronting for a "hidden dictatorship" of bankers and industrialists.[23]

If the confused Pioneers had any further objections, their older peers could have set them straight. Back in January 1919, when the great German leaders Rosa Luxemburg and Karl Liebknecht had tried to make a Communist revolution in Germany, the Social Democrats—who had just come to power—had had them killed using a right-wing paramilitary group. That atrocity became, in Communist lore, like a primal blood-crime, committed by Social Democrats. Thereafter Social Democrats everywhere could never be forgiven—their sin was indelible and inexpiable.

As the economic crisis deepened in the early 1930s, the American Communist movement grew quickly. Many people who had lost their jobs or been evicted from their homes were grateful when Communists moved their furniture back or organized demonstrations and strikes on their behalf—sometimes whole neighborhoods would participate.[24] The party's fiery rhetoric—about how the capitalist world was wracked by crises and plunging toward fascism and imperialist war—touched a nerve among the unemployed and the homeless alike: capitalism did indeed seem to be teetering. And then the Communists would tell them about the Soviet Union, the workers' fatherland. While signs at American factory gates said "no work," Soviet workers were marching confidently to work in factories. Planned, nationalized production was keeping people's bellies full, it seemed, while capitalism was leaving people hungry. Amid "raging waves of economic upheavals and military political catastrophes," said Stalin, "the Soviet Union stands apart like rock, continuing its work of socialist construction."[25] So it was that between 1930 and 1934, the Communist Party of the United States of America (CPUSA) recruited almost fifty thousand new members.

Rose Bookchin was still existing on her ex-husband's alimony payments, but around 1932 Nathan stopped sending them. Broke, Murray and Rose were now the ones evicted for nonpayment of rent, their furniture heaped on the sidewalk. They found a smaller apartment but couldn't pay the rent and got evicted again. They scuttled from one boardinghouse to another, sometimes a new one each month. Once in 1934 Murray had no shelter for three days, so he slept on the overpass bridge of the 149th Street subway station, along with other homeless people.

He put cardboard inside his shoes to cover the holes. His snap cap and leather jacket hung in tatters on his thin frame. He and Rose waited in breadlines outside churches, hoping for a bit of soup and bread, alongside gaunt Great War veterans, their medals pinned to their shabby jackets. Sometimes formerly wealthy men waited with them, quietly, wearing overcoats with now-worn velvet collars.[26]

Murray had to grow up fast, and he needed a job. Once again the Communists came through for him: around 1932 or 1933 they gave him work selling the *Daily Worker* on his home turf. So every twilight he would head over to the Simpson Street IRT station, don an apron, and meet a delivery truck, along with other boys. He would pick up a fifty-paper bundle, then head up Simpson Street to Crotona Park. In those days, the city parks had political identities, and Crotona was Communist territory—an excellent spot to sell this paper. In summer evenings starting around seven-thirty,

East Tremont residents (having no televisions yet) would mill around in the park, near Indian Lake, and talk politics.

They would talk about Adolf Hitler, and how the German working class absolutely must stop the Nazis in the coming elections. *Stalin will make sure the German Communists organize a revolution and keep Hitler from power,* someone might have said.

Even better, someone else might have jumped in, *the German Communists and Socialists should join forces, form an electoral alliance, to shut the Nazis out!*

No! It's out of the question! another might have objected. *We can't work with Socialists! Those bastards murdered Luxemburg and Liebknecht! Nazism and Social Democracy are twins! Stalin said so!*[27]

As the argument raged, people would call the *Daily Worker* boy over and buy a newspaper, sometimes using it to bolster their case. With the nickels and dimes Murray took in, he could buy food for himself and Rose the next day. But even more important to him, he could listen in on the arguments and discussions, and he absorbed all he could from the verbal fray.

In fact, he couldn't get enough of it. After he'd sold all his papers, he'd head over to East Tremont and Crotona, where the open-air street-corner meetings would just be getting underway. Communists, Socialists, and Trotskyists had designated corners at the intersection: their respective orators would set up their speaker stand, clear their throat, and then start declaiming about capitalism and revolution and fascism. Murray, moving from one to the other, hung on every word.

So eager was he to learn more about the Marxist ideology that underpinned his new family that he played hooky from public school and went into Manhattan to attend the Workers School. The CPUSA had established this Marxist academy in its headquarters building near Union Square (figure 1.1).[28] Taking a seat in an upstairs classroom, Murray would raptly sit through discourses called "Fundamentals of Communism" and "Political Economy." A year or two later, he took a class on *Das Kapital*. The instructor, to his astonishment, had memorized all of volume one, or so it seemed. If you called out a page number at random, he could recite the page contents word for word, Murray told me. Their proficiency and scholarship were breathtaking.

During the five years when he was its eager student, the Workers School gave him intensive and disciplined training in orthodox Marxism and Leninism. After class the students would go downstairs to the Cooperative Cafeteria, on the ground floor, and keep talking to any willing lecturers, like the American party leader William Z. Foster. "Cafeterias were the equivalent of European cafés," Murray recalled. "They were all over New York at that time—get a cup of coffee, sit around a table, twenty or thirty people, with one of the maestros, and we'd quiz them." They'd tell the students about the intraparty faction fights, or their trips to Moscow, where they'd met leading Russian Communists.[29]

FIGURE 1.1 The Communist Party headquarters (which Murray called the Daily Worker Building), home to the Workers School, ca. 1930

Charles Rivers Photographs Collection, Tamiment Library, New York University. Photographer: Charles Rivers.

The Pioneer meetings, Crotona Park, the street-corner oratory, the Workers School—they were all one big university for Murray. The passionate political discussions made life, "infused with radical politics," perpetually exciting.[30] Soon he would know enough to join the discussions himself.

* * *

When their homelessness finally became unbearable, Rose swallowed her pride and applied for New York City's municipal home relief program. She qualified, and the agency found an apartment on 177th Street for her and her son, and it even paid the rent. At least for now, Murray's living situation was secure.

When a Young Pioneer reached the age of fourteen, he or she could be co-opted into the teenage organization, the YCL. But by 1934, thirteen-year-old Murray showed such promise that his elders brought him into the YCL a year early. He considered it a great honor.

At YCL meetings, at the branch's headquarters on East Tremont Avenue, his Bolshevik education continued, with even greater intensity. The Young Communists acted like the Russians they all admired, even wearing the garb of commissars as depicted in Soviet movies: leather jackets and boots. They were in training to be commissars themselves, leaders of the revolution, members of its elite vanguard. Once the capitalist system entered its final crisis, their mission would be to lead the proletariat to socialism, and to create the proletarian dictatorship that would eliminate private property, socialize production, and distribute wealth according to need. Eager to fulfill their historic mission correctly, these zealous revolutionaries applied themselves to identifying the right strategy with painstaking care. The regular weekly YCL meeting wasn't enough for them: when it was over, they moved to a cafeteria to discuss, say, Lenin's *State and Revolution* and debate its meaning for the American situation, until two or three in the morning.

Murray and his comrades obsessively analyzed newspapers for signs of capitalism's terminal crisis—and in the year 1934 the signs were many. In February, Socialist workers in Vienna raised red flags and mounted an insurrection, fighting in the streets with rifles and machine guns. The Young Communists, sleepless, waited bleary-eyed through the nights by their radios for news of the uprising—which was soon, heartbreakingly, crushed. In April, workers at the Auto-Lite factory in Toledo, Ohio, went on strike and clashed with the National Guard in the "Battle of Toledo"—two were killed. In May, Communist teamsters in Minneapolis organized a trucking strike that turned into a general strike that shut down the entire city. In July, Communist longshoremen struck at the port of San Francisco. Other workers joined them, to the point that a general strike closed much of that city, too. In October, coal miners rose up and gained control of Asturias, Spain. Murray and his comrades, in feverish anticipation, were sure the Spanish miners would proclaim a Soviet republic.[31] But then a general named Francisco Franco appeared on the scene, leading the Spanish Foreign Legion northward; after three weeks, his troops quelled the uprising, killing thousands.

The Young Communists cast many wistful gazes on Germany, the country that Lenin had hoped would make the next socialist revolution. But contrary to all expectations, the KPD, the German Communist Party, was eerily quiet. It had done nothing to stop Hitler from coming to power in January 1933, and now nothing much was heard from the German comrades at all. Still, the Nazis surely didn't stand a chance against the powerful, well-organized German working class. Any day now the YCL-ers expected to read news of a German Communist-led strike or call to arms against Hitler.

Murray showed such promise as a leader and an ideologist that the YCL leadership tasked him with leading a Young Pioneer troop. Then they made him education director for his YCL branch. That meant that he set the agenda for the branch's educationals: in effect, he became its political commissar. He went to executive committee meetings and sat on the Bronx County committee.[32] Far from resenting all these meetings, he attended them eagerly, proud to be playing such an important role in the movement.

"Every day was an experience," he later recalled, for the world seemed on the brink of a profound upheaval. "The magic and romance of the October Revolution, the drama of it," suffused his life. How he chose to spend his time—seek out an open-air meeting, go to a demonstration, jump into a debate in the park—could, he knew, affect the course of world history. Even a young boy in the Bronx could help push the revolution forward, indeed was required to—and it would have been unthinkable to abjure that profound responsibility. Far from seeming onerous, the great task made life exhilarating.[33]

He understood, let it be said, that the Communist Party was no democracy. He and his comrades could debate issues, even at YCL meetings, but once the party adopted a position, and once the Young Communists were out in public, they were required to defend it, whether they agreed with it or not.

Murray disagreed, for example, with the party's demonization of Socialists as "social fascists." He had friends who were Socialists—members of the Young People's Socialist League (YPSL). Sometimes he went to Socialist rallies to observe them, quietly, and found them to be just as radical as the Communists—they waved the red flag, called themselves Marxists, and defended the Soviet Union.

But when the CPUSA commissars ordered him and his fellow commissars-in-training—on pain of expulsion—to disrupt Socialist open-air meetings, they all obeyed. They would go and listen and wait for the question period, then challenge the speakers about the murders of Luxemburg and Liebknecht. The harassment would sometimes lead to fistfights and break up meetings.

They did the same at Trotskyist meetings. But when they harassed these speakers as "social fascists," the Trotskyists argued back forcefully. *Your so-called comrade Stalin is executing millions of peasants in Ukraine!* they might counter. *He's forcing them to collectivize—and if they resist, he cuts off their food. It's mass murder by starvation!*

To which Murray and his comrades might reply, *Those kulaks—the peasants Stalin killed are actually rich and privileged. We can't let them stand in the way of building socialism!*

To which the Trotskyists might retort, *Stalin expels everyone who doesn't agree with him 100 percent! He even expelled Trotsky, Lenin's comrade in the October Revolution, and sent him into exile!*

Murray respected these dissident Communists—their arguments often made sense, and they seemed like decent people. But he knew, too, that the Communists were the exclusive vanguard, leaders of the march to a glorious future. By virtue of standing outside the vanguard, the Socialists and Trotskyists were "objectively counterrevolutionary," enemies of the international proletarian revolution. So it was possible to justify harassing them.

Yes, the Soviet Union was authoritarian, but sometimes people must abdicate claims to freedom in the name of moving history forward, he rationalized. As Engels had said, freedom is the recognition of necessity. Besides, once the final conflict was underway, the proletariat would need a centralized military-style organization to defeat the US Army, where taking orders would be a necessity.[34]

The YCL-ers had no doubt that they would ultimately defeat the American bourgeoisie and transform a nation of 125 million people into a socialist society—"the inexorable laws of history" were on their side.[35] After all, only a few thousand Bolsheviks had changed Russia, and one man—Lenin—had galvanized an entire city around the simple slogan "Peace, bread, land." Once they prevailed, the new society would actually be more democratic than the bourgeois republic it replaced. So broad would be the democracy that the very word would disappear from the popular vocabulary. For the sake of achieving that glorious outcome, the YCL-ers were willing to set aside democracy temporarily.

Until that utopian vision was fulfilled, however, a huge responsibility weighed on these teenagers' scrawny shoulders: it would be up to them to meet the challenge of history and lead the revolution.

No wonder he couldn't sit still in classes at Morris High School. And when he did show up, he turned classrooms into battlegrounds. When the American history teacher disparaged John Brown as a fanatic, Murray rose to fervently defend his abolitionist hero. When the European history class studied the French Revolution, the teacher would sigh, "All right, Bookchin, you can take over now and defend Marat!" Murray would stand up and extol the far-left Jacobins. Sometimes, he told me, he just wore the teacher down. "Go ahead, Bookchin," one teacher groaned, "present your Red point of view."[36]

But as often as not, Murray and his comrades skipped school and roamed the streets, looking for ways to spark an insurrection. When they came across a picket line of striking workers, they'd fall in, no questions asked.

Or they'd join an illegal demonstration, carrying banners and placards, swaggering and yelling their way to city hall. There "Cossacks"—police on horseback—were lined up along the street. At a signal, the Cossacks would charge toward the demonstrators, brandishing clubs. As the hooves thundered, the boys grabbed their placards and removed the wooden sticks, using them as makeshift clubs, and "whacked the hell out of the police," Murray told me. "Lots of times we got them off their horses, and we injured them enough that we squared off pretty well." Then Black Marias—police vans—would arrive and round them all up and take them to the Tombs. Shoved into a big cage, the arrestees would stink it up with a collective piss-in. They were quickly released.[37]

The CPUSA commissars, impressed once again by Murray's precocity, soon appointed him a street-corner orator, for the open-air meetings. They gave him the opening slot at the intersection of East Tremont and Crotona and assigned him the crucial initial task of attracting a crowd. Now, as soon as he finished selling his *Daily Worker* in the park, he'd rush over there and set up the YCL stand.

He'd mount the platform, clutch the railing, and try to look fierce and commanding. Then he'd shout at the top of his lungs, "Comrades and fellow workers—I herewith open this meeting of the Young Communist League!" And he was off. "We face today the possibility of a second world war that will wipe out Western civilization." To keep a group's attention, he learned, he had to be extremely expressive, gesticulating and clenching his fists.

A crowd would collect. Sometimes Socialists showed up to harass him with hostile questions about, say, the 1921 rebellion of the Kronstadt sailors against Bolshevik tyranny—the Red Army had quashed it with much bloodshed. Murray had to respond by asserting the party's rationalizations: that the sailors were petty bourgeois, or that their revolt had been financed by imperialist powers. If a Trotskyist challenged him on, say, the Stalinist belief that "socialism in one country" was really plausible, Murray would have to be ready to argue the Stalinist line that socialism existed in the Soviet Union.[38]

If he proved unable to answer a critic, the audience would laugh him down, and his public humiliation would reflect poorly on his whole YCL branch. So it was important to snap back decisively. And when he gave a good answer, they'd cheer, and his confidence soared. He got better at it, to the point that his listeners would hear him out. As training in public agitation, it was rigorous but effective—he learned to speak forcefully and dramatically, without notes. Speech-making soon became a pleasure. He couldn't wait for nightfall: for most of 1934, he orated on that street corner nearly every evening.

Meanwhile in France, on the night of February 6, far-right forces took to the streets and rioted against a leftist coalition government that had been in power in Paris for two years. As a result of this pressure, that government fell, and a more conservative one took its place, acceptable to the Far Right. Suddenly a fascist coup seemed like a possibility in France. A few days later, the French Communists did something strange, something that their German counterparts had not done a year earlier: they defied Stalin's "social fascist" demonization policy and joined forces with the Socialists to carry out a general strike. In fact, that June, French Socialists and Communists formed a unity pact.

Strangely, at least to the Bronx YCL-ers, the top Communists raised no objection. In fact, Stalin approved the collaboration. Hitler's Nazis were rearming Germany, he said, and they might one day attack the Soviet Union. During 1934–35 he had decided that the Soviet Union must form defensive alliances with capitalist governments to ward off Nazi aggression. On May 2, 1935, France and the Soviet Union signed a mutual assistance pact.

Stalin wanted to form more such alliances, with other capitalist governments, but the chances were slim as long as his Communist parties were trying to overthrow those very governments. So the word went out to Communist parties around the world: cease vilifying Socialists as "social fascists" and embrace them as allies. Communists should even, if possible, follow the French model and join forces with Socialists, even with Social Democrats, to form so-called Popular Front governments that would be friendly to the Soviet Union.

The French Popular Front alliance came to power in June. The ultrarevolutionary period was over.

Perhaps it was Gil Green, head of the American YCL, who in late 1935 informed Murray and the other young commissars-in-training that they were to stop attacking Socialists and start working with them.[39]

Murray was relieved at first—no longer would he be required to torment his Socialist friends at their meetings. He'd never really thought they were "social fascists" anyway. But he was dismayed that the party's new Popular Front line called for making alliances not only with Socialists (who were fellow revolutionaries) but with Social Democrats (who were reformists). Wasn't that class collaboration? What about socialist revolution?

The party commissars told him, in effect, to forget about revolution.[40]

Murray was dumbfounded. It contradicted everything they had hitherto taught him. The reversal was incomprehensible. Shocked, he stopped attending YCL meetings and went looking for a group that was still revolutionary. He found one in the Young Spartacus League, a group of Trotskyists.

They would surely have welcomed the talented young commissar. Surely, too, they explained to him that the Soviet Union's new Popular Front line, abjuring socialist revolution, should have come as no surprise. Stalin was a counterrevolutionary and had been for a long time—he had hijacked the Russian Revolution and diverted its momentum into the creation of a new tyranny that, far from liberating the workers, was oppressing them in new and increasingly atrocious ways.

Trotsky, by contrast, was still a Bolshevik revolutionary and wanted to put the revolution back on course. He intended to mount an uprising to overthrow Stalin, take control of the Soviet Union, and revive the spirit of true Bolshevism. He would succeed, they said, because unlike Stalin, he had the mighty proletariat as his social base.

Murray decided to join these like-minded comrades and gave them his name and address.

In the spring of 1936, perhaps while at the movies with his mother, Murray saw a black and white Movietone newsreel, showing footage of massive demonstrations in Spain. Thousands of workers were marching in the streets of Barcelona and Madrid, with clenched fists, wielding rifles and flags.

That February, the Spanish people had elected to power one of those Popular Front governments, a left-liberal coalition that included the Communists. But Spain's reactionary generals despised the Popular Front, so much so that, led by General Franco, they commenced a military rebellion against it and indeed against the Spanish Republic itself, in the name of the country's most reactionary military and religious traditions. Setting out from Spanish Morocco, they crossed Gibraltar and rolled through Spain, capturing numerous cities as they went. In some places, however, like Catalonia, resistance from popular militias beat them back.

A civil war had begun. To support the generals' revolt, Mussolini and Hitler sent troops and tanks. But only one country came to the aid of the beleaguered Spanish Republic: the Soviet Union, which sent tanks, fighter aircraft, guns, and military advisers. In fact, Stalin's Comintern was now encouraging Communists everywhere to join international brigades, to fight fascism in Spain.

Murray volunteered for the American brigade (named after Abraham Lincoln), but at fifteen he was too young to fight in Spain himself, and as much as he implored them, they refused. Determined to help the Spanish proletariat nonetheless, he raised money for arms and medical aid. But to do so he had to rejoin the YCL. Thus six months after he dropped out, he returned. "I didn't go back to support Stalinism," he said. The YCL just "gave me an avenue whereby I could express my support for the Spanish loyalists."[41] He'd go into the subways and launch into a speech, then hand out leaflets that urged, *Support the antifascist struggle!*

During the six months of Murray's absence, the Communist Party had undergone a dramatic transformation, thanks to the Popular Front policy. When he reentered his old branch headquarters on East Tremont Avenue, he was shocked to see that on the wall where a red flag had once hung, there was now an American flag. Images of Marx and Engels and Lenin had been replaced by portraits of Washington and Jefferson and Lincoln. When the commissars spoke, no more did they hail the coming proletarian revolution—now they voiced full-throated patriotic support for President Roosevelt. Another wall bore the slogan "Communism is twentieth-century Americanism." In November the party's political bureau formally adopted the New Deal's prolabor legislative program as its own.[42]

When it came to attracting new members to the Communist Party USA, the new Popular Front policy was a huge success. Liberals and labor unionists flocked to the Popular Front banner, suffused with admiration for the workers' paradise—the Soviet Union—and its fully employed workforce.

Meanwhile in Moscow, even as Stalin was anxiously courting Western liberals, he was also busy rewriting history. He and his henchmen were putting the still-living revolutionaries of 1917 on trial, charging them with conspiring to assassinate Stalin and, working with the Gestapo, turn the Soviet Union over to fascism. At a show trial, in August 1936, the prosecution offered no evidence to prove its case against them, because there was none—the verdict had been decided in advance. Yet the accused, standing on the dock, repented their supposed crimes in spectacular self-incriminations. They were found guilty and executed. Leon Trotsky, living in exile, was sentenced to death in absentia.

Murray understood that the charges were absurd, that the confessions had been extracted by torture. As for Trotsky, the more Stalin denounced him, the more Murray was intrigued by the onetime commander of the Red Army. He read his three-volume *History of the Russian Revolution*, in its eloquent literary translation by Max Eastman, and was floored. Contrary to everything the YCL had told him, Trotsky had formidable intellectual power, both as a historian and as a social theorist. Moreover, his account of the revolution showed that the YCL leadership had lied when it claimed the Bolsheviks had banned factions even before the 1917 revolution. On the contrary, before the seizure of power, dissent had been allowed.

At a second Moscow show trial, in January 1937, seventeen more old Bolsheviks were speciously charged, convicted, and executed. By now, in the Stalinists' universe, Trotsky had become the generic symbol of satanic evil. Stalinists routinely branded all their enemies as "Trotskyist-fascists" or "Trotskyist-imperialists" or "Trotskyist-Zionists."

In response, Trotsky himself, in exile since 1927 and now residing in Coyoacán, Mexico, assembled a dossier to refute Stalin's charges against him. An impartial international commission agreed to examine his case, headed by the eminent liberal American philosopher John Dewey. In April 1937 Dewey traveled to Mexico and spent a week interrogating Trotsky. He found his testimony straightforward and evidence-based. He declared Trotsky not guilty and even called the encounter "the most interesting single intellectual experience of my life."[43]

In Murray's eyes, Trotsky had become truly heroic. Still, he remained a YCL member in order to do support work on behalf of the Spanish proletariat. And when dealing with the commissars, he kept his head down.

One day the head of the YCL's Bronx County Committee noticed that at a meeting young Bookchin seemed strangely inattentive and asked him what was going on. Murray shrugged. "Comrade Bookchin," the man said, "I've gotten reports that you have views that aren't compatible with those of the movement. And I've heard you express some sympathies for the traitors being tried in Moscow, the fascists, Trotskyists. And I've heard that you have some differences with the Popular Front. You've been in the movement for a long time. We don't want to lose you. Would you like to see a comrade psychiatrist?" Murray stalled, saying he'd think about it.[44]

During the 1930s a new working class had emerged in the United States, employed in the burgeoning auto, steel, rubber, textile, glass, and electrical industries. These industrial workers, many of them immigrants, clamored to join labor unions, but the existing ones wouldn't admit them. In November 1935, United Mine Workers president John L. Lewis created the Committee for Industrial Organization (CIO), as an umbrella for industrial unions like the United Electrical Workers (UE), the United Auto Workers, and the United Steelworkers. Industrial workers stampeded into them, then participated in sit-down strikes, which were wildly popular and wildly effective.

The UE wanted to establish a foothold in the factories and mills and machine shops of northern New Jersey and needed organizers. So it turned to the people who had the most experience in such tasks: Socialists and Communists. UE organizers noticed Murray's oratorical talents and physical bravado. He was only sixteen but looked older, so in 1937 they recruited him. He took the ferry across the Hudson, then the train to Jersey City. He'd find a spot in front of a factory, then stand up and address the workers, extol the might of the insurgent proletariat, and urge them to go to the next UE meeting.[45]

New Jersey's political establishment, however, was determined to keep labor unions out of the state. Governor Harold Hoffman minced no words—he would eject CIO organizers "by bloodshed if necessary." Jersey City boss Frank Hague expressed the same attitude: "I am the law. I decide; I do; me!"[46] Hague forbade the CIO organizers to use public places for rallies and made sure they couldn't rent meeting halls—when workers, mobilized by Murray and his comrades, arrived at the local hall, they found the doors locked. Other times, when the organizers tried to distribute leaflets, company goons charged after them with axes and even shotguns,[47] and if they caught up with the terrified youths, they'd beat them and then drop them outside the county line, or stuff them onto a train and ship them back across the river. By such means did Boss Hague keep labor unions out of Jersey City.[48]

Unknown to Murray, thousands of miles away in Mexico, the exiled Trotsky was studying newspaper reports of these very events. Boss Hague was "an American fascist," the old Bolshevik concluded, and his tactics were "a rehearsal of a fascist overthrow." The only way to stop him, he thought, was to organize workers into defense committees to fight them physically. Otherwise "we will be crushed. . . . I believe that the terror in the United States will be the most terrible of all."[49]

Through it all, Murray was studying the Movietone newsreels for news about the Spanish Civil War. It wasn't just about fighting Franco anymore. The films showed Spanish people marching with clenched fists, singing "The Internationale," waving flags. Clearly something more than a civil war was going on.

Then, in early May 1937, the *Daily Worker* reported that street fighting had erupted in Barcelona, the Catalan capital. The Barcelona proletariat, said the Stalinist organ, had mounted a pro-fascist uprising. It was nonsense, Murray understood immediately. Tossing the paper aside, he picked up the bourgeois *New York Times*, which told a much more credible story. "Anarchists tonight were in control of a part of Barcelona," he read, "after a rising in which at least 100 persons have been killed." The Catalan authorities had mastered the city center, but "the Anarchists control the suburbs and outlying districts" (figure 1.2).[50] In other words, the Barcelona uprising wasn't fascist—it was anarchist.

Anarchists! Who knew they even existed anymore, let alone that they were capable of organizing? Who were these anarchists, anyway?

He found answers by devouring George Orwell's *Homage to Catalonia* (1938), Felix Morrow's *Revolution and Counterrevolution in Spain* (1938), and Franz Borkenau's *The Spanish Cockpit* (1937). Starting in the late nineteenth century, he learned, panish workers had organized themselves into a huge, militant anarchist (actually anarcho-syndicalist) trade union called the National Confederation of Labor

FIGURE 1.2 A Barcelona barricade in May 1937. Anarchists defend their revolution against the forces of Spain's Popular Front government, which ultimately suppressed it and obliterated the revolution's collectivist institutions. For Bookchin, the episode would remain a lifelong preoccupation
Courtesy Fundación Anselmo Lorenzo, Madrid, and Labadie Collection, University of Michigan.

(Confederación Nacional del Trabajo, or CNT). By July 1936, when Franco and the generals had tried to conquer Spain militarily, the CNT was quite strong, so much so that *cenetista* militias—organized along libertarian lines—had been able to defeat them in Catalonia and Aragon. Now finding themselves in political control of those areas, the CNT anarchists had collectivized factories and established workers' committees to run them, and in the countryside they had collectivized farms. The workers and peasants were in control.

For years Murray had been hoping that every upsurge reported in the press would turn into a social revolution. Now, finally, after all the aborted efforts, here was one that had succeeded: a proletarian revolution. But bizarrely enough, it had been carried out not under the red flag of Bolshevism but under the black-and-red flag of anarcho-syndicalism. The Communists were supposed to be in favor of proletarian revolution, but now they were in power, as part of Spain's Popular Front government, and instead of supporting this one, they actively sought to suppress it, indeed to dismantle the anarchist collectives and smash the anarchist militias, using the Popular Front's military forces, which they controlled thanks to Soviet aid. The whole scenario was topsy-turvy. In May 1937, as Murray read in the *Times*, when the Barcelona anarchists took to the streets to defend their revolution, the well-armed Communists vanquished them, dooming the anarchist revolution.[51]

Reading Orwell's *Homage to Catalonia,* Murray became enthralled with the Spanish anarchists. "I said, they cannot be wrong! Not people like that."[52] The more he learned about them and their revolution, the greater became his passion for them, and the more his hatred for the Stalinists grew. The anarchists' revolution that the Stalinists had extinguished, Murray concluded, had been nothing less than the greatest proletarian revolution in history.

In mid-1937 Stalin's Comintern dropped the Popular Front party line and adopted a new, even less revolutionary one, which it called the Democratic Front. Communist parties were now to accept as allies and members not just Socialists and Social Democrats but just about anyone. In East Tremont, a Bronx county leader visited Murray's YCL branch to announce the new policy. He took no questions, permitted no discussion, just dropped the bombshell and left.[53]

Murray's head was spinning. Only three years earlier, the CPUSA had been so ultrarevolutionary that it had excluded Socialists. Now, in the name of protecting the Soviet Union, the Young Communists were required to join forces with "even the vilest capitalist reactionary . . . even with J.P. Morgan." And now thanks to the Popular Front policy, new people were flooding into the YCL who knew little and cared less about Marxist ideology. They hadn't read *Capital* or any of the other texts, let alone master its arguments and memorize page numbers. YCL educationals deteriorated into travesties of their former selves. Instead of discussing Marxism, "a whole meeting will be taken up with such safe matters as selling tickets for some social affair."[54]

Ominously, the more mindless the party became, the less it seemed to tolerate deviations or dissidence, aping Stalin's dictatorship. More and more it vilified any dissent as Trotskyist. In the spring of 1938, as Stalin was demonizing Trotsky in yet another show trial, the CPUSA formally amended its constitution to prohibit its members from having a "personal or political relationship" with Trotskyists, who it said were "organized agents of international fascism."[55]

Once again, a county official beat the path to the East Tremont YCL, this time to announce the ban on Trotskyists. When he was finished, one YCL-er spoke up, asking how they could identify a Trotskyite at future meetings. Perhaps the official eyed the questioner suspiciously when he replied: "Nine out of ten times, anyone who asks a question is probably a Trotskyite." This was so brazen that other YCL members objected raucously. The party had to send out another official to cool them down.[56]

But the new line was serious: the Young Communists were ordered to spy on one another. Living at home with a Trotskyist family member was forbidden, and any YCL-er who did was required to move out.[57] Reading Trotskyist literature was excluded. Anyone suspected of associating with a Trotskyist was put on trial, subjected to vitriolic accusations, and forced to make a humiliating confession, in a facsimile of the Moscow show trials. Those found guilty—and they were many—were punished by expulsion.[58]

For Murray, as discontented as he was, expulsion was a frightening prospect, since he had no life outside the YCL. He hoped fervently that this new Democratic Front madness would prove short-lived, that the party would soon admit its mistake, drop the ally-with-capitalists line, and return to being a revolutionary party.[59]

But as the months passed, the inquisitorial mania, far from breaking, only intensified. The YCL leadership now detected Trotskyism even in a single word or in body language or a person's tone of voice. Members were banned from touching Trotskyite literature with their fingers, as if it carried some infectious disease.[60]

Around this time, Murray developed a crush on a girl at his high school, Dorothy Constas. For weeks he saved up his nickels to take her to see a movie and have an ice cream frappe. Dorothy was emotionally fragile and not much interested in politics, but her sister Helen was a Lovestoneite (a follower of the now-executed Bolshevik Nikolai Bukharin), which was almost as bad as being a Trotskyist. The YCL brought Murray up on charges, but since he was such a longtime member, one who had left the movement and then returned, they held back from expelling him.

Thereafter he consciously broke the rules and deliberately became a bad Communist: he went out of his way to associate with Trotskyists, as if to goad the commissars into ejecting him. He even invited his friend Boopy Miller to address his branch. Miller had previously been expelled from the YCL as a Trotskyist. His presence at the meeting caused an uproar. But still the YCL refrained from expelling Murray.[61]

Right-wing movements were now on the rise in the United States. The Catholic priest Charles Coughlin, in his regular radio broadcasts from Michigan, howled that atheists, Bolsheviks, and Jews were out to destroy the country. An "international conspiracy of Jewish bankers," Coughlin told his 30 to 50 million listeners, was responsible for the Depression and for the Russian Revolution. Coughlin supported Franco in Spain, gave Nazi salutes, defended the *Kristallnacht* pogrom, and reprinted Goebbels's speeches and the fraudulent "Protocols of the Elders of Zion" in his magazine, *Social Justice*.

New York was home to far-right groups as well. The Christian Front comprised several thousand members in the area, mainly Coughlin's Irish American listeners. The German-American Bund, an American Nazi organization, openly spouted anti-Semitic, antilabor diatribes on busy street corners. Gangs of Bundists, sometimes dressed in storm trooper uniforms, attacked lone workers and Jewish storekeepers in broad daylight. Neither they nor the Christian Fronters saw any difference between Jews and "Reds," and both groups' thugs broke up leftist street corner meetings with particular glee.[62]

Horrified, New York's Trotskyist groups created a Workers' Defense Guard (like the ones Trotsky thought should fight Boss Hague) "to protect workers' meetings, halls and institutions against hoodlum violence by the incipient fascists."[63] Murray and some of his comrades trained themselves in the cellar of his apartment building. A half-dozen strong, they would line up and drill; they would hold mock fistfights and practice using clubs, to fight the fascists if they showed up in East Tremont or anywhere else.[64]

In early 1939 the far-right groups called a mass meeting and rally to mobilize against labor unions, radicals, and Jews. It was to take place at Madison Square Garden on February 19, George Washington's birthday.⁶⁵ The Socialist Workers Party (SWP), the official Trotskyist party, called on all workers and antifascists—indeed, all New Yorkers—to assemble outside the Garden for a counterdemonstration. The SWP issued leaflets and plastered posters all over New York.⁶⁶

When Murray heard about it, he insisted to the YCL commissars that they must all go to the demonstration to stop the fascists, no question. But the commissars refused to participate—because the Trotskyists were organizing it. Appalled, Murray made his way to Madison Square Garden with a few comrades. There he surely witnessed some of the twenty thousand people entering the Garden for the despicable meeting.⁶⁷

Inside (as was later reported), the walls were hung with banners that read "Smash Jewish Communism!" American and Nazi flags were arrayed under a thirty-foot painting of George Washington. The head of the Bund inveighed against "Frank D. Rosenfeld" and the "Jew Deal." Any mention of Franco's military successes in Spain drew "great shouts of jubilation ringing from the topmost balcony." For two solid hours the speakers cursed "international Jewry." At the end, the audience sang "The Star Spangled Banner" with their arms extended in the full Nazi salute.⁶⁸

Meanwhile outside, under SWP banners reading "Workers Unite to Fight Fascism," the counterdemonstration assembled: they were Trotskyists, Socialists, and disaffected Stalinists like Murray; workers and African Americans; Italian antifascists and Irish republicans; Germans against Hitler; Great War veterans; and more.⁶⁹ A whopping fifty thousand people showed up, more than twice the number inside.

The speakers at the counterdemonstration called on the Workers Defense Guards to fight fascism. Murray and his comrades stood at the ready with their two-by-fours.⁷⁰ Organized squadrons of SWP as well as YPSLs (members of the Young People's Socialist League) arrived.

A few feet from the SWP speakers, at Eighth Avenue and Forty-eighth Street, a wall of police sat astride horses. At a signal, these Cossacks charged, trampling demonstrators, clubbing them, and arresting them. But after every sally, the crowd re-formed, determined to hold its own. "We didn't run from the police," Murray recalled, "we ran toward them as they came toward us." At the end, the police could do no more than clear a path between ranks of demonstrators, just wide enough that the Nazi supporters could emerge from the Garden. As the fascists walked by, they looked "panic stricken," Murray recalled. "I'll never forget their faces."⁷¹

Nor could he forget the Stalinists' refusal to participate.

A few months later, on August 24, 1935, Stalin concluded the ultimate nonaggression pact, with Adolf Hitler himself. Stalin promised not to get involved in the European war that Hitler was planning, and the two dictators divided up eastern Europe into "spheres of influence." The socialist fatherland was now an ally of the Third Reich—and presumably safe from attack by the Wehrmacht.

Appalled, Communists the world over abandoned the party in droves. Murray sauntered over to East Tremont and asked the commissars how they could possibly defend Stalin's pact with Hitler. One functionary assured Murray that the agreement "contained an escape clause." For example, "if Hitler invaded Poland, the Soviet Union would consider the nonaggression pact broken." Then, on September 1, Hitler invaded Poland. "How we waited ... for the announcement of that escape clause!" said Murray. "But it never came."[72] On the contrary, a few weeks later the Soviet Union, with the Nazis' approval, invaded eastern Poland and gobbled up large sections of it for itself.

Murray went back to East Tremont one last time and demanded, "Whatever happened to our antifascist policy?" The section leader who had previously suggested psychotherapy whirled around, pointed his finger at him, and said, "Bookchin, you're expelled!" Murray calmly asked, "Why did you wait so long?"[73]

He knew just where to go. After all, the Trotskyists weren't making alliances with fascists—they were actively fighting them.

NOTES

1. Murray Bookchin, "A Marxist Revolutionary Youth," interview by Janet Biehl, in Bookchin, *Anarchism, Marxism and the Future of the Left* (San Francisco: A.K. Press, 1999), 16. See also Erich Haberer, *Jews and Revolution in Nineteenth-Century Russia* (Cambridge: Cambridge University Press, 1995); Irving Howe, *World of Our Fathers* (New York: Harcourt Brace Jovanovich, 1976); and Michael Goldfarb, *Emancipation: How Liberating Europe's Jews from the Ghetto Led to Revolution and Renaissance* (New York: Simon & Schuster, 2008).

2. On the Jewish community in Yednitz (present-day Edinet, Moldova), see Mordehkai Reicher and Yosel Magen-Shitz, eds., *Yad l'Yedinitz: Memorial Book for the Jewish Community of Yedintzi, Bessarabia (Yedintsy, Moldova)*, online at http://bit.ly/qlpRPy.

3. Bookchin, "Marxist Revolutionary Youth," 16.

4. These are the dates from the Ellis Island records at http://www.jewishgen.org; Murray's birth certificate says Rose was born in 1897.

5. In Bookchin, "Marxist Revolutionary Youth," 16, Murray gives the date as 1906 or 1907, but the Ellis Island records at http://www.jewishgen.org say 1913. The immigrants are listed with names misspelled, as Ceitel Kaluska, from Jeddenits, Russia, born 1868; Ruchel Kalucki, from Jedinetz, born 1894; and Daniel Kalucki, from Jedinetz, born 1895. They entered Ellis Island on Oct. 13, 1913.

6. Bookchin, interview by Doug Richardson, unpublished ms., 1973, 2, Murray Bookchin Papers, Tamiment Library, New York University, hereafter MBPTL; and copy in author's collection. On this group, see Edgar B. Speer, "The Russian Workingmen's Association, Sometimes Called the Union of Russian Workers (What It Is and How It Operates)," A Bureau of Investigation Internal Report, Apr. 8, 1919, online at http://bit.ly/qfC7FD.

7. Kate Simon's evocative memoir, *Bronx Primitive: Portraits in a Childhood* (New York: Viking, 1982), describes the same neighborhood, just a few years before Murray was old enough to explore it. See also George Diamond, "I Remember Tremont: 1911–1918," in Lloyd Ultan and Gary Hermalyn, eds., *The Bronx in the Innocent Years, 1890–1925* (New York: Harper & Row, 1985); and Deborah Dash Moore, *At Home in America: Second Generation New York Jews* (New York: Columbia University Press, 1981), esp. 73–75.

8. She named him Mortimore, thinking it had grand British overtones. Both the name and the misspelling (of Mortimer) would embarrass him throughout his life. The name his childhood friends gave him, Murray, was the moniker that stuck.

9. Nathan would remarry and father another son, Robert. Murray would not meet his half-brother until he was in his thirties.

10. Clayton Hubert Thomas, *"Those Anarchist Bastards": Judge Webster Thayer and the Making of the Sacco-Vanzetti Case's Arch-Villain* (Cambridge, MA: Harvard University Press, 2006).

11. Zeitel's comments on Sacco-Vanzetti are variously recounted in Bookchin, "Marxist Revolutionary Youth," 21; "Interview Peter Einarssens mit Murray Bookchin, Okt. 1984," part 2, trans. Harald Simon, *Schwarzer Faden*, no. 27 (1988), 42; and Jutta Ditfurth, "Ein Streiter für Utopia," *Zeitmagazin*, Mar. 13, 1992, 50.

12. Murray Bookchin Video Biography, a multipart interview by Mark Saunders, Burlington, Vermont, May 1995, Spectacle, online at http://www.spectacle.co.uk (hereafter MBVB), part 21. See also Thomas Pawlick, "A Return to First Principles," *Harrowsmith*, no. 43 (1982), 30–40, and "Listening to the Earth," panel discussion with Bookchin, Gary Snyder, Morris Berman, and Ernst Callenbach, Planet Drum Foundation, Apr. 10, 1979; published as "Cities: Salvaging the Parts," *Planet Drum* 1, no. 3 (1981), 13.

13. Bookchin, "Marxist Revolutionary Youth," 23–24.

14. *New Pioneer*, Sept. 1932, 5; "Pioneering," "Dear Comrade Editor" section, *Pioneer*, Oct. 1931, 23.

15. On the Pioneers, see Paul C. Mishler, *Raising Reds: The Young Pioneers, Radical Summer Camps, and Communist Political Culture in the United States* (New York: Columbia University Press, 1999). Murray always referred to his Pioneer comrades as boys, but there were certainly Pioneer girls as well.

16. Bookchin, "Marxist Revolutionary Youth," 24, 28.

17. Sasha Small, "The Three L's—Lenin, Liebknecht, and Luxemburg," *New Pioneer*, Jan. 1933, 12–13; *New Pioneer*, Jan. 1932.

18. De Leon, "Karl Marx at Home and at Work," *New Pioneer*, Mar. 1933, 13.

19. Beth S. Wenger, *New York Jews and the Great Depression: Uncertain Promise* (New Haven, CT: Yale University Press, 1996), 108–10.

20. The parade is described in *Daily Worker*, May 2, 1934.

21. Or they sang "Rise Up, Fields and Workshops": "Rise up, fields and workshops, / Come out, workers, farmers! / To battle march onward, / March onward, world stormers! / Eyes sharp on your guns, / Red banners unfurled; / Advance, proletarians, / To take back our world!" See Socialist Song Book online at http://hengstrom.net/songbook/. Murray used to sing this song to me, as well as "The Youthful Guardsmen."

22. MBVB, part 1.

23. Harvey Klehr, *The Heyday of American Communism: The Depression Decade* (New York: Basic Books, 1984), 178.

24. Sidney Lens, *Unrepentant Radical* (Boston: Beacon Press, 1980), 36. See also Wenger, *New York Jews and Depression*, 110.

25. *Daily Worker*, Jan. 29, 1934.

26. Bookchin, "Economics as a Form of Social Control," *Liberation*, Jan. 1975, 7.

27. In July 1931 an official Communist publication declared that the KPD must throw "all its forces into the struggle against the Social Democrats." Jean Van Heijenoort, *With Trotsky in Exile: From Prinkipo to Coyoacan* (Cambridge, MA: Harvard University Press, 1978), 2.

28. Advertisements in the Communist press put the building variously at 50 East Thirteenth Street, between Broadway and Fourth Avenue, and at 35 East Twelfth Street.

29. MBVB, part 2.

30. Bookchin, "Marxist Revolutionary Youth," 27.

31. MBVB, part 1.

32. MBVB, part 4.

33. "Interview Peter Einarssens," part 2, p. 40; Bookchin, "Marxist Revolutionary Youth," 26.

34. Bookchin, "Marxist Revolutionary Youth," 36.

35. Ibid., 37.

36. MBVB, part 1. See also "Interview Peter Einarssens," part 2, p. 40.

37. Bookchin, "Marxist Revolutionary Youth," 31.

38. Ibid., 34–35.

39. Ibid., 38–39.

40. MBVB, part 9.

41. MBVB, part 3.

42. Harvey Klehr, *The Heyday of American Communism: The Depression Decade* (New York: Basic Books, 1984), 191, 222.

43. Alan M. Wald, *The New York Intellectuals: The Rise and Decline of the Anti-Stalinist Left* (Chapel Hill: University of North Carolina Press, 1987), 131.

44. MBVB, part 4.

45. MBVB, part 3.

46. Hoffman quoted in John T. Cumber, *A Social History of Economic Decline: Business, Politics, and Work in Trenton* (New Brunswick, NJ: Rutgers University Press, 1989), 121. On Hague, see Gerald Leinwand, *Mackerels in the Moonlight: Four Corrupt American Mayors* (Jefferson, NC: McFarland, 2004), 93–94; and http://jerseycityonline.com/frank_hague.htm.

47. Bookchin, "Marxist Revolutionary Youth," 42; MBVB, part 3. See also Bookchin, "Turning Up the Stones: A Reply to Clark's October 13 Message," http://bit.ly/qc1Yw6.

48. The CIO brought a lawsuit against Hague that in 1939 went to the US Supreme Court. In *Hague v. Committee for Industrial Organization*, the Court ruled that Hague had arbitrarily suppressed free expression and that his actions were unconstitutional. The decision established the citizens' right to use public space for the expression of political opinion.

49. Leon Trotsky, "Completing the Program and Putting It to Work," June 7, 1938, in *The Transitional Program for Socialist Revolution* (New York: Pathfinder Press, 1973), 139–40, 143.

50. "Anarchists in Open Revolt against Barcelona Regime," *New York Times*, May 5, 1937, 1.

51. William P. Carney, "Rift in Loyalists Is Seen by Rebels," *New York Times*, May 6, 1937, 6.

52. "Interview Peter Einarssens," part 1, p. 39.

53. Art Preis, "YCL Member Describes Internal Life of Stalinist Youth Group," *Militant*, May 31, 1941. The ex-YCL member whom Preis interviewed is unnamed, but in my opinion the interviewee was unmistakably Murray. In this and the next chapter, I draw on this article accordingly.

54. Ibid.

55. Klehr, *Heyday*, 221.

56. Preis, "YCL Member."

57. Ibid. According to historian Theodore Draper, "Suspects were often obliged to break with parents, wives, husbands, siblings, and children who were associated with or

merely sympathetic to other 'criminal' left-wing groups." Draper, "American Communism Revisited," *New York Review of Books*, May 9, 1985.

58. Preis, "YCL Member."
59. Ibid.
60. Ibid.
61. MBVB, part 17. Miller's real name was Milton Sachs.
62. *Militant*, Aug. 2, 1941; MBVB, part 4; James P. Cannon, *Socialism on Trial: The Official Court Record of James P. Cannon's Testimony in the Famous Minneapolis "Sedition" Trial*, 4th ed. (New York: Pioneer Publishers, 1965), 73.
63. Cannon, *Socialism on Trial*, 73.
64. MBVB, part 4.
65. *Militant*, Aug. 2, 1941. The old Madison Square Garden was located at Eighth Avenue and Fiftieth Street.
66. Ibid.; MBVB, part 4.
67. MBVB, part 4.
68. Felix Morrow, "All Races, Creeds Join Picket Line," *Militant*, Feb. 24, 1939.
69. Ibid.
70. MBVB, part 4.
71. Ibid.; *Militant*, Aug. 2, 1941; "Nazis Cheer New York Cops for Clubbing of Picket Lines" and "'History Is Written Not in Ink but in Blood!' Declares Jew-Baiting Swastika Leader," *Socialist Appeal, Official Organ of the Socialist Workers Party, Section of the Fourth International*, Feb. 24, 1939.
72. Preis, "YCL Member."
73. MBVB, part 4.

2

Labor Organizer

IN 1938 TROTSKYISTS worldwide, claiming to be the authentic heirs of the Bolshevik Revolution, had formed their own Fourth International, so as to rally the working class to a genuine proletarian revolution.

The following year, as Hitler invaded Poland, the Trotskyists predicted that this incipient new European war would follow the scenario of the Great War. Once again Germany was the aggressor. Once again the main belligerents were imperialist countries competing for hegemony. So close were the parallels, in fact, that the new war seemed to be a mere continuation of the first. That being the case, Trotskyists resolved to follow the playbook that had worked so well for the Bolsheviks in 1917. The Bolsheviks had opposed the first imperialist war, so the Trotskyists of 1939 would oppose the second. And just as the war-weary Russians had rallied to the antiwar Bolsheviks in 1917, so the sooner-or-later-war-weary proletariat would eventually rally to the Trotskyists, who would oppose this war. This was the "tried and tested program," as American Trotskyist leader James P. Cannon labeled it.

And just as the first imperialist war had led to revolution, so too would the second. Indeed, given that capitalism seemed to be teetering, the second imperialist war would likely end in multiple revolutions: the proletariat in the various Western countries would demolish capitalism; the Russian proletariat would overthrow Stalin's regime; and the German workers would overthrow the Nazis. The war, Trotsky predicted confidently in July 1939, would "provoke with absolute inevitability the world revolution and the collapse of the capitalist system." And the Fourth International's program, he affirmed, would "be the guide of millions, and these revolutionary millions will know how to storm earth and heaven."[1]

Radical intellectuals on several continents rallied to the banner of the Fourth, dazzled by acumen of the Russian Revolution's theoretical mastermind, the commander of the Red Army, and not least the author of the great *History*; he was the paragon of the activist intellectual, a man of letters who had also led troops against a reactionary enemy and prevailed. For the past decade Stalin had subjected him to a slanderous propaganda campaign and dispatched agents over three continents to murder him, yet his defiance had never faltered. His courage was breathtaking. For all these reasons, Bookchin would recall years later, "Trotsky won my deep admiration and ideological support."[2]

In early 1940, the Fourth International's largest branch was the American one, the Socialist Workers Party (SWP), with 2,500 members, and New York was its largest section; this was the group that had organized the counterdemonstration at Madison Square Garden. The New York SWP-ers actively supported Trotsky at his refuge down in Coyoacán, purchasing a house for him there, supplying bodyguards, and sending him money. They maintained an ongoing correspondence with him and journeyed to meet with him. "We were in very, very intimate contact with him after he came to Mexico," wrote Cannon.[3]

Meetings of the New York SWP, held at national party headquarters at 116 University Place, were attended by luminaries like the philosophers Sidney Hook and James Burnham, the journalists Dwight Macdonald and George Breitman, and many others. They made Trotskyism, for a time, "the leading American radical movement in terms of per capita brain power."[4] One of the brightest stars in the firmament of the New York SWP was Jean van Heijenoort, a Frenchman with a Dutch name who had been Trotsky's international corresponding secretary throughout the 1930s, traveling with him from one place of exile to another.[5] More recently, in 1940, Van, as he was called, had married a New Yorker and moved to the city, where he now frequented SWP meetings. With his revolutionary glamour, Van dazzled nineteen-year-old Bookchin. ("I knew Trotsky's secretary!" he would exult to me fifty years later.) Bookchin was also honored to befriend Felix Morrow, age thirty-four, the author of *Revolution and Counterrevolution in Spain*, and Al Goldman, forty-three, who had been Trotsky's attorney during the Dewey Commission inquiry. He found a same-age comrade in Dave Eisen, a budding journalist.

Stalin, with his murderous collectivizations, his show trials, and his gulags, was turning the dream of socialism into an abattoir, but Trotsky taught his followers around the world that Stalin was merely an aberration. As Bookchin's new friend Goldman remarked, Trotsky "wrote and explained, and we read and understood and continued the struggle."[6] Trotsky was the bearer of revolutionary hope. As the socialist author George Lichtheim observed, "Trotskyism stood for the utopian side of Communism: belief in an imminent world revolution."[7]

Not that Trotsky was above criticism, even to his New York supporters. In 1918–19, after the revolution, he and his fellow Bolsheviks had banned factions within their own party, suppressed all non-Bolshevik political parties, and formed the Cheka, the secret police. When the New York SWP-ers questioned him about these measures, he explained that they had been historically necessary.

Perhaps most troubling was the Red Army's brutal and bloody suppression of the dissident prodemocracy movement at the Kronstadt naval base in 1921. This brave, principled sailors' revolt had been smashed by none other than Leon Trotsky himself. When the Americans asked him how he could possibly justify this action, he responded: "How could a proletarian government be expected to give up an important fortress to reactionary peasant soldiers?"[8] Dwight Macdonald, unsatisfied with this response, challenged him, whereupon Trotsky shut down the discussion: "Everyone has the right to be stupid on occasion, but Comrade Macdonald abuses it."[9]

In fact, Trotsky had no more interest in democracy or civil liberties than had Lenin or Stalin. But the SWP-ers nonetheless gave the Old Man the benefit of the doubt. Their devotion both to him and to the coming neo-Bolshevik revolution was unbounded. "Our whole lives revolved around the party," one comrade, Sidney Lens, recalled.[10] And even though Trotsky himself was dogmatic and rigid, his New York followers were boisterously disputatious, dissecting even the tiniest differences of opinion among themselves in protracted discussions. Under a Trotskyist regime, they all would have been shot.

Bookchin felt at home in this hyperarticulate enclave, where political life and personal life once again converged. Much to his regret, however, he couldn't attend SWP meetings regularly—he had to go to work. In late 1939 he had dropped out of high school to take a job in industry, not only to earn a living but to help carry out Trotsky's injunction to organize the proletariat for the coming, inevitable revolution. So early on weekday mornings he commuted from the Bronx apartment, which he still shared with his mother, to Bayonne, New Jersey, where he toiled from eight to five in a foundry.[11]

There work conditions were brutal: the heat was intense and searing, the noise mind-numbing, and the air laden with hazardous substances. Bookchin's job was to pour molten metal from a hot furnace into molds—arduous, punishing work, requiring great physical strength and stamina. The heat scorched his face, and the heavy load strained his five-foot-five frame. During those years, he told me, the work was so hard that he was always underweight. But the SWP exhorted its members not only to perform the work but to excel at it—after all, in order to gain the respect and confidence of their fellow workers, eventually to recruit them, they must be "the best workers on the job."[12] Bookchin would not let the revolution down by slacking.

He did manage to get to SWP meetings sometimes, although not as often as he would have liked. His grueling job left him exhausted at day's end, barely able to stay awake on the train home, let alone sit through a meeting. Often he could muster just enough energy to deliver his mother's daily insulin injection—Rose was by now a diabetic—before collapsing into bed, for a few hours' sleep, then rise in the dark to repeat the ordeal the next day.

* * *

In 1940 the United Electrical Workers, the CIO union for which Bookchin had tried to organize in Jersey City, set up a local in Bayonne, and Bookchin and his fellow foundry workers were proud to join. The local elected him shop steward, which meant that he handled their grievances about pay and working conditions. Most of his fellow workers were African Americans, and he felt a special urgency to defend them against racial discrimination. A black worker named Alec became a particular friend. At one point the company was about to fire Alec, and the white UE president, "very racist," did nothing to prevent it, but Bookchin defended him and saved his job.[13]

Even as he was pouring metal and handling grievances, Bookchin was also trying to enroll new members for the SWP. As a party member himself, he had the revolutionary task of actively propagating Trotskyist ideas to workers—attempting to persuade them to attend SWP lectures and classes and if possible to join. That was no mean feat in the UE, a union whose leadership was militantly Stalinist. Nonetheless, away from the Stalinists' earshot, Bookchin talked up the SWP to any foundry workers who would listen, during lunch hours, after work, and maybe even at grievance meetings.

If they showed any interest, he would patiently explain the basic principles of Marxism. He tried to persuade them to form a study group, where theory, not just practical issues of wages and working conditions, could be discussed. But most of the foundry workers, he found, merely "tolerated" that kind of talk at best, "perhaps out of friendship or perhaps even out of curiosity."[14] He had at least one success: Archie Lieberman, a UE organizer, joined the SWP.[15] But for the most part, the Bayonne foundrymen felt no urgent need to prepare themselves for a revolution or to join the neo-Bolshevik cause.

Back in New York, the Trotskyist intellectuals were arguing over a new development. In November 1939 Stalin had invaded Finland, a small peaceable country that posed no threat to the Soviet Union. Outraged, Max Shachtman, a leading Trotskyist who had cofounded the SWP along with James P. Cannon, condemned the invasion as an act of imperialist aggression. Most of the SWP intellectuals agreed with him, and they certainly expected that Trotsky would concur.

Down in Coyoacán, however, Trotsky heard of their objections—and shot back his disapproval: the invasion, he said, was not imperialistic. The Soviet Union, despite Stalin's atrocities and his hideous regime, was still the home of the Bolshevik Revolution. It was still a socialist state—Lenin's nationalization of industry remained in place. It was therefore by definition incapable of imperialism. And it was therefore by definition superior to bourgeois-democratic Finland. So the invasion was justified—and so was Stalin's invasion of eastern Poland, for that matter. Trotsky insisted that his followers endorse both.

Max Shachtman refused to approve either of those acts of naked aggression. The Soviet Union had nothing at all to do with socialism anymore, he retorted to Trotsky. By no stretch of the imagination was it a workers' state. It was a prison camp, a charnel house, ruled by latter-day tsars, "a modern despotism of immense proportions drenched in blood."[16] It was entitled to no support whatsoever from anyone.

But Trotsky tolerated no deviations in his followers, and in the New York group, it was James P. Cannon who carried out the task of enforcing orthodoxy. At an SWP convention in April 1940, Cannon insisted that the assembled Trotskyists endorse the two Soviet invasions. Disgusted, Shachtman walked out, taking most of the stellar intellectuals along with him. Once outside, they formed a new, separate Trotskyist party, which they called the Workers Party.[17]

Some of the New Yorkers, however, chose to remain in Cannon's SWP. Among them were Al Goldman, Felix Morrow, Jean van Heijenoort, Dave Eisen—and Murray Bookchin. Their reasoning is unknown to me, but surely one purpose for staying would have been to affirm their loyalty to the hero of the Russian Revolution; to this group of friends and comrades, that transcendent fact still meant a great deal.

A few months later, in August 1940, Trotsky was sitting at his desk in Mexico penning a diatribe against the imperialist war, when a Stalinist agent crept into the room and plunged an ice ax into his brain. A few hours later, he died in a hospital. Stalin had finally achieved his goal of killing off the entire Bolshevik revolutionary leadership from 1917.

When Bookchin heard the news of the death of Leon Trotsky, he was undoubtedly heartbroken. The whole Fourth International plunged into deep mourning. But the second imperialist war was underway, so even in its grief the Fourth recommitted itself to carrying out Trotsky's program, indeed with ever more fervor: to transforming the imperialist war into an international socialist revolution.

The Stalinists, for their part, were taking an entirely different position. In June 1941, Hitler tossed aside his pact with Stalin and did what the Soviet dictator had so long feared: he invaded the Soviet Union. Stalin then joined the Allied cause. The Comintern parties, including the CPUSA, were delighted to find themselves finally on the antifascist side of the war, and overnight they became enthusiastic supporters of the Allied war effort, favoring US involvement as well.

In late 1941, as war raged in Europe, Bookchin, age twenty, took the curious step of joining the merchant marine. He had a romantic notion of seeing the world from the prow of a ship, he later told me, like the self-educated socialist Jack London, whom he much admired. But it seems to have been a strange moment for tourism, and I have come to suspect he had another reason as well. Trotsky's old international secretary, Jean van Heijenoort, still carried on an extensive correspondence with Trotskyists around the world. Once the war made international postal delivery impossible, Van recruited comrades to join the merchant marine and form a network of couriers.[18] Perhaps Bookchin, who admired Van, joined in order to participate in that network of mail carriers.

For whatever reason, Bookchin obtained his seaman's certificate on December 3, 1941, and prepared to ship out on a coastal freighter the next day. That night some of his friends took him out drinking—to give him a hearty send-off, they said. The muscatel flowed, and convivial Bookchin polished off glass after glass, while his friends surreptitiously dumped theirs. He got roaring drunk—and the next day, as his friends intended, he was too hung over to report for duty. He missed the boat.

Four days later Imperial Japan bombed Pearl Harbor, and the United States entered World War II. On December 11, Hitler declared war on the United States. "Wolf packs" of Nazi U-boats along the Atlantic seaboard fired torpedoes at any merchant ship whose silhouette appeared, in their periscopes, against the glittering shoreline. Burned-out carcasses of American ships soon littered Atlantic beaches. Had Bookchin gone to sea on December 4, his ship would likely have become one of those easy targets. His friends had undoubtedly saved his life. He put his seaman's certificate in a frame, and for the rest of his days, wherever he lived, it would hang on the wall over his desk, "as a reminder to myself not to do anything stupid."

Once the United States entered the war, most Americans embraced the Allied cause enthusiastically—50 million American men were drafted, with minimal resistance. Among the most ardent supporters of the war were the American Stalinists, who tried to make up in energy what they lacked in numbers. The Stalinist leaders of the UE, for instance, full-throatedly supported the war effort to defend the socialist fatherland.

But the Trotskyists remained firm in their definition of the war as imperialist rather than antifascist, and in their commitment to leading an international proletarian revolution against ruling class in all the major warring states. Trotskyists were not conscientious objectors, opposed to all war as such—on the contrary, Trotsky had encouraged his followers to join their respective armies, to learn to fight in preparation for the revolutionary conflict and to recruit new comrades from among their fellow soldiers. Their position, rather, was to oppose the imperialist war as such (following the "tried and true" program) and refuse to take sides in it.[19] Murray was exempted from the draft because his diabetic mother depended on him for her daily insulin injections, but he made himself useful as a civilian, channeling his energy into instigating the revolution in Bayonne.

Once the United States entered the war, American factories converted quickly to war production and sought to intensify it: they demanded that workers toil for longer hours and faster. But the workers, now unionized in the CIO, were eager to keep the pace of production and working conditions reasonable. Their only muscle was the strike, and in 1941 more American workers went on strike than in any year since 1919. They saw no contradiction between patriotism and an insistence on decent working conditions.

Government and industry fought back, appealing to American workers to patriotically sacrifice their right to strike for the duration of the war. In December 1941 the Roosevelt administration asked them to voluntarily take a "no-strike pledge." The leadership of the CIO unions, who were mostly Stalinists, were delighted to comply and took the pledge—no mere workers' discontents, in their view, should be permitted to obstruct the war effort. And the UE leaders were among the happiest. By 1940–41, the UE was "the main Communist fortress in the labor movement." Its leadership permitted no strikes at all and demanded stepped-up war production from its members.[20]

The Trotskyists, however, had no intention of complying with the no-strike pledge, not by a long shot. Fomenting strikes was basic to their program of instigating and leading proletarian revolution against all ruling classes. So they rejected the pledge, which they considered class collaboration, and continued to promote revolutionary labor militancy.

The UE leaders, annoyed, wanted to tame obstreperous locals like Bayonne and to that end invoked the no-strike pledge, denying the right of the locals to strike. As Murray's fellow Bayonne organizer Archie Lieberman put it, the UE Stalinists were "the worst strikebreakers."[21] The Bayonne rank and file, however, scorned the no-strike pledge and elected Bookchin to serve on the District 4 council to uphold the right to strike. There he clashed with the Stalinists, who had a practice of positioning their comrades strategically at meetings and attempting to "overawe dissenters" like him "with vituperation and character assassination." When that failed to faze Bookchin, they offered to pay him for his heretofore-unpaid work as a shop steward. He declined to be bought off.[22]

Instead, he threw himself into exhorting his fellow foundry workers to mount a political strike against the imperialist war. He talked to them about Communism; they listened and nodded. Then when he paused to take a breath, they told him about their grievances concerning working conditions and wages. They had no interest in overthrowing capitalism—they wanted to strike, but only for immediate practical goals. No matter how much he tried, he couldn't awaken a revolutionary spirit in them. Far from joining the SWP, "they always drifted away. . . . And that was very, very shaking to me."[23]

At least he could find solace at SWP headquarters. Doubtless after a hard day of futile agitating, it felt good to sit down and relax at his political home on University Place.

Sometime in 1941 or 1942, as he sank into a chair, Bookchin might have noticed a new man in the SWP meeting room. Small in stature, he looked a bit like Richard Wagner, an effect enhanced by the heavy German accent and a certain flamboyance. The man talked animatedly to the comrades clustered around him, explaining the political situation in Europe, and fascinating them with tales of his own narrow escapes from the Gestapo.

But probably not even the charismatic German could keep Bookchin's weary eyes open for long. Let's allow him to rest as we catch up with the life of Josef Weber (whose name Murray would always pronounce "VAY-bur," the German way).

Born in 1901 in the industrial Ruhr, young Weber had been an incorrigible rebel who (like Bookchin) dropped out of secondary school but later educated himself, an autodidact. Elated by the Bolshevik Revolution, he had joined the German Communist Party, or KPD ("while Rosa Luxemburg was still alive!" Murray would later marvel). But in the 1920s, he realized that Stalin was creating a new tyranny, and by 1929, when Stalin sent Trotsky into exile from the USSR, Weber had become one of Trotsky's ardent international supporters. Meanwhile he earned his living as a bandmaster, composer, and pianist, harboring a special love for Wagner's operas.

In late February 1933, just after Hitler became chancellor of Germany, the Nazis outlawed leftist parties. Police arrested every KPD member they could find and sent them all to concentration camps. The several hundred German Trotskyists escaped the initial wave of persecution because of their relative obscurity—apparently the Nazis weren't familiar with Trotskyists at first. Some of them bravely tried to organize the proletariat to rise up in revolt against Hitler, handing out flyers near factories.[24] But most German Trotskyists, like Weber himself, fled the country, finding homes in various European cities. Taking the name International Communists of Germany (IKD), the scattered émigrés formed a coordinating committee called the Auslandskomite (AK). "The AK of the IKD" (as Murray would always call it) established a newspaper, *Unser Wort* (Our Word), to report on the German situation to the rest of the world and to appeal for help. Published first in Prague and then in Paris, most issues of *Unser Wort* led off with an article by Trotsky, analyzing developments and prescribing strategies.

In late 1933 Josef Weber decamped to Paris, where he began writing for *Unser Wort* under the cover name "Johre." He quickly rose to become the paper's coeditor. Trotsky, who at that time was living in France, endorsed the publication of *Unser Wort* as "a great achievement" and its strategy as "correct."[25]

From their vantage point in Paris, the IKD émigrés watched in horror as the Nazis terrorized the German proletariat into submission and criminalized labor agitation. The Gestapo arrested anyone who tried to distribute leaflets or, after realizing that Trotskyists were Communists too, *Unser Wort*.[26] Once they decimated the Left, the Nazis proceeded to ban all other non-Nazi political parties, cultural institutions, and social organizations as well.

Only one sector of German society seemed to be putting up a fight against Nazi domination: the churches. As a good Marxist, Josef Weber was astonished and puzzled by the very existence of church resistance, but he also wondered, "Can or must we support it?"[27] In late 1934 he and his fellow German Trotskyist exiles assembled in Switzerland for a meeting, where he told them that since the proletariat had proved unable to fight the Nazis, the IKD must fully support the "only oppositional tendency in Germany that could potentially achieve mass support": the churches. Indeed, he said, the IKD must support any movement that contested the fascist state, whatever it might be.[28]

His comrades' response was outrage. Marxists rejected organized religion, they reminded him. The factory, not the church, was the proper arena for struggle! Some walked out, accusing the IKD and Weber of betraying the very fundamentals of Marxism. But those who remained in the IKD entrusted the leadership of the AK to Weber and his close comrade Otto Schüssler, who now became its central figures.[29]

As a man, it must be said, Weber was notable for his self-confidence, especially when "my mostly unusual opinions . . . proved themselves more and more to be well founded."[30] To his great credit, he was open-minded enough to discard Trotskyist orthodoxy when it no longer corresponded to reality. But much to his frustration, his friends "were by no means always in agreement with my views," and to get his way, he complained, he sometimes had to "drag them along after me." Hence some remembered him as "rigid and intolerant." He was given to supercilious harshness, dismissing those who disagreed with him as "backward elements and newcomers"—their observations were "absurd" and laden with "all the usual petty bourgeois gossip and dirt."[31] That is, he was prone to insult. In the AK of the IKD, vitriolic faction fights became a way of life.

But as he debated or insulted his critics, Weber could always righteously point out that he had the support of the one who mattered most. "I agree, as you may know, with Johre on the disputed questions," Leon Trotsky wrote of him in a 1936 letter.[32] In 1937 Trotsky praised the "really excellent" articles that came "from Johre's pen." Asked to take sides in another quarrel in the IKD, Trotsky said, "I haven't the slightest hesitation. . . . We have to support [Johre's] group with all our might."[33]

Still, the incessant factional fights took their toll on the IKD, and most of its members finally gave up in frustration with this contentious man and departed the group (when they were not tracked down and assassinated by Stalin's henchmen, as Walter Held was). Weber and Schüssler remained, persisting because of "our better policy and understanding," as Weber assured himself. But by the late 1930s, the IKD over which the two men presided was tiny and marginalized, a "more or less phantom organization."[34]

Politics aside, Weber enjoyed living in the City of Light. He savored the cuisine, the art, "operetta, comedy and satire, irony and serene catholic sentimentality, pathos and passion, eroticism and the luxury industries." Nowhere else "has civilization borne a more beautiful flower on its broad vulgar stem" than in France, he remarked.[35] He fell in love with ironic, picaresque French novels, like Diderot's *Jacques le Fataliste*, in which a nervy protagonist exuberantly defies conventional shibboleths with panache. It was, surely, the kind of orthodoxy-smashing role that he pictured himself playing.

In early September 1939, at the outbreak of war, France designated all Germans living within its borders as "enemies of the French state" and detained them in internment camps.[36] Weber was among them and was probably still in such a camp on May 10, 1940, when the Reichswehr smashed through the countryside and reached Paris. On June 14 jackbooted Nazis goose-stepped down the Champs-Élysées. The beautiful flower of civilization capitulated to Hitler's thugs on June 25.

The Nazis ordered the new Vichy regime to hand over all antifascist expatriates living there. Political radicals, labor leaders, artists, authors, musicians, economists, and scientists, most of them Jews, faced the horror of imminent removal by the Gestapo. By the thousands they fled south, bringing only what they could carry. Weber, who recognized the imminent danger from the internment camp, somehow escaped and joined the southward exodus.

Marseilles, on the Mediterranean coast, was the last stop. By the time Weber arrived there, the city was jammed with penniless refugees, desperate to leave France.

(The dissident leftist author Victor Serge mentions seeing "the German emigration" there, which must be a reference to Weber.) Official US policy was to do nothing to help them, lest it rile Hitler, but a rogue official in the US consulate, seeing that the very heart and soul of European culture was urgently in need of rescue, set up a secret operation, issuing exit visas and retrofitting cargo ships to evacuate as many of the refugees as possible.[37] But the ships were too small and few to accommodate all of them. Weber talked his way on to "one of the last boats out of Marseilles" (as Murray would later put it to me).

Perhaps that boat was the old freighter *Capitaine Paul Lemerle*, which left in March, carrying three hundred refugees, forty of them "comrades," in Serge's term.[38] Upon reaching Martinique after a wretched month-long voyage, the refugees were interned in a former leper colony on the island for several months, while the Vichy government tried to figure out what to do with them.

Perhaps as Josef Weber languished in the Martinique leper colony, he brooded on something that one of his heroes, Rosa Luxemburg, had written back in 1915, during the first imperialist war. The alternatives that civilization faced, she had said in the so-called *Junius Pamphlet*, were "socialism or barbarism." If for some reason the international proletariat failed to rise up and create socialism, then capitalism would persist and degrade the world into barbarism. Now, as the Nazis were overrunning Europe, Weber might well have reflected, the world was discovering what barbarism was. It was midnight arrests and internment camps and swastikas and Gestapo, it was prison camps, jackboots, and executions. Luxemburg's words seemed prescient.

Finally, the Vichy regime allowed the refugees to depart from Martinique. And somehow, in late 1941 or early 1942, Josef Weber made his way to New York, where he looked up his American comrades at SWP headquarters.

Upon his arrival at University Place, his fellow Trotskyists welcomed him—he was surely an impressive figure, having eluded the Nazis twice. He would have explained to them that the IKD, which he represented, was "one of the oldest and most stable organizations of the Fourth [International]. . . . Under conditions and difficulties about which [you] do not have the slightest notion, we issued a paper [*Unser Wort*] in the emigration and up to the outbreak of the war, published brochures, books and documents." And he could boast of receiving the ultimate accolade: "Leon Trotsky greatly esteemed our work and never corrected us in a single political question."[39] And he was cultivated and charming to boot.

But as awed as the Americans surely were, they were also hungry for news about the coming proletarian revolution. Where in Europe was the proletariat resisting Hitler? they pressed him. Weber had written—apparently on Martinique—an article on that very subject, he told them, and handed them a manuscript called "Three Theses."

The European workers' movement, the article said, was scarcely breathing. The Nazis had smashed all the labor unions and left-wing parties; they had murdered, imprisoned, and exiled the proletarian leaders; they had prohibited even the expression of revolutionary ideas. As a result, "there is no longer an independent . . . proletarian political or workers' movement" in the Nazi-overrun countries. All that remains "are individuals and weak and uneven groups." Resistance groups do exist, but they are not exclusively workers' groups: "all classes and strata" participate in them: farmers and priests, officers and merchants, students and professors also resist fascism. They are fighting not for socialism but for national liberation and bourgeois democratic government, with freedoms of assembly and expression. Trotskyists, Weber concluded, must not ignore this reality—they must support this all-class, national, prodemocratic struggle, because it is a struggle against fascism.[40]

The Americans reacted to the article "as if they had suddenly been doused with cold water." It called into question "not only the policy and programme of the Fourth International but the validity of Trotskyism itself."[41]

Felix Morrow dared to venture faint praise of "Three Theses," but Jean van Heijenoort overruled him.[42] Europe's national liberation movements are not our potential allies, Van scolded him—they are obstacles to socialism. "The more I read your documents," he berated Weber, "the more I am against them. We will . . . see if we have to part company."[43] After that chastisement, Morrow fell into line.[44] Chairman Cannon, for his part, pronounced the article heresy.[45]

The journal *Fourth International* published Trotskyist condemnations of Weber's ideas. Weber wrote replies to them, but the editors refused to publish them. Only after fourteen months was the original troublesome article, "Three Theses," finally published, in the December 1942 issue; but even then it was accompanied by an official response, authored by Goldman and Morrow, explaining to readers that Weber's article was factually wrong: "The liberation struggle has actually unfolded under the leadership of workers' organizations and workers' groups" and was determined to achieve socialism. Morrow and Goldman pronounced it "embarrassing" to have to explain to the German comrade "the ABC's of Marxism," and they closed the piece with a clincher—some quotes from Trotsky.[46]

These indignant regurgitations of orthodoxy must have grated on Weber. After all, he was in the right: the resistance movements in occupied countries definitely comprised people from across the political spectrum. They derailed trains, bombed tracks, and blew up ammunition depots in a struggle not for socialism but for national liberation and democracy.

Van, however, remained intransigent. "The senior schoolmaster," Weber mused, "rejects the whole notion of a 'democratic revolution.'" But Weber did praise Morrow and Goldman for writing the reply—"They at least honestly wanted to discuss."[47] And his challenge to orthodoxy seems to have shaken them up, for doubts were beginning to percolate in their minds.

If Bookchin missed a lot of SWP meetings in 1943–44, he may be forgiven. At the Bayonne foundries, grievances were simmering along with the molten metal.

The union leadership, as we have seen, had pledged not to strike. But as the war went on, corporate profits and executive salaries were soaring, while workers' wages did not even keep up with the cost of living. Then in 1943 wages were frozen, even as industry sped up assembly lines and demanded longer work hours.[48] Workers' grievances accumulated, but the no-strike pledge crippled that one muscle that the unions could flex in opposition.

To press their issues, the workers were limited to bargaining with management, through shop stewards like Bookchin. Shop stewards during the war "were much more militant and aggressive than the national [union] leaders," according to the UAW, and were "truly more representative of the workers than the union officials."[49] In March 1943 a UE local was battling management in a foundry in Bayonne, the Bergen Point Iron Works (which may or may not have been the foundry where Bookchin worked), over issues of "wage adjustment, union security, checkoff, management rights, seniority, vacation, dismissal, arbitration, holidays, sick leave, transfer, and grievance procedures."[50]

But business and government had no tolerance for worker militancy. A rebellious worker would be warned "to find another job or relax his militancy for a while," Bookchin later recalled. Those who failed to heed the warning could be fired. Rebellious shop stewards, for their part, would be "called into 'personnel' offices or receive visits from the cops."[51]

When shop steward bargaining broke down, workers mounted "wildcat" strikes (strikes without union authorization). Beginning in the spring of 1943, a wave of wildcats swept through heavy industry. Was the long-awaited revolution arriving at last? No—most of the strikes were of short duration, usually a few days.[52] The wildcats led to mediation, but not to revolution.

In New York, SWP members' hopes rose again in the fall of 1943, when they learned that Italian industrial workers were striking at important factories in Milan and Turin and forming workers' councils—that is, soviets. Trotskyists rejoiced at "the first day of the proletarian revolution in Italy, the first day of the coming European revolution." But Weber calmly assured them that the Italian strikes meant nothing of the sort.[53]

Nonetheless, in October, an SWP party plenum, infused with orthodoxy, saluted the Italian workers for demonstrating that "the workers in alliance with the peasants and colonial peoples will prove capable of overthrowing capitalism."[54] Weber had wanted to attend the plenum, but Van had slyly tricked him into staying away. Outraged that he'd been excluded, Weber rebuked Van that one would have to be "blind" not to see that "the broad masses of Europe are 'national' in . . . their demand for independence."

Cannon suggested that "the German emigration" had a "certain psychology" and was "a little bit screwy."

Weber shot back that the SWP had failed to conduct the kind of open discussion that could "make possible a correct orientation for the international movement." If the Fourth had followed his advice, early on, and supported national liberation, it could have placed itself "at the head of the movement at least propagandistically and agitationally" and might have won "a substantial influence upon the consciousness of the masses." But instead it had let itself be blinded by orthodoxy. Gallingly, the

Stalinist parties had caught on to reality and were now playing a prominent role in resistance movements—and were winning prestige as a result. *That could have been us*, Weber must have seethed inwardly.

By now, Goldman and Morrow were conceding that Weber had a point. Even stalwart Van finally climbed down and admitted that the French resistance included not only workers but "large strata of the petty bourgeoisie," and that its immediate objective was not a socialist society but "the overthrowing of the German yoke," while its broader aims were "democratic and patriotic."

Seeing that he was making progress, Weber recounted the International's mistreatment of himself as well. "Do you believe," he asked them, "that the best way of promoting the European revolution" consists of "gagging and discrediting" European exiles? Who did they want to make the revolution with, "if not those rare specimens who have survived the European catastrophe physically and politically?" He had them.

Goldman and Morrow admitted their mistake. As for Van, he was becoming "a sort of 'champion' of the national question"—but alas without, Weber complained, crediting himself.[55]

In November 1943, Morrow, Goldman, and Van, joined by Bookchin and Dave Eisen and several others, took the daring step of forming a dissident faction within the SWP, called the SWP Minority (also known to history as the Goldman-Morrow faction). Its members were those who agreed with Weber's analysis.

Ever since Max Shachtman's departure in 1940, SWP chairman James Cannon had dreaded the emergence of another faction—he wanted no more splits. Yet now he had one on his hands. Furious, he insisted that SWP members must maintain an undeviating commitment to orthodoxy—that is, to Trotsky's 1938 Transitional Program. Nothing therein was to be changed, not a comma.

Al Goldman complained that Cannon preferred a homogeneous "monolith" to a revolutionary party. During the war years, he said, "anyone who presented any new idea"—like Weber—"was looked upon as a disturber of the peace."[56] Bookchin agreed that Cannon was authoritarian: when it came to dissident views, he had discovered to his chagrin, the Trotskyists "were no different from the Stalinists."[57]

By 1944, Bookchin had toiled for five years in the hellish foundry. He'd done his best to organize his fellow workers. He'd handled their grievances and fought the UE's Stalinist leadership on their behalf. He'd tried to teach them Marxism.[58] But they were no closer to becoming a revolutionary proletariat than when he started. So he hung up his apron and goggles and left the foundry.

The auto industry seemed to hold more promise for a revolutionary agenda—its proletariat was feisty and assertive, and the United Auto Workers (UAW) was the country's most militant union, a spearhead of labor activism. It was not controlled by Stalinists. At that moment it was waging a bitter struggle at General Motors, "the most hard-bitten and reactionary corporation in the world," as Bookchin recalled. Since the war began, GM's corporate profits had doubled and executive salaries had skyrocketed; meanwhile, the workers were being pushed to work faster and longer for the same pay. The UAW believed that management was taking advantage of the workers' no-strike pledge. So the rank and file revolted, more persistently and broadly than in any other industry, in wildcat strikes.[59]

After the frustrations of Bayonne, the class conflict raging at GM was doubtless irresistible to Bookchin. So he took a job in a GM machine shop on Eleventh Avenue in Manhattan[60]—and got a UAW card.

The shorter commute and the less onerous work meant that he could spend more time at University Place talking politics with his SWP Minority friends. They expressed their perplexity that no proletarian revolution was in the offing. The war was turning in the Allies' favor, and it was clearly only a matter of time before they—not the working class—defeated Hitler. What were they, as Trotskyists, to make of this unexpected turn of events?

To address this question, Josef Weber, who had in the past been right about so many things, came out with a new article, long and densely theoretical, called "Capitalist Barbarism or Socialism." Published in October 1944 in Max Shachtman's magazine, its title (surely not by coincidence) alluded to that memorable formulation by Rosa Luxemburg: "socialism or barbarism." In 1915 Luxemburg (figure 2.1) had written that the world faced a choice between "the victory of socialism" by the international proletariat, or "the triumph of imperialism and the collapse of all civilization."[61]

FIGURE 2.1 Rosa Luxemburg, whose maxim "socialism or barbarism" was a touchstone for both Weber and Bookchin

In the absence of a revolutionary transition to socialism, Weber asserted in his new article, the postwar world would indeed revert to barbarism. Once the Allies defeated the Axis, the victorious imperialists would dismantle Germany's and Japan's economic and political institutions and drive both countries back to precapitalist levels, reducing them to agricultural hinterlands. Then Russia and especially the United States would strip Germans and Japanese of their human rights and resettle them in prisons and ghettos and concentration camps, transforming them into a slave labor force. The imperialist overseers would do the same to every other country that they wished to exploit. Barbarism, in other words, would look very much like the Third Reich, its labor and population policies extended as a "world phenomenon." Weber called this vision of decline the "theory of retrogressive movement," or the retrogression thesis.[62]

The article's bombastic prose style, ex cathedra assertions, and esoteric Marxist jargon make for tough reading today. But in 1944 Bookchin found it electrifying. Eager to talk to its author, he made his way to Weber's Bronx apartment.

Surely Weber (age forty-three) was delighted to see the young proletarian intellectual (age twenty-three) standing in the doorway before him. Perhaps as they sat down together, the young man told him about the militant but entirely nonrevolutionary proletariat he had encountered in Bayonne. Surely Weber detailed his shabby treatment at the hands of the orthodox Trotskyists. Perhaps he explained to Bookchin about Luxemburg's "socialism or barbarism" formulation. (*I was in the KPD when she was still alive!*)

The soon-to-be-victorious Allies were already scheming to "retrogress" Germany and Japan after the war, he must have explained to Bookchin. Only a month before, at the Quebec Conference of September 16, Roosevelt and Churchill had agreed to a plan, developed by Treasury secretary Hans Morgenthau, to eliminate "the war-making industries in the Ruhr and in the Saar" and to convert Germany "into a country primarily agricultural and pastoral in its character."[63] Stalin, too, was plotting a postwar demolition of Germany's industrial capacity, and he intended to demand huge war reparations, which millions of Germans would have to toil in slave labor for more than a decade to pay.[64]

That was how monopoly capital worked, Weber might have explained: the United States, by eliminating its major capitalist rival—Germany—would thereby extend the reach of its own imperialist system, colonizing Germans, making them dependent on itself for manufactured goods.

In 1944, it should be remembered, no one had any idea what would happen after the Allied victory; for all anyone knew, the world would sink back into economic depression. Weber's grim scenario would have seemed all too plausible to Bookchin: in his own short lifetime, he had already experienced homelessness and dislocation, tribunals and expulsions, antiunion goon squads, and treachery. Indeed, Weber's retrogression thesis seemed to him a stroke of genius, on a par with Trotsky's 1938 Transitional Program. Weber himself seemed a likely candidate for Trotsky's successor—perhaps he would even become the next important Bolshevik leader.

A few months after Weber and Bookchin's first conversation, the Allies took Berlin, and in August 1945 the Japanese surrendered to General MacArthur. No proletariat in any nation rose up to break its shackles and usher in the socialist order. On the contrary, in the capitalist countries, the workers had ardently supported their respective national war efforts. The Germans had fought to defend Hitler "all the way to the bunker" (as he put it decades later, still amazed by the fact). And the Russians, perhaps terrorized into submission (or perhaps not), had remained quiet during what Stalin called the "great patriotic war." The unshakable laws of history that Trotsky had invoked to predict that the war would end in socialist revolution had turned out to be nothing more than his own wishful thinking.

To be sure, just after V-J day, as soon as the no-strike pledge passed into history, the American working class roared to life. Industrial workers from coast to coast went out on strike, demanding full employment and wage increases. By October 1945 the strike wave was gigantic: 43,000 oil workers, 200,000 coal miners, 44,000 lumber workers, 70,000 truck drivers, and 40,000 machinists all laid down their tools. In November, 300,000 UAW workers—among them Bookchin—struck at General Motors. Within twelve months of V-J day, more than 5 million American workers had gone on strike (although not all at the same time). It was the largest strike the country had ever seen. But after 113 days, the UAW, having exhausted its limited strike fund, ended its strike, accepting a small wage increase and some fringe benefits and contract changes.[65] Workers in the other industries soon returned to work as well, having achieved modest gains. All in all, the captains of industry had prevailed.

Then in August, amid the chaos of demobilization, Bookchin was finally drafted into the US Army. For years he had had deferments for taking care of his mother; it's unclear to me what changed in 1946. In any case, he didn't mind being drafted: "I was still a Bolshevik. I believed that we should be trained for armed insurrection," and besides, the army was full of workers. Far from being a conscientious objector, he said, "I was a conscientious soldier."[66]

Private Bookchin, a member of Company A of the 526th Armored Infantry Battalion, was stationed at Fort Knox, Kentucky, the army's center for mechanized cavalry.[67] Trained as a tanker, he and his unit practiced maneuvers. Bombs blew up around foxholes, and "friendly" bullets whizzed by. He also served in a more familiar role, as an auto parts clerk for the motor pool.[68]

While he was away, his friends in New York kept him up to date, by letters, on political developments. At a November 1946 convention, the SWP had charged Felix Morrow and Dave Eisen with disloyalty, for the sin of questioning Trotskyist orthodoxy and for forming the Minority faction. The SWP-ers in attendance had vilified them. Chairman Cannon had refused to allow Morrow and Eisen even to respond to the charges, and the convention's vote to expel them had been almost unanimous.

Bookchin saw clearly that when he got back to New York, he would not be returning to the SWP.[69]

In early 1947 the army terminated the wartime draft in favor of an all-volunteer army, and released all postwar draftees. On June 14, after ten months of service, Bookchin was honorably discharged. He returned to the apartment he shared with his mother, at 710 East 175th Street, and to work at GM. The UAW had participated in the huge strike wave before he was drafted; likely Bookchin expected the UAW to soon mount another strike. But in 1947 the company was rethinking its way of handling its 400,000 blue-collar workers and decided on a new strategy: it would co-opt them. In the spring of 1948, GM offered the UAW a contract with a guaranteed annual wage, benefits for sick leave, health insurance, and vacations, as well as improved working conditions. In exchange, the UAW was to guarantee that its members would not strike for two to four years. The UAW accepted the contract.

Many of Bookchin's fellow workers surely favored the deal, for all the benefits it would provide to them. But Bookchin saw it as a decisive collapse of revolutionary will: a truly revolutionary union would have rejected the contract and forced annual wage negotiations. But the UAW not only accepted it, it went on to phase out union shop stewards and replace them with company-employed grievance men. Worse, it allowed the company to buy off the heads of some UAW locals by paying their salaries. These union leaders would actually become company employees.

The GM-UAW settlement of 1948 demonstrated to Bookchin once and for all that industrial proletariat was not revolutionary. Industrial workers might sometimes be class conscious and even militant, but only in an effort to improve their lot in the existing system, not to change it. Now, as workers removed their union buttons and eyed homes in suburbia, they were identifying not with their class but with the very company that exploited their labor. The change was of epochal proportions, Bookchin thought: the proletariat was being brought "into complicity with capitalism."[70]

Having been a Marxist since the age of nine, the realization came as a shock. For if the proletariat was not revolutionary, then proletarian socialism was an illusion, and Marxism was based on a fallacy. Bookchin left General Motors, dazed and uprooted.

Drawing twenty dollars a week as a veteran, he had the liberty to ponder the dizzying change as he drifted around New York. From Union Square, he could see the *Daily Worker* building, where he'd spent years studying Marxism, and the Trotskyist headquarters on University Place, but both were now out of bounds. Fortunately, he was not alone in his political dislocation. Other refugees from Marxist movements were in the same plight—they were facing the failure of Marxism, just as western Europe appeared to be collapsing economically. To commiserate and, as always, to discuss what to do and where to go next, they congregated in the low-priced restaurants and cafeterias of Fourteenth Street.

Here at Child's or Schrafft's, you could get a blue plate special for two bits. Even better was Horn & Hardart, the automat, where individual servings of macaroni and cheese, or Salisbury steak, or pie waited behind little glass windows. You dropped coins into a slot, and the food came out hot, on real dishes. Another nickel got you a cup of coffee. Here in the cafeterias Bookchin could sit down at a table with other lost souls and talk freely, as they could no longer do at the old headquarters nearby.

Here Bookchin met Dwight Macdonald, who was now editing an independent Left magazine called *Politics*. "The validity of Marxism as a political doctrine," Macdonald might have said at his table, "stands or falls on its assertion that the proletariat is the historical force which will bring about socialism." Since the proletariat had not fulfilled this role, "the rock of Historical Process on which Marx built his house has turned out to be sand."[71]

Macdonald was rejecting Marxism, but Felix Morrow, Bookchin learned, had given up on radical politics altogether—he'd taken a regular job in publishing.[72] Even Jean van Heijenoort, Trotsky's old secretary, threw in the towel in 1948. It had been one hundred years since Marx and Engels wrote *The Communist Manifesto*, Van announced, but the movement had nothing to show for it. He ceased political activity altogether and returned to his first love, mathematics, eventually becoming a respected professor.[73]

Bookchin must have been relieved to sit down with his old friend Dave Eisen. Neither of them wanted to follow in Macdonald's or Morrow's or Van's footsteps and leave revolutionary politics. They surely discussed the latest brainstorm from Josef Weber—he had suggested that all the old SWP Minority members leave the SWP and form a new, independent group with him. They could start a new magazine, to figure out the new direction collectively.

Weber had been right many times, Bookchin knew, and now he was turning out to be right about retrogression. Newsreels were showing skeletal bodies of Jews in concentration camps, stacks of unburied dead, gas chambers, still-smoldering crematoria. Bookchin had been reading the transcripts of the Nuremberg trials—the forced labor, starvation, tortures, and enslavement. The Stalinist massacres at Katyn forest and Babi Yar, the US bombing of Hiroshima and Nagasaki—it was all a descent into barbarism. It was retrogression.

Weber himself didn't care for the cafeterias of Fourteenth Street. He had showed up there once and sat down to a steaming bowl of soup—only to find a cockroach amid the boiled noodles. He'd flown into a rage, calling the cafeterias "hellish inventions," and had stormed out, never to return. "A certain percentage . . . of all food eaten by New Yorkers," he judged, "consists of roaches or their excrement."[74] New York, it seems, wasn't Paris. So Murray and Dave would have to seek him out on his own turf.[75]

When they arrived at Weber's Bronx apartment, his new wife, Millie, might have opened the door to them. He had recently married her, resolving his immigration status.

They all sat down amid the German books and papers and surely talked about their great shared dilemma. "Everyone understands 'something is wrong,'" Weber might have said. "That 'something' is the failure of the socialist movement to lead society under conditions most favorable to it: war and its aftermath." And Marxism—as "the theory and praxis of the 'proletarian revolution' and the 'dictatorship of the proletariat'"—was dead, "absolutely dead."[76] On that they all agreed.

But socialism? To give up on socialism would be to accept barbarism. And as a devotee of Rosa Luxemburg, Weber could never do that.

In 1939, Weber might have continued, Trotsky, just before he was killed, had said something very important. He had said that if somehow the unthinkable should happen and the second imperialist war ended without a revolution, "then we should doubtless have to pose the question of revising our conception of the present epoch and its driving forces."[77] In other words, Trotsky had said, if the revolution didn't come, we would have to rethink the socialist project.

Rethink socialism with me, Weber might have said. *So we don't end up with the barbarism of retrogression.*

He'd been circulating his 1944 article-manifesto, "Capitalist Barbarism or Socialism," to his friends in Germany. Those who agreed with the retrogression thesis were starting a new magazine called *Dinge der Zeit*. The first issue had just been published—dated June 1947, the month of Bookchin's discharge. Weber's plan was for the group to publish an English-language sister edition, to be called *Contemporary Issues*. The new journal wouldn't be like the old Trotskyist organs, suppressing discussion, insisting on one party line. *Contemporary Issues* would have no "dictatorship of the editorial office."[78] Instead, it would let ideas and theory develop from facts. It would actively encourage readers to participate, to raise objections and doubts, to challenge its editors and writers. No topic would be off limits.

Murray and Dave, in the end, agreed to work with Weber and his international comrades. Together they would choose socialism over barbarism.

NOTES

1. Leon Trotsky, "The USSR in War" (Sept. 25, 1939), in *In Defense of Marxism: The Social and Political Contradictions of the Soviet Union* (1942; New York: Pathfinder Press, 1973), 50; Leon Trotsky, *Writings, 1938–39*, ed. Naomi Allen and George Breitman (New York: Pathfinder, 1974), 87, 232.

2. Bookchin, "A Marxist Revolutionary Youth," interview by Janet Biehl, in Bookchin, *Anarchism, Marxism and the Future of the Left* (San Francisco: A.K. Press, 1999), 44.

3. James P. Cannon, *Socialism on Trial: The Official Court Record of James P. Cannon's Testimony in the Famous Minneapolis "Sedition" Trial*, 4th ed. (New York: Pioneer Publishers, 1965), 69, 68.

4. William L. O'Neill, *The Last Romantic: A Life of Max Eastman* (1978; reprint New Brunswick, NJ: Transaction, 1991), 191–92. Among their European comrades and sympathizers were Victor Serge, a onetime Bolshevik who had been imprisoned by Stalin and now lived in Paris; the Ukrainian Boris Souvarine, expelled by the Comintern in 1924, and author of an important critical biography of Stalin; and C. L. R. James, author of the seminal 1938 work on Toussaint L'Ouverture, *Black Jacobins*. In the 1990s, Murray spoke to me of all of them with admiration.

5. Anita Burdman Feferman, *Politics, Logic, and Love: The Life of Jean van Heijenoort* (Wellesley, MA: A.K. Peters, 1993), 116.

6. Albert Goldman, "Trotsky's Message—Socialism Is the Only Road for Humanity. Extracts from Albert Goldman's Speech at the New York Trotsky Memorial Meeting, Aug. 21, 1942," *Militant*, Aug. 27, 1942.

7. George Lichtheim, *A Short History of Socialism* (Glasgow: Fontana Collins, 1975), 282.

8. Quoted in Leszek Kolakowski, *Main Currents of Marxism*, vol. 3, *The Breakdown*, trans. P. S. Falla (New York: Oxford University Press, 1978), 195.

9. Quoted in Michael Wreszin, *A Rebel in Defense of Tradition: The Life and Politics of Dwight Macdonald* (New York: Basic Books, 1994), 83.

10. Sidney Lens, *Unrepentant Radical* (Boston: Beacon Press, 1980), 40–41.

11. Bayonne foundries in 1939–43 included the Bayonne Steel Casting Company, the Bergen Point Iron Works, the Bergen Point Brass Foundry, and Babcock and Wilcox. I do not know which of them, if any, was Murray's employer.

12. Cannon, *Socialism on Trial*, 37, 38–39.

13. MBVB, part 13.

14. Ibid.

15. Archie Lieberman, "The Lessons of Working Class History," *Against the Current*, no. 57 (July 1995), 42, and *Against the Current*, no. 59 (Nov.–Dec. 1995); David Finkel, "Remembering Archie Lieberman," *Against the Current*, no. 103 (Mar.–Apr. 2003), 41.

16. Quoted in Kevin Coogan, introduction to Dwight Macdonald, *The Root Is Man* (1946; New York: Autonomedia, 1995).

17. On the Shachtman-Cannon schism, see Alan M. Wald, *The New York Intellectuals: The Rise and Decline of the Anti-Stalinist Left from the 1930s to the 1980s* (Chapel Hill: University of North Carolina Press, 1987), chap. 6.

18. Murray never mentioned this reason to me, but after his death, when I found out about the network of couriers, it seemed to me entirely plausible. Feferman, *Politics, Logic, and Love*, 189.

19. Socialist Workers Party, "Resolution on Proletarian Military Policy," Sept. 1940, quoted in Robert J. Alexander, *International Trotskyism, 1929–1985: A Documented Analysis of the Movement* (Durham, NC: Duke University Press, 1991), 815.

20. Bert Cochran, *Labor and Communism: The Conflict That Shaped American Unions* (Princeton, NJ: Princeton University Press, 1977), 279; "Bayonne Locals Push on Action to Win the War," *UE News*, Dec. 20, 1941; "400 Delegates of 158 Shops Discuss Problems of War," *UE News*, Jan. 31, 1942.

21. Quoted in Finkel, "Remembering Archie Lieberman."

22. Cochran, *Labor and Communism*, 285–86; Bookchin, interview by Doug Richardson, unpublished ms., 1973, MBPTL, copy in author's collection.

23. MBVB, part 13.

24. Margaret Dewar, *The Quiet Revolutionary* (Chicago: Bookmarks, 1989), 163.

25. Weber's fellow editors were Otto Schüssler (cover name Oskar Fischer) and Heinz Epe (cover name Walter Held). Trotsky's praise is in "A Real Achievement," Jan. 24, 1934, in Leon Trotsky, *Writings, 1933–34*, ed. George Breitman (New York: Pathfinder Press, 1972), 228–30.

26. Dewar, *Quiet Revolutionary*, 168; Evelyn Lend, *The Underground Struggle in Germany* (New York: League for Industrial Democracy, 1958), 45.

27. Josef Weber (as Johre), "Reale Vorgänge und Lehren," *Unser Wort*, no. 20 (June 1934), 2.

28. Josef Weber (as Lux Adorno), "The Kirchenkampf and the Politics of the Working Class," *Unser Wort*, no. 58 (June 1935). For this important article, Weber adopted a pseudonym from Theodor Adorno, whom he greatly admired. See also Wolfgang Alles, "German Trotskyism in the 1930s," *Revolutionary History* 2, no. 3 (Autumn 1989), 7, online at http://bit.ly/oiypya; and "Unsere Reichskonferenz: Ein Überblick," *Unser Wort*, no. 54 (Mar. 1935). The 1934 conference was in Dietikon, Switzerland.

29. Dewar, *Quiet Revolutionary*, 158; Alexander, *International Trotskyism, 1929–1985*, 422, 420.

30. Josef Weber (as Ernst Zander), "Some Comments on the Organizational Question" (written Jan. 1, 1951), *Contemporary Issues* 11, no. 44 (Sept.–Oct. 1962), 267.

31. Dewar, *Quiet Revolutionary*, 196; Weber (as Zander), "Some Comments," 268–69.

32. Leon Trotsky, "Results of the Open Letter" (written Jan. 18, 1936), in George Breitman, ed., *Writings of Leon Trotsky, Supplement (1934–40)* (New York: Pathfinder Press, 1979), 637. See also Siegfried Kissin, "My Political Experiences in the Trotskyist Movement," ed. Ted Crawford, in *Revolutionary History* 13, no. 1 (n.d.), http://bit.ly/pdk-GcW. See also Pierre Broué, "Otto Schüssler: A Biographical Sketch," trans. Ted Crawford, in *Cahiers Leon Trotsky*, no. 1 (Jan. 1979); http://bit.ly/pRYYRn; and "Franz Meyer," http://bit.ly/ukVNDv.

33. Leon Trotsky, "The International Conference Must Be Postponed" (written Sept. 26, 1937), in Breitman, *Writings of Leon Trotsky, Supplement (1934–40)*, 742.

34. Pierre Broué, "Walter Held," trans. Ted Crawford, in *Cahiers Leon Trotsky*, no. 1 (Jan. 1979), http://bit.ly/16udIgI; Weber (as Zander), "Some Comments," 268–69; Broué, "Otto Schüssler."

35. Josef Weber (as International Communists of Germany, IKD), "Capitalist Barbarism or Socialism: On the Development of Declining Capitalism, and on the Situation, Tasks and Perspectives of the Labor Movement," *New International*, supplement (October 1944), 349.

36. Ibid., 350 n. 18. Weber was sent to a camp in Loir-et-Cher—perhaps it was Marolles. See also "Conditions in the French Detention and Internment Camps," July 31, 1941, *Jewish Virtual Library*, http://bit.ly/tTcpKH.

37. Sheila Isenberg, *A Hero of Our Own: The Story of Varian Fry* (New York: Random House, 2001), 3; Victor Serge, *Memoirs of a Revolutionary: 1901–1941*, trans. Peter Sedgwick (London: Oxford University Press, 1963), 362. The official was Hiram Bingham. See Isenberg, *Hero*; and Andy Marino, *A Quiet American: The Secret War of Varian Fry* (New York: Macmillan, 2000). See also Eric Jennings, "Last Exist from Vichy France: The Martinique Escape Route and the Ambiguities of Emigration," *Journal of Modern History* 74, no. 2 (June 2002), 289–324. Murray told me that Weber told him that he (Weber) left Marseilles on "one of the last boats," that it had something to do with the State Department, and that the boat stopped in Martinique.

38. Serge, *Memoirs*, 366.

39. Josef Weber (as IKD), "The SWP and the European Revolutions," *New International*, Dec. 1944, 414. In this period Weber signed his writings as "IKD" or "the AK of the IKD."

40. Josef Weber (as International Communists of Germany, IKD), "Three Theses on the European Situation and the Political Tasks" (written Oct. 19, 1941), *Fourth International* 3, no. 10 (Dec. 1942), 370–72; online at http://bit.ly/o8Q41o, 3–5. Perhaps he wrote it in Martinique.

41. Weber (as IKD), "SWP and European Revolutions," 412; Sam Levy, "The Proletarian Military Policy Revisited," *Revolutionary History* 1, no. 3 (Autumn 1988).

42. Rodolphe Prager, "The Fourth International during the Second World War," *Revolutionary History* 1, no. 3 (Autumn 1988).

43. Daniel Logan, quoted in Weber (as IKD), "SWP and European Revolutions," 413; "Daniel Logan" is a pseudonym for Jean van Heijenoort; see *Lubitz' TrotskyanaNet*, http://bit.ly/vP1DD7.

44. Prager, "Fourth International."

45. Albert Goldman, *The Question of Unity between the Workers Party and the Socialist Workers Party* (Long Island City, NY: Workers Party Press, Jan. 1947).

46. Felix Morrow [and Albert Goldman], "Our Differences with the 'Three Theses,'" *Fourth International* 3, no. 10 (Dec. 1942), 372–74, online at http://bit.ly/sH4y2s. The only byline was Morrow, but in "Capitalist Barbarism" Weber says the piece was written by "Morrow and Morrison," Morrison being a pseudonym for Goldman, per *Lubitz' TrotskyanaNet*, http://bit.ly/vP1DD7. See also Weber (as IKD), "SWP and European Revolutions," 414.

47. Weber (as IKD), "SWP and European Revolutions," 415.

48. On wartime labor struggles, see James B. Atleson, *Labor and the Wartime State* (Urbana: University of Illinois Press, 1998); and Nelson Lichtenstein, *Labor's War at Home* (New York: Cambridge University Press, 1982).

49. UAW shop steward handbook, quoted in Irving Howe and B. J. Widick, *The UAW and Walter Reuther* (New York: Random House, 1949), 119.

50. *Bergen Point Iron Works and United Electrical, Radio, & Machine Workers of America, Affiliated with the Congress of Industrial Organizations*, 539 (1943), online at http://bit.ly/pM8MoM.

51. Bookchin (as Harry Ludd), "The Fate of American Civil Liberties," *Contemporary Issues* 4, no. 16 (Nov.–Dec. 1953), 234.

52. "By the end of 1945, 3.5 million workers had engaged in 4,750 work stoppages, costing employers 38 million workdays." Atleson, *Wartime State*, 132, 141–42.

53. "Italian Workers Elect Own Factory Committees," *Militant*, Sept. 21, 1943; quotation in Peter Jenkins, *Where Trotskyism Got Lost: World War Two and the Prospect for Revolution in Europe* (Nottingham, UK: Spokesman Books, 1977), online at http://bit.ly/r9YB2G; Weber (as IKD), "SWP and European Revolutions," 413–14.

54. SWP, "Perspectives and Tasks of the Coming European Revolution," Resolution Adopted by the Fifteenth Anniversary Plenum of the Socialist Workers Party, Nov. 2, 1943, *Fourth International* 4, no. 11 (Dec. 1943), 329–34.

55. For the story recounted in the preceding paragraphs, see Weber (as IKD), "SWP and European Revolutions," 412–14. There Weber refers to "Daniel Logan," a pseudonym for Jean van Heijenoort.

56. Goldman, *Question of Unity*. On the Goldman-Morrow faction, see Wald, *New York Intellectuals*, 253–57.

57. Bookchin, interview by Jeff Riggenbach, *Reason*, Oct. 1979, 37.

58. Lieberman, "Lessons of Working-Class History," 42; Finkel, "Remembering Archie Lieberman."

59. MBVB, part 21. Kevin Boyle, "Autoworkers at War: Patriotism and Protest in the American Automobile Industry, 1939–1945," in Robert Asher and Ronald Edsforth, eds., *Autowork* (Albany: SUNY Press, 1955), 118–19; George Lipsitz, *Rainbow at Midnight: Labor and Culture in the 1940s* (Urbana: University of Illinois Press, 1994); and Atleson, *Wartime State*, 144.

60. The shop is described in Christopher Gray, "The Car Is Still King on 11th Avenue," *New York Times*, July 9, 2006.

61. Weber (as IKD), "Capitalist Barbarism," 349. For Luxemburg's "socialism or barbarism" formulation, see her *Junius Pamphlet* (1915), chap. 1, online at http://bit.ly/njEQXV. The idea of a choice between "socialism or barbarism" actually dates back to Marx and Engels. In the *Communist Manifesto* they wrote that the class struggle would end "either in the revolutionary reconstitution of society at large, or in the common ruin of the contending classes." Karl Marx and Friedrich Engels, *Manifesto of the Communist Party*, chap. 1, online at http://bit.ly/qrTcnx.

62. Weber (as IKD), "Capitalist Barbarism," 331, 333–34.

63. Robert Dallek, *Franklin D. Roosevelt and American Foreign Policy, 1932–1945* (Oxford: Oxford University Press, 1995), 475.

64. "Everything that Germany possesses 'above the minimum necessary to survive,' has to contribute to the reparations fund for compensating the allied nations," read one Russian planning report, issued in July 1944. Quoted in Robert Gellately, *Stalin's Curse* (New York: Knopf, 2013).

65. Jeremy Brecher, *Strike!* (Boston: South End Press, 1977), 228–30; John Barnard, *American Vanguard: The UAW during the Reuther Years, 1935–70* (Detroit: Wayne State University Press, 2005), 215; Lipsitz, *Rainbow at Midnight*, 108; James Matles and James Higgins, *Them and Us: Struggles of a Rank and File Union* (Englewood Cliffs, NJ: Prentice-Hall, 1974), 141–42, 146; and Art Preis, *Labor's Giant Step: Twenty Years of the CIO* (New York: Pioneer, 1964).

66. MBVB, part 14.

67. John E. Kleber, ed., *The Kentucky Encyclopedia* (Lexington: University Press of Kentucky, 1992), 346.

68. MBVB, part 14; honorable discharge, copy in author's collection.

69. Dave Eisen to Leon Brownstein, Nov. 2, 1946; Eisen to Brownstein, Dec. 5, 1946; IKD Faction to WP, "High Road or No Road"; Eisen to Barney Cohen, October 22, 1946, all courtesy Dave Eisen.

70. Bookchin, "Postwar Period," interview by Doug Morris, in Bookchin, *Anarchism, Marxism*, 47–48; MBVB, part 31. In the 1990s Murray sometimes referred to a great strike of 1948, but his memory played tricks: the great strike took place in 1946, and the historic compact was in 1948.

71. Dwight Macdonald, "The Root Is Man" (1946), reprinted in *The Memoirs of a Revolutionist: Essays in Political Criticism* (New York: Farrar, Straus & Cudahy, 1957), 267.

72. Wald, *New York Intellectuals*, 286–89.

73. Feferman, *Politics, Logic, and Love*, 215–17.

74. Josef Weber (as Erik Erikson), "Critical Review (New York eriksonized)," *Contemporary Issues*, no. 7 (Autumn 1950), 229–35.

75. Weber and his wife lived in an apartment south of Fordham Road near the river; Jack Grossman told me it might have been on Jerome Avenue.

76. Josef Weber (as IKD Faction) to the WP, "The High Road or No Road," written Apr. 18, 1947, *New International* (Aug. 1947); Josef Weber (as Wilhelm Lunen), "The Problem of Social Consciousness in Our Time," *Contemporary Issues* 8, no. 31 (Oct.–Nov. 1957), 505.

77. Leon Trotsky, "The USSR in War" (Sept. 25, 1939), in *In Defense of Marxism: The Social and Political Contradictions of the Soviet Union* (1942; New York: Pathfinder Press, 1973), 50.

Weber wrote about this passage in 1947 in IKD Faction to WP, "High Road or No Road." Curiously, Macdonald invoked the same quote in his 1946 essay "The Root Is Man," in Macdonald, *The Root Is Man*, 32.

78. Josef Weber (unsigned), "The Great Utopia: Outlines for a Plan of Organization and Activity of a Democratic Movement," *Contemporary Issues* 2, no. 5 (Winter 1950), 16.

3

Rethinker

IN 1947–50 SOME two dozen people read "Capitalist Barbarism or Socialism," rallied to its retrogression thesis, and joined the new group to publish *Contemporary Issues* (*CI*). Most were former Trotskyists; some came from Labor Zionism; and all wanted a new socialist movement to emerge from the magazine's discussions, one that would offer "a democratic solution to the crisis of mankind."[1]

Josef Weber named that movement, in advance of its birth, the Movement for a Democracy of Content. The meaning of "democracy of content" would never be entirely clear, except that it differed from "formal democracy," a mere set of rules and procedures for decision-making. A democracy of content would involve ends, not just means; ethics, not just instrumental procedures. It would become the "living model for the transformation of the whole of society."[2]

How would the movement be structured? Under capitalism, Weber warned, most organizations, even the best intentioned, "alienate themselves from their original aim" and develop a bureaucracy that becomes an end in itself. Such creeping normalization "ruins" movements and "bolsters up the system." The Movement for a Democracy of Content, as Weber had it in mind, would eschew any such "leadership-apparatus" in favor of a simple and clear democratic structure, so that even if "thousands of unprepared people" entered, it would "remain transparent to and controllable by all." The movement would certainly not form a traditional political party, because "all parties are no good!"[3]

Members of the *CI* group would take on no specialized roles—every member would be simultaneously editor, theorist, propagandist, correspondent, organizer, secretary, proofreader, and typist. Two friends in London organized the printing and paid the bills, but no *CI* members would be paid for their movement work, since paychecks would lead to "a new commerce in commodities"—and the movement must remain outside the capitalist economy.[4]

Weber attracted members to *CI* with the content of his ideas but also with his personal charm—an unusual trait in so committed a revolutionary. Weber was "almost always smiling, full of fun," wrote one *CI*-er, "fond of telling jokes, easy to talk to, affectionate, 'physical.' The absolute opposite of what I expected a German intellectual to be.... He not only didn't brood or exhibit melancholy, he thought very little of those who did."[5] He had a piano in the living room, which he played often and well.

Weber embraced what he called Pantagruelism, derived from the writings of the sixteenth-century French novelist Rabelais, who had written *Gargantua and Pantagruel* as an act of rebellion against medieval Catholicism. This work depicts life in the Abbey of Thélème, an anarchic utopia whose residents are not ruled by laws but live "according to their own free will and pleasure," following only one rule: "Do What Thou Wilt." That arrangement works, in Rabelais's novels, because "in honest companies" people "have naturally an instinct ... that prompteth them unto virtuous actions, and withdraws them from vice."[6]

Pantagruelism thus involves a passionate love of life, a libertine zest, and an exuberant faith in human goodness. Pantagruelists follow their instincts, seeking satisfaction for the body in physical pleasure and for the mind in the delights of curiosity and study. Pantagruelism was also potentially an engine of the socialist revolutionary project as a principle of rebellion: the disastrous nature of capitalism was obvious, and the "awakening Jacques sooner or later must ask: 'Why should we continue to be your slaves, and why should you wish to remain our masters ... if there is more than enough for all of us?'"[7]

Weber deeply admired Denis Diderot's novel *Jacques le Fataliste* (1773–75), which he considered a Pantagruelist masterpiece. The protagonist, Jacques, is a man of uninhibited sexuality who considers even monogamy to be "thoroughly hypocritical and untenable." For Jacques, the adulterer is the true moralist, for "every healthy man and woman is an 'adulterer,' if not in flesh, the more frequently in thought."[8]

But Weber put aside wit and sensuality when it came to Richard Wagner, whom he revered: his physical resemblance to the nineteenth-century German composer extended to a personal identification. In 1944 he had even presumed to write a revised scenario for *The Ring of the Nibelungen*, which he self-commended as superior to the original: "Wagner intended the same thing as the writer of these lines, only ... he was, alas, just Richard Wagner and not the writer of these lines." Yes, as Murray told me, Weber had "a bloated conception of his own genius." He had a strong concept, too, of his own crucial role in revolutionary politics, and the Movement for a Democracy of Content agreed: said the London group, "Jupp [Weber] is our absolute spiritual leader."[9]

As strong as his self-image was, Weber was physically frail. By 1947 he had had several heart attacks. Soon after the second issue of *CI* was published, he had another one and spent two months in a hospital. Still, his frailty evoked sympathy and brought out the group's solicitude and protectiveness toward him.[10]

Almost every evening Bookchin would visit Weber's apartment to help him with his research and writing, learn from him, and revel in the companionship of a towering intellect with great political experience on two continents. As *CI* began publishing, Weber's typewriter churned out a stream of articles: manifestos, editorials, analyses, replies to critics, replies to readers. He would write in German first, then read off a translation to Bookchin, who transcribed it. "I helped him clarify his thoughts," he told me. Bookchin would then edit it, revise it, and type up the final version. His mentor's writing, Bookchin thought, was "luminescent."[11]

CI was, in effect, Bookchin's graduate school. He soaked up everything Weber had to offer, especially concerning Hegelian philosophy and socialist theory and German revolutionary history. Weber taught him, for example, about the interwar Marxian crisis theorists, who had tried to prove that capitalism must inevitably collapse. "Even his errors were brilliant," Murray once told me.

Their intellectual and political relationship soon became personal as well: at twenty-six, Murray had finally found a father figure. He abjectly adored Weber. And when he began writing his own articles for *CI*, he wrote for the purpose of advancing Weber's work.

As a writer and theorist, Weber's immense self-esteem tended to undermine his intellectual rigor. He happily issued ex cathedra assertions and opinions but neglected the hard work of amassing the factual evidence upon which they might be based. So Bookchin stepped in to do the research for him. So when Weber's bombast interfered with his rigor, Bookchin supplied the missing documentation. He learned to be a writer by acting as Weber's bulldog.

The *CI*-ers met every Friday evening to talk politics. The meetings were of "intense intellectual interest," remembered Annette Jacobson. Weber and Bookchin offered "brilliant illumination of past and current world events and topical issues." Murray had a "phenomenal knowledge of history," recalled Reni Bob. When he disagreed with someone, recalled Jacobson, he was "magnanimous and reasonable. . . . He would expound, might illustrate tangentially, and like Joe, offer enlightenment in a kind way."[12]

Members clashed sometimes—"there was often heated argument at meetings," recalled Bob. So Weber laid down a rule: group discourse had to be civil. The Democracy of Content stood for "non-barbarism, i.e., for a civilized discussion even with opponents." Discussion should be constructive, avoiding personal sharpness. Least of all should members engage in what Weber called "venom and gall," or gratuitous vituperation. "I take great care, in questions of judgment," Weber explained, "not to allow feelings to be roused at all."[13]

The *CI*-ers socialized together. Bookchin's closest friends were Dave Eisen, his old SWP buddy, and Jack and Mina Grossman. In 1951 Bookchin married Beatrice

Appelstein, whom he brought into *CI* and who became "a good comrade," Murray told me.

The group collectively edited the magazine, which appeared regularly. To build the Movement for a Democracy of Content, "we tried to approach people based on their immediate interests," Jack Grossman told me. "All efforts to retard the tendency to barbarism were worthwhile."[14] Weber's career in Europe, with the International Communists of Germany (IKD) underground and in exile, had taught him to be secretive, and he retained that trait in New York; at his insistence, the articles that he and other *CI*-ers wrote were published in the magazine under cryptonyms. Weber's articles appeared under the names Ernst Zander, William Lunen, and Erik Erikson; Bookchin wrote as M. S. Shiloh, Robert Keller, Harry Ludd, and Lewis Herber.

The first issues (1948–50) focused intensively on retrogression in all its aspects. Led by Weber, the *CI*-ers considered Stalinist Russia to be the acme of retrogression. (*CI*-ers called it "Russia" rather than "the Soviet Union.") Stalin's totalitarian state "has not the slightest connection with Socialism or 'Communism'"—the workers were in control of nothing. Rather, it was "state capitalist," a combination of capitalist organization and state ownership. As such, it waged a "ceaseless civil war against its own population." Stalin's regime was "more belligerent, more rapacious, murderous and ruthless than the most savage of other imperialisms."[15]

In 1950 Bookchin wrote his first published article, called "State Capitalism in Russia" (bylined M. S. Shiloh), to substantiate Weber's assertion and affirm his conclusions. Russia, he wrote, embodied "capitalist retrogression . . . in every feature." Workers, technicians, and engineers were enslaved to the state, toiling in "the bleak hells of Siberia, in mines and on wastes where life is scarcely maintainable and quickly passes out of existence." The system's "absolute foundation" was arms production, which was also "the life fluid" of capitalism.[16]

Over and over again, Weber would make an observation, and then Bookchin would scour newspapers and books to document it. Like an eager graduate student, he did Weber's homework for him with energy and enthusiasm. Weber had said that Stalinist Russia was "the most violent, bloody and reactionary inquisition and barbarism that ever existed in history," surpassing even Hitler's Germany. So Bookchin echoed that Stalinist Russia today "assum[es] the functions of Hitler during the thirties."[17] Weber laid down ideas, and Bookchin echoed them, sometimes to the point of recklessness: when Weber wrote that Stalin killed more Jews than Hitler ("Stalin again takes the lead"), Bookchin echoed that "German fascism showed more hesitation in the application of repressive measures than Russian fascism."[18]

Bookchin later regretted that he had "downplayed the level of popular anti-Semitism in the Third Reich" and excessively blamed Stalin for the annihilation of Europe's Jews. Still, in the 1950s he was right to raise the alarm about the threat of Russian anti-Semitism, correctly noting that "the Kremlin has singled out the Jewish people for discrimination and liquidation." Stalin was, in fact, in those years accusing prominent Russian Jewish doctors of plotting against him. Decades later, documents from the Stalin archive would confirm that plans like those Bookchin described in the early 1950s had been underway: had Stalin not died when he did, Russian Jews would have undergone another Holocaust.[19]

Besides Stalin's Russia, retrogression's other great bastion was the United States, the very acme of "Capitalist Barbarism." Some might have thought that the United States was a lesser evil than Stalinist Russia, but "the Devil can't be driven out with Beelzebub," Weber said. Indeed, the incipient Cold War was a mere sham; the two regimes were actually collaborating to impose retrogression, to reduce Germany "to colonial status," and to drive the world toward an international slave order, ruled by a few monopoly capitalists. Soon it would bring about "general crises, annihilation of nations and expulsion of peoples in 'peacetime,' deportations, . . . civil war, colonial oppression"—in short, "the decisive ruin of mankind."[20]

Bookchin did the research on American retrogression as well. "The germs of . . . totalitarianism," he wrote, were "rotting the core of American democratic life" and "roll[ing] out the carpet for American fascism." McCarthyism was one symptom. "Congressional committees and 'loyalty boards,'" he wrote in 1953, have constituted themselves "into a permanent, quasi-court system with fantastic prerogatives, . . . and a court procedure that [is] completely alien to the constitutional guarantees of the country." The House Un-American Activities Committee was "duplicating the step-by-step march towards totalitarianism in Europe."[21]

Weber was much impressed by Bookchin's work. So sometime in the early 1950s, he bestowed on him a great honor: he designated him as his heir, the Engels to his Marx. Murray must have been elated.

Although Weber and the *CI* group rejected Marxism, they did not throw it entirely into the dustbin. They especially admired the dialectical philosophy that underlay it and sought to rescue it for *CI*'s rethinking. Weber introduced the *CI* group to the writings of the Frankfurt School, a kind of Marxist think tank (formally called the Institute for Social Research, or ISR) that had fled Germany in 1935 for New York. Weber held the ISR intellectuals in high esteem—they "rank in knowledge of the social process far above any of their contemporary colleagues."[22]

The Frankfurt Schoolers, like the *CI*-ers, had been grappling with the era's compelling questions: Why had the proletariat failed to fulfill its historic role and create a socialist revolution? Why had fascism arisen instead? Their most basic and most intriguing answer—as recounted in Herbert Marcuse's *Reason and Revolution* (published 1941), and Max Horkheimer's *Eclipse of Reason* (1947)—was that reason had been degraded into an instrument for domination.

In premodern times, they argued, reason had been infused with many excellent qualities: reflection, speculation, discernment, judgment, and critique. It had been concerned with content, not form; with ends, not means; and above all with ethics. In a word, it had been dialectical. But the Enlightenment had separated means from ends, form from content, and reduced reason to procedure, utility, and calculation, a tool for manipulation and domination. This process of amoral instrumentalization had culminated in the cold efficiency and technical precision of Nazi Germany's mass exterminations. After the war, the nexus of instrumental reason had shifted to the United States, as was manifested in its prevailing philosophical currents, positivism and empiricism. To Weber and Bookchin, instrumental reason, as described by the Frankfurt School, was yet another symptom of retrogression.

Both were fascinated by the concept of dialectical reason. As a philosophy of organic development, it describes the processes of separation and incorporation that propel a development forward: that which exists, that which contradicts it, and the new transformative product of their interaction that preserves what is valuable in both. It looks both forward and back. But dialectical reason, ethically charged as it was, could also be a tool to judge and critique the existing society against an ethical standard. Using standards deriving from ethics-infused dialectical reason, one could also affirm the potentialities within the existing corrupt society for the creation of a rational society.[23]

In this connection, dialectical philosophy had yet another, more personal appeal for Bookchin and Weber: it was a study in alienation. It gave the two men—both of whom had been dislocated by history, albeit in different ways—a framework to criticize the barbaric, retrogressive American *what is* and thereby discern the possibility of a rational *what could be*. By upholding ethics against instrumentalism, dialectical critique was part of the revolutionary process. Simply by engaging in critical discourse, the CI-ers were helping to drive revolutionary change forward.

That said, Weber's interpretation of dialectics turned out to be somewhat idiosyncratic, even crude. He used it to construct a literal analogy between capitalism and organic growth. Like a plant or an animal, he said, capitalism had been born, it grew up, it became mature, and now it was rotting, and one day it would die. Retrogression was the "putrefaction" of capitalism, a quasi-biological inevitability.[24]

But where Weber used dialectics to brood about decline and rot, Bookchin picked up on the Frankfurt School's more sophisticated idea of using ethically charged dialectical reason for critique. Capitalism, deploying instrumental reason, measures value in terms of potential profits and reduces objects to commodities. It strips social life of meaning and content, reducing people to disconnected, competitive individuals. A "competitive industrial spirit" now permeates "nearly every aspect of American life."[25] Similarly, the bureaucratic state reduces people, once members of a community, to atoms, to units—mere statistics in a dominating system.

But writers who offered a dialectical critique could inspire an ethical revolt against capitalism and the state, against commodification and bureaucracy. By showing them an alternative to *what is*—the ethical *what could be*—they could spur people to rebel against their own dehumanization. Knowing they had a choice, people could and would choose ethics over instrumentalism; cooperation over competition; morality over manipulation; content over form; and face-to-face interaction over bureaucracy. Where Marxian socialism had claimed to be scientific, the new socialism would be "drenched in an ethical dimension."[26]

Weber scoffed at his protégé's ideas about ethics. You can't depend on "virtue" to bring about social change, he said. That was "only another commodity among innumerable sham-products." The idea that "we need a morality" is a "deceptive abstraction." All that mattered was the "materialistic point of view."[27]

Perhaps to address this objection, Bookchin acknowledged that people were not going to make an ethical revolt out of the blue: some crisis in capitalism, some systemic breakdown, would also have to propel them. "That a crisis is inevitable no longer is doubted," he wrote. "The only question is when it will come and, above all, how it will be 'managed.'" The nature of the crisis was still unclear, but whenever it came, people would and could choose to dispense with the "irrational society" of capitalism in favor of an ethical, "rational society," one "based on use instead of profit, on co-operation instead of competition, on reason instead of demoniacal blindness."[28]

Even as Weber and the *CI* group were pondering retrogression, the American, British, and French authorities occupying Germany were having second thoughts about the Morgenthau Plan, the decision to dismantle Germany's industrial infrastructure and reduce the country to farmland. Deindustrialization was turning out to be too harsh a program on the German people—and if their hunger and poverty became too extreme, they might turn Communist. So in late 1946–47 the Truman administration set aside the Morgenthau Plan and instead affirmed that "an orderly, prosperous Europe requires the economic contributions of a stable and productive Germany."[29]

The Marshall Plan was, in effect, a reversal of the Morgenthau Plan. Over the next years, the United States would pour billions of dollars into Europe, not to ruralize it but to rebuild its industrial plant. That material assistance would help West Germany grow at the astonishing rate of 8 percent per year throughout the 1950s. By 1960 its gross national product (GNP) would be second in the world, behind only that of the United States.[30] Instead of retrogressing, West Germany would undergo an "economic miracle." As for Japan, instead of turning the Japanese into slaves, the United States imposed a democratic constitution on the country.

Domestically, the United States did not relapse back into economic depression. On the contrary, it entered a period of unprecedented prosperity. Americans were buying cars and moving to the suburbs. By 1950 Bookchin understood that American industry was facing, not a contraction, but a seemingly limitless expansion. Weber's retrogression thesis turned out to be thoroughly and profoundly false.

Weber, however, could not or would not bring himself to admit it. And in deference to their mentor, *CI*-ers held their tongues—uncomfortably. Sometime around 1951 Bookchin and another *CI*-er, Leon Brownstein, got up the nerve to confront him. As tactfully as they could, they asked Weber how his thesis could still be viable when the economy was so clearly booming. Weber wagged his finger at them and said, *Wait! You'll see.*

The retrogression thesis had been *CI*'s unifying idea, and with its failure, *CI* members had every reason to leave the group—it might well have disbanded. At this time, owing in great part to the McCarthyite red scare, the American Left as a whole was shriveling. Once-numerous radical bookstores were shuttered, leftist publishers went out of business, and public forums vanished. Onetime Socialists and Communists, like Felix Morrow, found careers in mainstream institutions.

But the *CI* group persisted. In fact, its members not only stayed together, they began meeting twice a week instead of once, adding Saturdays to the regular Friday meetings. The members would continue to meet to talk about politics and philosophy, and edit the magazine, for the rest of the decade.

Why? One reason was surely their respect for Weber. The worldly European socialist, born in the country of Karl Marx, still had a charismatic pull on them, and Bookchin found himself unable to resist that revolutionary prestige. Even after Weber ceased to be his mentor, Bookchin deferred to him reflexively as a result of the psychological bond. Many of the group members too were emotionally attached to him, regarding him as their second father, as Reni Bob recalled, or "maybe first, in the case of Bookchin, who didn't have a father."[31]

These earnest young people also felt strongly connected to one another. "I believe the group stayed together out of comradeship," recalled Annette Jacobson, "and a bonding because of common personal values and philosophic principles." Over sixty years later, she still recalled fondly "the atmosphere of respect, friendship, and amazing camaraderie."[32]

Finally, retrogression or no, the *CI*-ers shared the conviction that the only alternative to socialism was barbarism—and that barbarism, no matter how economically comfortable, was intolerable. Capitalist America might be prosperous, but it still was instrumental, and it still commodified all aspects of life. Weber had laid down the marker in 1950: "If we fail to transform the capitalist mode of production into a socialist mode, barbarism . . . assumes the sharply delineated outlines for the doom of all modern society."[33] Even if everyone else in the United States conformed to the culture of barbarism, the *CI*-ers would stand tall and refuse.

So they continued with their common project, the magazine, in which they studied possible sources of new social conflict. They viewed every social problem as a manifestation of capitalist irrationality, and they sought to tie all solutions to specific ills to a general opposition to the system as a whole.

Not only did they stay with *CI*, they made personal sacrifices to do so. Instead of going to graduate or professional schools, these articulate, by-now-well-read people took sometimes menial jobs so that they could remain part of the group. Bookchin got a job at the welfare department as a bookkeeper. Murray Bob worked for a shipping company. Chet Manes (who had been in the SWP Minority) worked as a janitor. Jack Grossman was a structural steel detailer. None of them had much money.

As for Weber, he initially was funded by Vincent Swart, the *CI* project's London-based financier, but at some point that funding dried up. *CI*-er Chet Manes took a second job, as a night janitor in a school near his home, so he could turn those paychecks over to Weber. He didn't mind—it was a privilege, an honor to support his mentor—he was just doing what Engels had done for Marx.

* * *

As Bookchin and Weber rethought the revolutionary project, they sorted through their old Marxist ideas to salvage whatever might still be illuminating or useful. One thing they rescued, as we have seen, was dialectical philosophy; another was Marx's view of technology ("the means of production"). Marx had considered it capitalism's historical mission to develop technology to the point where it could provide for humanity's material needs.[34] In recent centuries, technological advances, from the steam engine to the factory system to steel production, had been so immense that onerous toil could, in principle, be eliminated. Machines could perform the grueling physical work that people had once done. Once technology reached the point of automation, capitalism would have no further basis for existence, and the apparatus could pass into the hands of the workers, who could use it to create socialism.

Even before the war, Trotsky had thought the United States, having reached the highest stage of technological development, was "ripe for socialist revolution and socialism, more ripe than any other country in the world."[35] Now, in the postwar era, even further technological advances were underway—the preconditions were more than ripe. But all this technology was still in the wrong hands—and so it was being used, not for socialism, but to intensify want and exploitation. Transfer it to the people's hands, and they could use it to eliminate toil and gain the leisure time to develop their creative sides. Bookchin attended the RCA Institute on the GI Bill, where he not only got his high school equivalency degree but studied electronic engineering; these latter classes reinforced his view that machines could ultimately replace most human toil. To a man who had once done hard labor in a foundry, that notion was positively utopian.

In the late 1940s, several thinkers were arguing that the new technologies were opening up utopian possibilities. Max Horkheimer and Erich Fromm, both of the Frankfurt School, agreed that "the present potentialities of social development" surpassed the dreams and visions of all previous utopias.[36] The anarchist Paul Goodman's 1947 book *Communitas* offered plans for utopias in cities.

In this vein, Weber, hoping to provide a new platform for the *CI* group, wrote a manifesto that he called "The Great Utopia." For millennia, he argued, humanity's dreams of achieving paradise had remained at the level of fantasy, because people had been required to toil to meet their basic material needs. Now technology had progressed to the point that it could provide for those needs—in superfluity. Hence the potentiality exists for humankind to finally be liberated from the burden of onerous physical labor and to be free to devote most of their time to creativity and enjoyment. At present this technology is being used to manufacture endless consumer goods to satiate artificially contrived needs.[37] To channel its use to utopian ends, society must eliminate wasteful production and instead produce goods of quality that meet real needs—that is, society must produce not for profit but for use.

Since the material conditions for such a rational society exist, said Weber, "there is nothing in principle to obstruct the solution to the social question." Humanity must simply decide to do it. It must make that choice "by democratic majority decision." The vehicle for doing so was the open, transparent Movement for a Democracy of Content. He appealed to readers to join the movement, in ever greater numbers, "until the party is the people and the people is the party."[38] Once enough people joined, they could seize control of technology and economic life.

To attract the public, Weber thought, *CI* needed to develop a specific practical plan for converting the wasteful present-day economy to a popularly controlled utopia. The *CI* group, he said, must undertake a detailed inventory of all of America's economic and social resources, then distinguish what was socially useful from what was socially wasteful. The group would then be able to demonstrate concretely that after eliminating what was wasteful, society could be organized rationally to satisfy everyone's basic needs. Once *CI* placed this plan before the public, citizens would recognize it as a good idea and rush to join.

Developing this "World Plan," as Weber came to call it, would be a huge task—it would first require analyzing the US national budget, as a reflection of the national wealth. But making it happen became his priority. Not that he would undertake the job himself—he never did care to do his own research. The whole group, or perhaps a few members, could do it. But as much as Weber urged them, the group found the task so huge as to be daunting. No one, it seems, volunteered.

Weber, who had crossed an ocean to escape the Nazis, had no love for his new place of refuge. He had adored living in Paris, but he actively despised New York. Not least of his discontents was the food. The Coca-Cola and the hot dogs, even the mustard, were nauseating: "My stomach revolts." Most foodstuffs were bland and processed: in New York "I have never eaten a true vegetable or fruit or chicken or pork-chop."[39] The canned peas and carrots were particularly loathsome instances of American culinary retrogression.

In 1948 two books appeared that, to Weber, began to explain what was wrong with American food. Fairfield Osborn's *Our Plundered Planet*, published in 1948, documented that since the war's end, a new chemical industry had emerged. Industrialized agriculture was now routinely spraying pesticides onto crops and spreading fertilizers and herbicides into soil. These petrochemicals were being touted as miraculous—"Better things for better living through chemistry," as DuPont put it. But the chemicals were causing an upswing in cancer rates, Osborn asserted.[40] The other book, William Vogt's *The Road to Survival*, contended that the rapacious capitalist system, in its quest for profit, was rendering North America, one of the earth's wealthiest and most fertile regions, sterile. It was exhausting topsoil and leveling forests and exterminating fauna. Both books were pleas for conservation and better human management of the natural environment.

You could see the problem with industrially produced foodstuffs, Weber thought, in "the bad taste of many American fruits, vegetables and meats." Chemical fertilizers not only reduced their nutritional value but (as Osborn argued) were "responsible for the appalling increase" in the incidence of "heart maladies . . . cancer and other modern plagues."[41] So enamored was Weber of this thesis that he asked *CI*-er Phil Macdougal to review the two books for the magazine.

Macdougal did so, writing that the picture that Osborne and Vogt painted was indeed "frightening"—the loss of arable land and topsoil constituted "the most immediate single threat to civilization." But he disagreed with Weber's view that chemicals were so toxic that they should be banned altogether. Inorganic fertilizers could actually be a boon, Macdougal said—they replenished played-out soil and increased its bounty. He favored their "proper use."[42]

Weber had hoped for an endorsement of his own thinking and was annoyed by the review.[43] As for the rest of the *CI* group, they were perplexed. Osborne and Vogt made a compelling and persuasive case, but they had trouble accepting it because Weber himself was undermining it. He insisted on associating it with the ideas of one Bernard Aschner, an unconventional Viennese practitioner whom he admired. Aschner thought cancer could be caused not only by chemicals but by physical blows, love bites, hysterectomy, and even personal appearance. ("We know that cancer families are usually black-haired.") The group saw right away that Aschner was a quack, a mystical crackpot. Weber's using him as a source distorted whole the issue of environmental destruction.[44]

To achieve some clarity, Weber called for an investigation into the matter. His protégé and heir Bookchin, who was as puzzled as anyone, rose to the challenge, as he had so often done before.[45] Just at that moment, in the fall of 1950, a US congressional committee was holding hearings on the subject of chemicals in food. The committee's final report, issued in 1952, concluded that chemical additives in food "raised a serious problem as far as the public health is concerned."[46]

As he combed through the transcript, Bookchin learned that parathion, DDT, chlordane, diethylstilbestrol—in other words, fertilizers, pesticides, preservatives, and flavoring and coloring agents—have potentially devastating effects on the human body. These chemicals, he wrote in a long article, produce "abnormalities in the organism as a whole" that can disrupt "cellular structure," from which processes cancer may emerge. "The Problem of Chemicals in Food" was published in *CI* in 1952.[47]

Weber must have been pleased that his protégé and heir had vindicated him, about the carcinogens if not about Aschner.[48] But this time Bookchin had not simply done Weber's homework—he realized that this subject had enormous implications, that the adulteration of food was part of "the misuse of industry as a whole." Capitalism was reshaping agriculture. To maximize profits, industrial agriculture was cultivating crops on a large scale, in monocultures, which required pesticides because they were vulnerable to infestations and fertilizers because they degraded soil. Chemicalization was part of the instrumentalization of food production, which commodified both farming and gustation. Capitalism as a system, it turned out, was harmful to human health and well-being. The very concept, Bookchin recognized, was explosive.[49]

True to its original promises, the pages of *CI* remained open to discussion. Letters from readers were welcomed and published, even hostile ones. Weber took responsibility for answering them, albeit in verbose, rambling replies.[50] Sometimes he violated the rules of civility that he had laid down at the outset. When someone criticized "The Great Utopia" as "old stuff and warmed-up wisdom," he lashed back with a bitterly sarcastic piece called "The Dog behind the Stove," written partly in verse: "Look at my cur! He barked, the lout: / 'Utopia?—That's old shit'! / But when I kick him on its snout— / He yelps devoid of wit." He appeared not to care that his arrogant, contemptuous style might have a chilling effect on future letter writers. In fact, he relished being able to "finish somebody off 'artistically' and . . . in a satirical manner." After all, "misplaced modestly cannot prevent us" (meaning himself) from affirming that "we are extremely interesting."[51]

In 1953 Weber had had another heart attack, which left him less able to manage his temper. Now whenever someone irritated him, he confessed to a friend, that person was in "danger of being killed on the spot . . . the heart stops to beat, and I'm for a moment simply 'not there.' Then I feel the blood roar in my head and all control is lost." Once his heart was stirred up, "it remains in an irritated condition for quite some time, affecting me day and night and requiring considerable efforts to get it back to what I can call 'normal.'"[52]

It didn't help matters that New York's perpetual din kept him awake at night. While Weber tossed and turned, "the Elevated trains make noise two hundred yards away; . . . motor cars, lorries, buses, aeroplanes make noise; everywhere radios make noise, and again and again one is awakened by voices and sounds like shots." With its clattering and thundering subways and its wailing sirens, New York ground its residents down. It was "unwholesome, ugly, uninspired, nerve-racking, crushing, hostile to the sense, hostile to life, a catastrophe"—the very embodiment of retrogression, a concept he seemed increasingly willing to rescue by diluting its meaning.[53]

Most irritating of all, the CI group was dragging its heels on undertaking the statistical survey of the American economy, the all-important "World Plan." Their delay in "giving the matter flesh" was causing the blood to roar in his head.[54] Sometime in 1953, he asked his protégé and heir to do the job. Probably Bookchin would have preferred to continue researching chemicals in food, but how could he turn down a personal, urgent appeal from his beloved mentor? So he agreed, putting his own project aside.

But what an enormous task it was—the ultimate research assignment. Getting down to work, he first made a list of basic human needs—food, clothing, shelter, furnishings, medical care, utilities, water, transportation, energy, and so on—then went to the library every day to pore over statistics on the various economic sectors that supplied goods and services to meet those needs. He researched the capacities of factories and distribution facilities.

Everywhere he looked he found colossal waste. Raw materials were being squandered. Corporations were producing commodities, in staggering quantities, that were useless or destructive or both. To boost consumer consumption of them, managers mobilized a huge labor force—"an over-sized army of clerks, accountants, bookkeepers, typists, managers, executives, sales men, brokers, dealers, engineers, foremen, psychologists, supervisors, advertising specialists and artists." The biggest wastrel of all was the US government itself, which was rolling in unnecessary expenditures. Overall, Bookchin concluded, the present social order, a "thoroughly artificial system," could be maintained only because of "an exhausting parasitism that, octopus-like, slowly strangles all strata of the population."[55] He wrote up his research, and as he finished each chapter of the World Plan, he presented it to the *CI* group, who discussed it and praised it.[56] The entire document was slated for publication in 1954.

On March 1 of that year, the United States tested a hydrogen bomb at Bikini Atoll in the South Pacific. The force of the blast was a thousand times more powerful than the bombs dropped on Hiroshima and Nagasaki. The wind shifted eastward, blowing the radioactive plume toward a Japanese fishing boat, the *Fortunate Dragon*. Crew members' skin was blackened and blistered. In September the vessel's radio operator died. The fallout reached Japan, where some dubbed the incident a second Hiroshima.

If anything was a contemporary issue, the *CI* group agreed, it was thermonuclear weapons. Bookchin churned out a pamphlet called "Stop the Bomb," demanding an immediate end to the tests, since the fallout they produced could render "large parts of our marine food supply radioactive and unhealthful" all over the world. Bookchin's leaflet broke new ground by opposing not only these terrifyingly destructive weapons but also the "peaceful" uses of the atom that the Eisenhower administration was also urging at the time, to generate energy. Atomic power plants, too, he pointed out, would produce radioactivity, which could contaminate the food supply and damage health. Radiation, whether generated by bombs or by power plants, was the most lethal of all chemicals in food.[57]

CI distributed twenty thousand copies of the leaflet around New York—and the response exceeded their wildest expectations. Letters poured in, expressing "pleasure at meeting others who oppose this insanity" and offering to help the campaign.[58] The prominent pacifist A. J. Muste told Bookchin personally that the "Stop the Bomb" leaflet transformed his thinking about nuclear power.[59] The leaflet was widely reprinted. The large Japanese daily *Asahi Shimbun* published excerpts on page one. It was a "surprise," wrote a reader, "for us Japanese . . . to know that there are many people also in America who want to stop the A-Bomb or the H-Bomb." Japanese scientists and intellectuals wrote letters to *CI*, explaining that "even vegetables in this country now are more or less dangerously radioactive, as the earth itself is fully radioactive."[60]

Atmospheric nuclear testing became one of *CI*'s most important topics; the magazine published many articles on the subject between 1954 and 1960.

Given the alarming health crises that were emerging, Bookchin found it difficult to concentrate on the World Plan. Not only was the statistical work tedious, but the project had come to seem strangely trite. That American capitalism could produce ample resources to go around seemed obvious, even without statistical evidence. And irritatingly, the project was quantitative in nature, instrumental, as if achieving utopia were simply a matter of accounting. But you could not simply redistribute wealth to satisfy material needs, Bookchin thought—you would also need an ethical revolt, something Weber refused to admit. The Plan seemed more and more like a monumental waste of time.

Doggedly loyal to Weber nonetheless, Bookchin wrote a stirring introduction, arguing that the US economy was "brimming and overflowing with the means of life." If Americans chose to get rid of the wasteful corporate and governmental bureaucracies, they could achieve "the fruition of a rational system."[61] But when he presented the introduction at a group meeting, Weber objected to it, perhaps because Bookchin interjected some lines about ethics. In any case, having prodded his protégé and heir into writing the World Plan, Weber vetoed, for publication in *CI*, the introduction he had written. The blood in his head must have been roaring something fierce.

Bookchin had had enough—he dropped the World Plan project. The issue of chemicals in food and nuclear weapons had sparked his thinking in new and creative directions.

NOTES

1. Josef Weber (unsigned), Editorial, *Contemporary Issues* (hereafter *CI*) 1, no. 2 (Winter 1948), 81; Weber (unsigned), Editorial, *CI* 2, no. 5 (Winter 1950), 2; Weber (as Ernst Zander), "Germany and World Development," *CI* 1, no. 1 (Summer 1948), 11.

2. Josef Weber (unsigned), Editorial, *CI* 1, no. 1 (Summer 1948), 1; Weber (as Ernst Zander), "Majority and Minority," *CI* 3, no. 10 (Winter 1952), 112–13; Weber (unsigned), "The Great Utopia: Outlines for a Plan of Organization and Activity of a Democratic Movement," *CI* 2, no. 5 (Winter 1950), 18.

3. Weber (unsigned), "Great Utopia," 12, 14; Weber (unsigned), Editorial, *CI* 1, no. 2 (Winter 1948), 81.

4. Weber (unsigned), "Great Utopia," 16, 18.

5. Murray Bob, "Autobiography of the First 18 Years," courtesy Reni Bob.

6. François Rabelais, *Gargantua and Pantagruel* (1532–64; reprint Hong Kong: Forgotten Books, 1952), 137.

7. Josef Weber (as Wilhelm Lunen), "Appeal for an English Edition of Diderot's 'Jack the Fatalist,'" *CI* 4, no. 15 (July–Aug. 1953), 184, 175, 200.

8. Weber (as Lunen), "Appeal for an English Edition," 181, 149, 160.

9. Josef Weber (as Wilhelm Lunen), "The Ring of the Nibelung, or The Art and Freedom of Interpretation at the Middle of the 20th Century" (written May 1944), *CI* 5, no. 19 (Aug.–Sept. 1954), 177; quote is from Marcel van der Linden, "The Prehistory of Post-scarcity Anarchism: Josef Weber and the Movement for a Democracy of Content (1947–64)," *Anarchist Studies* 9 (2001), 137.

10. Bob, "Autobiography of the First 18 Years"; Josef Weber (as Ernst Zander), "Some Comments on the Organizational Question" (written Jan. 1, 1951) *CI* 11, no. 44 (Sept.–Oct. 1962), 271.

11. "Interview Peter Einarssens mit Murray Bookchin, Okt. 1984," trans. Harald Simon, *Schwarzer Faden*, no. 26 (1987), p. 41.

12. Annette Jacobson to author, June 7 and 17, 2008; Reni Bob to author, June 11, 2008.

13. Weber (as Zander), "Some Comments," 267, 277, 286, 294; Weber (as Ernst Zander), "An Unfinished Letter" (written Oct. 7, 1956), *CI* 10, no. 38 (May–June 1960), 93, 95, 98–99.

14. Jack Grossman to author, Sept. 5, 2008.

15. Josef Weber (as Ernst Zander), "The Campaign against Remilitarization in Germany: A Letter and a Reply," *CI* 6, no. 23 (May–June 1955), 180. Following Weber's lead, CI-ers consistently referred to the Soviet Union as "Russia."

16. Bookchin (as M. S. Shiloh), "State Capitalism in Russia," *CI* 2, no. 7 (Autumn 1950), 207, 222, 223.

17. Josef Weber (as Ernst Zander), "Concerning Germany and World Development," *CI* 1, no. 1 (1948), 7; Weber (as Zander), "Campaign against Remilitarization," 180; Bookchin (as Shiloh), "State Capitalism," 221.

18. Weber (as Zander), "Germany and World Development," 17 n. 7; Bookchin (as M. S. Shiloh), "A Social Study in Genocide," *CI* 3, no. 10 (Winter 1952), 132.

19. Bookchin to Jacob Suhl, Feb. 20, 1993, MBPTL and author's collection; Bookchin (as M. S. Shiloh), "Anti-Semitism in Eastern Europe," *CI* 4, no. 13 (Nov.–Dec. 1952), 42. On the Stalin archives, see Jonathan Brent, *Inside the Stalin Archives: Discovering the New Russia* (New York: Atlas, 2008).

20. Josef Weber (as Ernst Zander), "On the 'Inner Limit' of Capitalism" (written 1950), *CI* 10, no. 40 (1961), 272; Weber (unsigned), "Great Utopia," 8; Weber (as Ernst Zander) "War as a Way Out?" *CI* 2, no. 7 (1950), 158; Weber (as Zander), "Germany and World Development," 3, 12.

21. Bookchin (as Harry Ludd), "The Fate of American Civil Liberties," *CI* 4, no. 16 (Nov.–Dec. 1953), 206, 213, 230, 237.

22. Josef Weber (as Wilhelm Lunen), "The Problem of Social Consciousness in Our Time," *CI* 8, no. 31 (Oct.–Nov. 1957), 526.

23. Herbert Marcuse, *Reason and Revolution: Hegel and the Rise of Social Theory* (Boston: Beacon Press, 1941), 6. A late self-description of *CI* would read: "The writers and supporters of *CI* have sought to place existing institutions, social forms, cultural values, the entire human experience of our time, before the bar of reason. The basic questions they have asked are: Is an institution or social change rational? Does it satisfy the needs of man? Does it lead to the realization of human potentialities?" Editors, "Why We Publish," *Contemporary Issues: A Magazine for a Rational Society* (June–July 1963), 1–3.

24. Weber (as Lunen), "Problem of Social Consciousness," 526, 508, 480.

25. Bookchin (as Ludd), "Fate of American Civil Liberties," 207.

26. MBVB, part 37.

27. Weber (as Lunen), "Problem of Social Consciousness," 489, 497, 480.

28. Bookchin (as Robert Keller), "Year One of the Eisenhower Crusade," *CI* 5, no. 18 (June–July 1954), 101; Bookchin (as Lewis Herber), "A Follow-up on the Problem of Chemicals in Food," *CI* 6, no. 21 (Jan.–Feb. 1955), 56–57.

29. US occupation directive JCS 1779.

30. Paul Hockenos, *Joschka Fischer and the Making of the Berlin Republic* (New York: Oxford University Press, 2008), 18, 22.

31. Reni Bob to author, June 11, 2008.

32. Annette Jacobson to author, June 7, 2008.

33. Weber (unsigned), "Great Utopia," 4.

34. Murray had been familiar with this idea since childhood. *New Pioneer* had written in 1932, "Machinery, if in control of the workers, would cut down the hours of labor and give every one a chance to enjoy life." But that machinery "is now in the hands of the

bosses and is used by them . . . to kill 20,000 American workers in industry every year." See Bert Grant, "Science and Nature for Johnny Rebel: The Chemical Front," *New Pioneer*, Nov. 1932, 15.

35. Leon Trotsky, "Completing the Program and Putting It to Work," June 7, 1938, in *The Transitional Program for Socialist Revolution* (New York: Pathfinder Press, 1973), 72, 142. See also Leon Trotsky, "The Political Backwardness of the American Workers," May 19, 1938, in *Transitional Program*, 125.

36. Max Horkheimer, *Eclipse of Reason* (London: Continuum, 1947), v; Erich Fromm, *Man for Himself: An Inquiry into the Psychology of Ethics* (London: Routledge, 1947), 4.

37. Weber (unsigned), "Great Utopia," 4, 9.

38. Ibid., 3, 6.

39. Josef Weber (as Erik Erikson), "Critical Review (New York eriksonized)," *CI*, no. 7 (Autumn 1950), 230–35; Josef Weber (as Ernst Zander), "A Fragment on Chemicals in Food and Other Questions" (written ca. 1952) *CI* 10, no. 39 (Aug.–Sept. 1960), 229.

40. Weber quoted from them in his 1949 "Interim Balance Sheet: The Bankruptcy of Power Politics," *CI* 1, no. 4 (Autumn 1949), 273, 301n.

41. Weber (as Zander), "Fragment on Chemicals in Food," 222; Weber (unsigned), "Great Utopia," 6–8.

42. Phil Macdougal (as Stephen D. Banner), "Humanity's Resources and the New Malthusianism," *CI* 1, no. 3 (Summer 1948), 233, 242.

43. Weber (as Zander), "Fragment on Chemicals in Food," 223.

44. Weber makes his case and cites Aschner in his unsigned "Great Utopia," 7 n. 4. See Bernard Aschner, *The Art of Healing* (London: Research Books, 1947), 184, 290. On the subsequent discussion, see Weber (as Zander), "Fragment on Chemicals in Food," 218–19, in which Macdougal is quoted on 224.

45. Weber (as Zander), "Fragment on Chemicals in Food," 221.

46. Alfred Larry Branen, *Food Additives* (n.p.: CRC Press, 2002), 201. The House Select Committee to Investigate the Use of Chemicals in Foods and Cosmetics (known as the Delaney Committee) held hearings from September 14 to December 15, 1950.

47. Bookchin (as Lewis Herber), "The Problem of Chemicals in Food," *CI*, no. 12 (June–Aug. 1952), 238.

48. Weber (as Zander), "Fragment on Chemicals in Food," 221.

49. Bookchin (as Herber), "Problem of Chemicals," 209–11, 240–41; Bookchin (as Herber), "A Follow-up on the Problem of Chemicals in Food," *CI* 6, no. 21 (Jan.–Feb. 1955), 56.

50. Weber (as Zander), "Campaign against Remilitarization ," 194.

51. Josef Weber (as Erik Erikson), "The Dog behind the Stove," *CI*, no. 9 (Summer 1951), 49–50; Weber (as Zander), "Some Comments," 268; Weber (as Zander), "Campaign against Remilitarization," 213.

52. Weber to Vili, May 6, 1954, courtesy Reni Bob.

53. Weber (as Erikson), "Critical Review (New York eriksonized)," 230–35.

54. Weber (as Zander), "An Unfinished Letter," 111–12.

55. Bookchin (as Keller), "Year One of the Eisenhower Crusade," 91, 99. I am inferring what he wrote in the World Plan from other writings from the time; the "World Plan" itself is not available.

56. It was to be published in *CI* in 1954 under his pseudonym Robert Keller, as "Basis for Utopia: The Outlines of an Economic Plan for a New Society." Bookchin (as Keller), "Year One of the Eisenhower Crusade," 103. But the World Plan was never published, except for the

introduction Murray wrote, which was published in *CI* after Weber's death and incorrectly attributed to Weber.

57. Bookchin (unsigned), "Stop the Bomb: An Appeal to the Reason of the American People," *CI* pamphlet, MBPTL, copy in author's collection.

58. "A Note on the Campaign against the Hydrogen Bomb," *CI*, no. 20 (Oct.–Nov. 1954), 324.

59. Bookchin, "Postwar Period," interview by Doug Morris, in Bookchin, *Anarchism, Marxism and the Future of the Left* (San Francisco: A.K. Press, 1999), 51.

60. "Reactions to 'Stop the Bomb,'" *CI*, no. 22 (Mar.–Apr. 1955), 147, 150, 153; "A Note on the Campaign Against the Hydrogen Bomb," *CI*, no. 20 (Oct.–Nov. 1954), 325. The translated leaflet appeared in the June 6, 1954, issue of *Asahi Shimbun*.

61. "A World Plan for the Solution of Modern Society's Economic and Social Crisis. Introduction," *CI* 11, no. 41 (May–June 1961), 32–33. This article is incorrectly attributed to Weber (as Ernst Zander).

4

Eco-decentralist

BOOKCHIN DID NOT share Weber's animus toward New York. His childhood in East Tremont had been idyllic, with its close-knit ethnic community and beautiful park. The neighborhood was still congenial, as he saw every day when he went to give his mother her insulin injection. The deli was still selling pickles and whitefish and knishes, and he could still hear Yiddish murmured as he passed strangers on the sidewalk. His old YCL comrades were long gone, but their parents still lived there—and why would they move, when the buildings were rent controlled?

Rose Bookchin, nearly blind from diabetes, still lived in the four-story brick building on 175th Street. In early December 1952 a city agency announced that a six-lane highway was to be constructed through East Tremont, called the Cross Bronx Expressway. Apartment buildings that stood on the right-of-way would be torn down. A letter from the city—signed by Robert Moses—informed the residents that they had ninety days to leave.

Rose's building wasn't on the list, but thousands of her neighbors were forced, tearfully, to leave. The abandoned buildings were boarded up and vandalized; then came the wrecking ball. Now when Murray visited Rose, he could look eastward and see the earth-moving machines and bulldozers coming ever closer. Construction crews used dynamite to level the hills on which East Tremont stood. The staccato of jackhammers and dynamite surely plagued Rose. East Tremonters called the dust and grit that got in their pores "fallout."[1]

The kosher butcher, the greengrocer, and the other shopkeepers boarded up their storefronts and joined the exodus. Construction of the Cross Bronx Expressway destroyed what was left of Murray's childhood utopia.

For about fifteen years now, Robert Moses, New York's veritable dictator of public works, had been ripping up working-class neighborhoods to build highways, tunnels, and bridges. Many other American cities, too, were being remade on behalf of the automobile and some concept of efficiency. Starting in 1949, the federal "urban renewal" program had been designating old neighborhoods—mainly immigrant and working-class communities—as blighted, congested slums. That label became a warrant for sending in bulldozers to clear the way for erecting functional towers of glass and steel in anonymous, antiseptic concrete plazas.[2]

Why these incomprehensible acts of civic violence? How could perfectly good, humane communities be destroyed on behalf of inhumane high-rises? For insight into the nature and workings of cities, Bookchin first turned to the writings of his oldest teachers, Marx and Engels. As he pored through their books, several passages leaped out at him. "The whole economic history of society can be summed up," Marx had written (in a rather uncharacteristic passage), by the development of "the antithesis between town and country." And Engels had observed that this town-country antithesis had arisen as "a direct necessity" for industrial and agricultural production.[3] In other words, this urban problem was connected to the problem of industrial agriculture, and both were connected to capitalism.

Lewis Mumford's *The Culture of Cities* (1938) gave Bookchin further insight.[4] In medieval times, Mumford had written, small, close-knit European cities had been human in scale, with irregular streets and low-slung houses. They were attractive, communal, and traversable on foot. In their many green spaces, people could gossip, trade, pray, and politic face to face. Over the centuries, however, as kings created bureaucracies and standing armies and centralized authority in the nation-state, these humane small cities had given gave way to baroque, imperial cities, whose layout consisted of straight lines and rigorous visual axes. Thereafter, said Mumford, as capitalism and authority corrupted civilization, urban history continued to deteriorate, culminating in today's gigantic cities, organized for power and money.

Bookchin was inspired by both Marx and Mumford to write his own narrative of urban decline. The high point, in his view, was not the medieval communes but the small cities of ancient Attica, in the first millennium BCE. These poleis, notably Athens, had existed in balance with the surrounding countryside; their inhabitants "had firm ties to the soil and were independent in their economic position," which gave them a strong, self-reliant character. Economically, the ancient Athenians produced simple goods to meet their basic needs and nothing more. From this arrangement had arisen a remarkable political culture, with democratic assemblies and an "exceptionally high degree of public participation."[5]

Much later, with the rise of modernity in Europe, cities became commercial and industrial enterprises. Civic and communal life deteriorated, as buying and selling displaced other social roles. Products of workmanship became objects of exchange, or commodities, while traditional social relations yielded to relations of exchange. As the market moved to the center of social life, the quest for profits became the overriding endeavor.

Cities, as the venues for this historic transmutation, became dehumanized—and by the 1950s, their pathology had become extreme. They were too big, gigantic, *megalopoleis*. Housing was scarce and shoddy. Traffic congestion had reached the point of dysfunction. The subways were overcrowded and unreliable. Office work was monotonous and sedentary; stifled by tedium, urban workers had come to resemble machines, "enslaved, insecure, and one-sided." City dwellers encountered one another in passing, with mutual indifference or mistrust. The giant city was "a mere aggregate of dispirited [people] scattered among cold, featureless structures."[6]

No wonder Weber couldn't sleep. Nerves that were battered and raw from the noise were further assailed by advertising. And the automobile was everywhere. An expressway like the Cross Bronx reduced people "to mere byproducts of the highway and motor car." New Yorkers were being forced to surrender "residential space, parks, avenues, and air to a steel vehicle that looks more like a missile than a means of human transportation." As a result of the automobile, "urban air is seriously, in some cases dangerously, polluted," taking a toll on human health.[7]

Meanwhile, the separation of town and country necessitated the use of ever more chemicals in food production: not only pesticides and fertilizers and herbicides (to maintain monocultures) but also preservatives (to prevent deterioration during shipping) and food colorings (to create the appearance of freshness). "As long as cities are separated from the countryside," Bookchin wrote, "food will necessarily include deleterious chemicals to meet problems of storage, transportation, and mass manufacture—not to mention profit."[8] The separation of town and country was intimately connected to the problem of chemicals in food.

Bookchin concluded that "the possibilities of the city are exhausted" and "can never be revitalized." The city had reached its limits. The megalopolis had become a fetter on civilization, and humanity, in order to advance, must burst that chain.[9]

But what pattern of settlement would replace the gigantic city? Lewis Mumford, when writing of alternatives, had been inspired by the work of the English garden-city advocate Ebenezer Howard, the Scottish urban planner Patrick Geddes, and the Russian anarchist Peter Kropotkin, all of whom around the turn of the century had proposed creating green settlements, small, human-scale communities that were surrounded by open swaths of countryside dedicated to recreation and to agriculture that produced food for local consumption, that farmed "on soils enriched by urban refuse," and that made use of urban market gardens.[10] Following these predecessors, Mumford proposed that new cities should have greenbelts and parklands, combining "the hygienic advantages of the open suburbs with the social advantages of the big city."[11]

Bookchin thought that creating new green cities in hitherto rural areas would be excellent but went further to propose decentralization—breaking up the giant metropolises into "small, highly-integrated, free communities of [people] whose social relations are blemished neither by property nor production for exchange." Humanistic in scale and appearance, the new small cities would be integrated with the surrounding landscape. Their inhabitants would have easy access to farmlands, where they could raise crops and savor rural recreation.[12]

This ecological, decentralized society would have no need for chemicals in food. In fields that were small in scale, crops could be rotated, requiring no fertilizers, and crop diversity (as opposed to monocultures) would render pesticides unnecessary. The short distance between farm and marketplace would eliminate the need for preservatives and colorings. Agriculture would remain mechanized to reduce toil, but with the absence of chemicals, it could once again become organic, a concept that Bookchin absorbed from Sir Albert Howard's 1940 book, *An Agricultural Testament*.

The integration of town and country would enhance social solidarity as well as intimacy between people and land. People could develop a robust character like that of the ancient Athenians, creative and civic-minded, governing their small communities in equilibrium with the natural world. Communities would produce "goods solely to meet human needs and promote man's welfare." Profit-seeking would yield to social responsibility.[13]

If all these considerations were making eco-decentralization desirable and necessary, technological advances were making it possible. Rather than depending on dangerous and centralizing nuclear power, communities "could make maximum use of [their] own energy resources, such as wind power, solar energy, and hydroelectric power." Thanks to miniaturization and electronics, giant factories could be dismantled, replaced by small-scale, automated production facilities: "The smoky steel town . . . is an anachronism. Excellent steel can be made and rolled with installations that occupy about two or three city blocks." And since the machines would do most of the work, people would have to work only a few hours a day, allowing for "the self-assertion of . . . a spiritually independent and free personality." The pace of life would be more relaxed, set not by production schedules but by bodily and daily cycles. People could actually enjoy life, having "unrestricted access to the countryside as well as the town, to soil as well as to pavement, to flora and fauna as well as to libraries and theaters."[14]

They would not suffer social or intellectual privation. Thanks to the telephone, people "can now communicate with one another over a distance of thousands of miles in a matter of seconds." And thanks to trains and planes, "we can travel to the most remote areas of the world in a few hours." As a result of modern mass communication and transportation, "the obstacles created by space and time are essentially gone."[15]

By all these means, eco-decentralization could open "magnificent vistas for individual and social development." It would not only promote human health and fitness but lead to "a long-range balance between man and the natural world." The human personality could expand. Today's desensitized urban robots would be released from their insecurities, greed, and competitiveness and in their place develop self-confidence, a sense of moral responsibility, and cooperation. Genius could once again flourish, as it had in ancient Athens.[16]

Above all, people in the eco-decentralist society would have the free time to participate civically: to govern themselves. At some point in the early 1950s, Bookchin came across H. D. F. Kitto's *The Greeks*, which describes ancient Athens in ways that resemble Bookchin's ideal society. In the ancient polis, wrote Kitto, "town and country were closely-knit," and cities were built to a human scale. Production was for use, and wasteful consumption did not exist: "Three-quarters of the things which we slave for the Greeks simply did without." And needs for energy were met by the sun.[17]

As a result, ancient Athenians—or rather, their male citizens—had abundant leisure time, which the climate allowed them to spend outdoors. (Women, slaves, and resident aliens did not, alas, enjoy these privileges.) The typical Athenian man, said Kitto, "was able to sharpen his wits and improve his manners through constant intercourse with his fellows. . . . Talk was the breath of life to the Greek"—rather like Crotona Park and Union Square in the 1930s, Bookchin might have thought.

The political institution that organized Athenian sociability was the *ekklesia*, the democratic assembly, in which all male citizens were legal equals, with equal rights to debate and vote on issues of communal concern and equal rights to hold political office. Such an assembly (which would of course be modified to include all adult residents, including women and ethnic minorities) seemed to Bookchin to be the proper governing institution for the eco-decentralist society. It would be, he wrote, "democratic in content not only in form."[18]

Once again Bookchin had done Weber's spadework for him. Weber had complained about tasteless food; Bookchin theorized an ecological society that would produce wholesome, organic food to delight his palate. Weber had complained about New York's noise and cockroaches; Bookchin was ready to break up the city for him and disperse its pieces around the countryside. Weber had called for a utopia where machines eliminated toil; Bookchin thought out that utopia's social fabric. Weber had called for organizational transparency and freedom of expression; Bookchin identified its institutional form—the democratic citizens' assembly. In short, Bookchin provided the "content" for the Movement for a Democracy of Content. He and he alone had carried out the Trotsky-inspired "rethinking" that the CI group had been formed to undertake.

But sometime in the mid-1950s, Weber began to seem strangely distant from Bookchin's work—even uninterested. Perhaps he was annoyed that his protégé and heir had dropped the World Plan. When Bookchin pressed a copy of Kitto on Weber and urged him to read about ancient Athenian assemblies, Weber dismissed the whole notion of assembly democracy as "dilettantism."[19] And when Bookchin explained his eco-decentralist ideas, Weber complained that the group was "confused" over "theoretical questions": "We have de facto no theory."[20]

Despite Weber's rudeness to him, Bookchin tried to keep himself intellectually interesting to his mentor. He wrote up his ideas on town and country in an article called "The Limits of the City," in which he sketched the city's evolution over the millennia, from the village and the Athenian polis to the imperial city, the bourgeois city, and finally the megalopolis. His narrative treated each kind of city as a phase or moment of a larger urban process that unfolded dialectically. He submitted the article for publication in *CI*.

On the day the group discussed the article, they said it was excellent—at first. Then Weber weighed in. The article was too historical, he pronounced. All those pages on the history of Attica were nothing more than useless scholarship.

Bookchin defended the emphasis on history. It was necessary to show how the Athenian polis had come into existence so that readers could understand it.

Nonsense, Weber might have said. History "can 'teach' us only a few things (which have by now become truisms at best) and most of it is . . . a bourgeois fetish."[21]

But Joe, Bookchin might have protested, *much of Marx's* Capital *is about the history of capitalism. You can't really know something's nature unless you know its history, its development, how it emerged from what came before. That's the dialectic.*

Bah, Weber might have replied. *The present is what matters. Suppose you were talking about slavery. You wouldn't have to describe the history of slavery in order to oppose it—it would be enough* "to describe accurately what modern slavery is."[22]

To which Bookchin might have argued that a utopian must be interested in studying history at least for its turning points, the moments in history where things might have taken a different and better turn. What if Rosa Luxemburg and Karl Liebknecht had survived the attack of January 1919 and given the German working class real leadership? What if the Stalinists and the Social Democrats had joined forces electorally in 1933 and prevented Hitler from becoming chancellor? These were crossroads, moments of potentiality where humanity could have gone in a different direction. It's crucial to recognize them and learn from them, Bookchin would have insisted.

The "what ifs" of history are irrelevant, Weber might have snapped: "In history, it is exclusively a matter of what has actually happened, not of what might have occurred under different circumstances and conditions."[23] To disparage such counterfactuals, he quoted from his beloved Diderot's *Jacques le Fataliste*: "If, if, if—if the sea boiled, you'd have a lot of cooked fish!"[24]

Under the force of Weber's arguments, group members who had initially praised "The Limits of the City" reversed themselves. They echoed Weber's condemnation of the article as "too historical" and even "academic."[25] Its historical section contributes "nothing to our understanding of the present state" of the cities, said one. "Either we deal with contemporary issues, or we cease to be *Contemporary Issues*."[26]

Bookchin's good friend Dave Eisen rose to defend the article as "a striking interpretation of the significance of the city in civilization."[27] The debate over this article was strangely bitter, but in the end, Weber vetoed its publication.

It surely consoled Bookchin when news arrived that "The Problem of Chemicals in Food" was to be published as a book in Germany. The translator, Götz Ohly, said he hoped it would "serve as a warning signal lest this lunacy"—chemicals in food—"affect Germany as well."[28] For Bookchin, it must have been a much-needed validation.

When Stalin died in 1953, the group had rejoiced. In February 1956 the Soviet premier, Nikita Khrushchev, denounced Stalin's dictatorship and admitted that the 1930s Moscow show trials had been frame-ups. Half a year later, on October 23, Hungarian workers rose up in revolt against the Kremlin. They went on a general strike and formed soviets—not authoritarian ones but democratic ones, as in 1917. Hungary seemed poised to free itself from Russian control.

The Hungarian insurgents appealed to the West for help, broadcasting desperate appeals by radio for arms and ammunition. For nearly a decade, Western diplomats had been assuring eastern Europeans that should they rise up against their Russian suzerains, the West would lavish them with aid. So they had every reason to expect their appeal to be heard.

But Western governments reneged and sent no aid. Horrified, *CI* formed the Emergency Committee for Arms to Hungary to demand that the US government air-drop weapons to the insurgents. Bookchin was the principal spokesperson. In early November he wrote a leaflet admonishing Westerners not to allow the brave Hungarian rebels to perish, by "forc[ing] them to fight tanks with pistols and armored cars with pitchforks." It was one of those crossroads moments that could change history: arming the rebels could help topple the Russian dictatorship.[29] Hungarian expatriates in New York and Boston arranged a speaking tour of New England for Bookchin, who orated to thousands.

But by November 4, Soviet tanks had crushed the rebellion. *CI* mourned the demise of the bold uprising and the deaths of the heroic insurgents. The revolt had been particularly inspiring and illuminating in the context of the otherwise quiescent 1950s. The young Hungarians, Bookchin mused, "had grown up entirely under a Stalinist regime and had never known other social dispensations." Yet they had taken up arms in insurrection. If they could do it, "then maybe we had reason to hope that revolutionaries could emerge closer to home as well."[30]

The *CI* group leavened their hard work with dashes of Pantagruelism. The Bookchins and the Grossmans vacationed together in the Great Smoky Mountains. In 1956 *CI*-ers with theatrical inclinations formed the Grub Street Players, who performed obscure plays by Cervantes and Boccaccio in the living rooms where the group met. The most elaborate production was an eighteenth-century farce called *The Cornish Squire*, performed on December 28, 1957. As they had hoped, Weber loved it.[31]

But there was no denying it: the *CI* group was growing depressed. By 1958, they had been working together for ten years. They had produced several dozen issues of the magazine, and they had discussed politics and culture intensively twice a week. "Stop the Bomb" and "Arms for Hungary" had gained them attention and respect. But the magazine had not become a collection point for social resistance—the New York group was still only about twenty strong. No Movement for a Democracy of Content had emerged. "In four years of serious political agitation," Jack Grossman pointed out, referring to 1954–58, "we have not succeeded in making one *CI*-er out of the many people with whom we were in touch."[32] The result had been "a perceptible demoralization . . . inbreeding with the same old faces . . . the same predictable attitudes."[33]

Readers, for their part, complained that they couldn't figure out where to place *CI* on the ideological spectrum. "Who are you really, and what do you want?" they asked. *CI*'s ideological openness, which had been intended to generate discussion and creativity, made for an incoherent presentation.[34]

Weber, increasingly irate, blamed the *CI*-ers for the lack of progress. They didn't "think before they speak," he complained to his friends in London and Germany and South Africa; they didn't "listen carefully and register what has been said," and so engaged in "a lot of valueless talk and outright nonsense."[35] He claimed the group would fall apart without him. Least of all did he admit that his own difficult personality might have been an obstacle to movement building. As Murray told me, people tended to drop away from the group rather than challenge Weber on anything.

Weber's insulting remarks about *CI*-ers are particularly insensitive in the light of their loyalty to him, indeed their financial support for him. Certainly they were not the Frankfurt School (which was what Murray, in later years, suspected Weber had wanted them to become). But they were committed activists who were defying the spirit of their times to try to build a broad democratic movement. If they failed, it was not for lack of diligent effort and self-sacrificing dedication on their part.

But in the prosperous 1950s, most Americans weren't interested in hearing social critique, let alone in reducing their consumption. "In the absence of a radical atmosphere in the country," Grossman conceded grimly, "it is doubtful that many people will be won to the revolutionary perspective of our movement."[36]

As Weber got sicker, he lost interest in searching out possibilities for radical political change in contemporary issues. Instead, he issued sweeping condemnations of the present, which deserved "greater contempt" than any other historical epoch. Even though all the necessary conditions for utopia were in place, he groused, "social life remains . . . a sea of blood, dirt, baseness, irrationality and misery." He denounced whole fields of intellectual inquiry, like psychology, which was merely a replacement for religion "as an attempt to make people feel guilty." Scientists were morally bankrupt: they took comfortable jobs in industry and allowed their work to be used for "H-bombs, guided missiles, gases and bacteria for warfare, jet-fighters, insecticides, chemicals and so on."[37]

Even his onetime dialectical heroes Max Horkheimer and Theodor Adorno had sold out, producing "useless scholasticism" for "officialdom" and sometimes even stooping to issue "ordinary official propaganda." They had succumbed to the positivism they had once criticized and were now ruled "by the law of ignorance."[38]

Under capitalism, Weber concluded, "true social consciousness" was obtainable "only by individuals" who refused to "blindly or exclusively follow . . . narrow economic and political interests"—that is, people like himself. He, after all, had never taken a paycheck from "officialdom," the way the Frankfurt Schoolers had, and thus he remained independent and free of taint. (That he took paychecks from his protégé Chet Manes didn't seem to matter.) Against the money-grubbing society, he compared himself to the philosopher Spinoza, a "man of character" who had "stood up for [his] convictions" and refused to "sell his soul."[39]

Weber's conceit became sickening, Murray told me. People who weren't willing to flatter him or feed his vanity, it seemed, no longer had any place in the group.

Chet Manes, who had taken a second job in order to support Weber financially, never flagged in his adoration of his mentor. But for years he had resented Weber's designation of Bookchin as his heir. Manes thought he himself deserved that honor. As he brooded over the injustice, his annoyance grew. Finally he made a decision: he would no longer hand his second paycheck over to Weber.

That left Weber without an income. The CI-ers would have happily pooled their resources to support him, they agreed, but their own incomes were meager, and many of them now had families. They encouraged Weber to apply for disability payments—he could surely qualify, with his heart condition. But Weber refused, saying he could not endure the humiliation of taking public assistance. Evidently he preferred Chet Manes's paychecks. So he withdrew his designation of Bookchin as his heir.

Thereafter, Weber's rudeness to Bookchin intensified. Murray would ask a question, and Weber would reply curtly, if at all, and walk away. No matter what Murray wrote, Weber criticized it. He vetoed publication of yet another paper by Bookchin, this one called "The Decline of the Proletariat," on the nonrevolutionary nature of the working class—a favorite theme of Weber's, in earlier times, which perhaps had been Bookchin's point in writing it.

Then Weber started bad-mouthing Bookchin behind his back, to his friends overseas. He was a mere "journalist"—look at all that plodding research on those congressional hearings. He was a mere "agitator"—look at those futile speeches on behalf of the Hungarian uprising. And he was an amateur—look at that interminable historical article on Athenian democracy! Such "dilettanti" as Murray, he complained in a letter, "will always attempt to overbridge their insecurity by 'explaining' everything which they don't understand."[40] And he still hadn't finished the goddamn World Plan.

He stepped up his attacks, accusing Bookchin of being "used up" and "a disastrous human failure" who would "never write again." Unnervingly to Murray, things he hadn't told anyone except his wife Beatrice were suddenly coming out of Weber's mouth.

Bookchin was crushed. Why had Weber turned against him? The best things CI had produced—"Stop the Bomb," "Arms for Hungary," and eco-decentralism—had been his contributions. What had he ever done to Weber, except love him too much?

In the late 1950s, a CI-er named Jack Schwartz, who was a mathematician, developed a model to test the validity of Marx's labor theory of value. He ran the numbers—and the theory failed. He concluded that the economic analysis in *Capital* was incorrect.[41] He wrote a paper and in September 1956 presented it to one of the Saturday night meetings.

The blood must have roared in Weber's head as he listened. Yes, Marx had been wrong about proletarian revolution—but his economic analysis, Weber believed, remained unsurpassed. He rebuked Schwartz, saying he didn't understand *Capital*. *Explain it, then!* Schwartz insisted. *You said you wanted dialogue and argumentation.*

Weber dismissed the whole discussion as "impossible" and the level of insight as appallingly low.[42]

But Schwartz refused to be brushed off and kept the argument going for weeks. Group members took sides, no one willing to concede. Weber proposed that they agree to disagree, but Schwartz refused, saying that the group must admit that he was right, and that Marx's theory of value was false. If they did not, then "I will split."[43]

Weber seethed: Schwartz "can't find rest until he has conquered . . . fame for himself."[44] But he had to be answered. Who would do it?

Bookchin must have seen his chance to win back Weber's approval. He volunteered, stepping back into his old role as Weber's researcher.[45]

He was neither a mathematician nor an economist, but by now he was accustomed to self-education. For Weber's sake, he spent the winter of 1956–57 poring over *Capital* and making notes. At length he drafted a reply, arguing that Schwartz's error was to ignore the social dimension of Marx's labor theory of value—he occupied himself with relations between things, not between people. Bookchin presented his paper to the group that April. Jack Grossman applauded it as "one of the finest expositions of the significance of labor in political economy ever produced."[46]

But Weber denounced it as shallow and unacceptable, and he ripped into Bookchin for being "like the dog behind the stove." It was because of Bookchin, he said maliciously, that *CI* was degenerating.

Murray was speechless. He still loved his old mentor, even with all his flaws. Despite his cruelty, he couldn't bear to fight him or even criticize him.

Bookchin's relationship with Weber was further strained by Bookchin's belief, shared by Millie Weber (according to Murray), that Weber and Beatrice had become romantically attracted to each other. He faced the prospect of losing both his father figure and his wife.[47]

As a refugee from Nazism, Weber learned, he was entitled to compensation from the West German government, enough to live on. In 1958 he abruptly left New York and returned to his native country. The *CI* group, shocked by his treatment of Murray, gave him no send-off.

But even from across the Atlantic, Weber did not let up, writing letters to the *CI* group attacking Bookchin. In October 1958 he mocked supposedly joyless revolutionaries who "use their wives for their own ends." "Let, then, the wife . . . find a satisfactory sexual relation (especially with a man who is not a pig)." On and on he ranted, accusing this unnamed revolutionary of suffering from "saturnalia of neurosis" and "pathological perversion."[48]

That letter shocked even Weber's devotees in the New York group and fomented "hatred against Joe" that was "difficult to imagine," wrote one of them.[49] Dave Eisen protested Weber's "venomous personal attack." He begged Weber "to empty your pen of vitriol and conduct the discussion henceforth . . . upon another plane."[50]

Weber responded by laying it on even thicker. By now he was posturing as Jacques le Fataliste, Diderot's hero who had extolled adultery as the highest form of morality. Prohibitions of adultery, Weber complained, always lead "to submission to authorities like the state, the church, or (at the end) even the husband." He even tried to associate monogamy with capitalism: "Under capitalism's artificially maintained economy of scarcity," he wrote, "physical and moral compulsion" enforce "sexual privation," while "sex-guilt supports an economy of scarcity."[51] Without compunction, he invoked cherished political ideas to justify his personal agenda.

Bookchin would later tell me he thought Weber was trying to take *CI* down with him. He was right, I found in my research. "I decided to let it [*CI*] die of its own stupidity," Weber remarked in a 1959 letter, dismissing his years with these friends, admirers, supporters, and disciples as "years of endless effort with the result of an incredible mess."[52] He had taught and inspired them, then tried to destroy them. The group continued to meet, if nothing else in sheer defiance of his prediction that *CI* couldn't survive without him.

Then, in the summer of 1959, news arrived that on July 16, Josef Weber, in Germany, had died of a heart attack at the age of fifty-eight. But by now the *CI* group that had once revered him as a successor to Trotsky had come to despise him.

His early death rescued Bookchin from that toxic relationship. Sorting out all the wild and bitter emotions would take years. But in the end, Bookchin would generously decide that he preferred to remember Weber as he had been at his best, before illness warped his mind, and to value him as his mentor. But for now he had to stanch the bleeding.

NOTES

1. Robert A. Caro, *The Power Broker: Robert Moses and the Fall of New York* (New York: Alfred A. Knopf, 1974), 855, 859, 860, 887.

2. See Anthony Flint, *Wrestling with Moses: How Jane Jacobs Took on New York's Master Builder and Transformed the American City* (New York: Random House, 2009).

3. Karl Marx, *Capital* (1867; reprint Chicago: Charles H. Kerr, 1906), 1:387; Frederick Engels, *Herr Eugen Dühring's Revolution in Science (Anti-Dühring)* (1877; reprint New York: International Publishers, 1939), 323.

4. Lewis Mumford, *The Culture of Cities* (New York: Harcourt Brace, 1938).

5. Murray Bookchin, *The Limits of the City* (New York: Harper & Row, 1974), 24, 27–28. The original full manuscript of "The Limits of the City," written in 1959–60, no longer exists, to my knowledge. A truncated version was published as "The Limits of the City," *CI*, no. 39 (Aug.–Sept. 1960). But only in 1979 was the full article published, as the first chapters of the book *The Limits of the City*. It's impossible to determine how much of the 1959–60 material was revised for the 1974 book. I take Murray at his word that the first chapters of the 1974 book are more or less what he wrote in 1959–60.

6. Bookchin, *Limits of the City*, 28; Bookchin (as Herber), "Limits of the City," 197.

7. Bookchin (as Herber), "Limits of the City," 198, 204–5.

8. Bookchin (as Lewis Herber), "A Follow-up on the Problem of Chemicals in Food," *CI* 6, no. 21 (Jan.–Feb. 1955), 56.

9. Bookchin (as Herber), "Limits of the City," 215. "Limit" is a dialectical concept. See, for example, Sidney Hook, *From Hegel to Marx* (New York: Humanities Press, 1962), 69–70.

10. Peter Kropotkin, *Fields, Factories, and Workshops* (1912), discussed in Lewis Mumford, "The Natural History of Urbanization," in William L. Thomas Jr., ed., *Man's Role in Changing the Face of the Earth* (Chicago: University of Chicago Press, 1956), 395–96. Wayne Hayes told me that this article had an important impact on Murray.

11. Mumford, *Culture of Cities*, 396.

12. Bookchin (as Herber), "Limits of the City," 215–16.

13. Bookchin (as Herber), "Follow-up on Chemicals in Food," 56–57. To my knowledge, Bookchin did not use the term *eco-decentralist*; it's my shorthand for his ideas from this period. He would develop these ideas further in the next years in Bookchin (as Herber), *Our Synthetic Environment* (New York: Alfred A. Knopf, 1962), 237–45.

14. Bookchin (as Herber), *Our Synthetic Environment*, 209, 241, 242; Bookchin (as Herber), "Follow-up on Chemicals," 56–57.

15. Bookchin (as Herber), *Our Synthetic Environment*, 241.

16. Ibid., 244; Bookchin (as Herber), "Follow-up on Chemicals," 56.

17. H. D. F. Kitto, *The Greeks* (Chicago: Aldine, 1951), 30, 36–37.

18. Bookchin (as Herber), "Follow-up on Chemicals," 57.

19. Josef Weber (as Ernst Zander), "The Campaign against Remilitarization in Germany," *CI* 7, no. 27 (May–June 1956), 207–8.

20. Josef Weber (as Ernst Zander), "An Unfinished Letter" (written Oct. 7, 1956), *CI* 10, no. 38 (May–June 1960), 92, 112.

21. Weber to Dave Eisen, Jan. 3, 1959, MBPTL, copy in author's collection.

22. Weber is paraphrased in Dave Eisen to Vincent, Oct. 11, 1948, courtesy Dave Eisen.

23. Josef Weber (unsigned), "The Great Utopia: Outlines for a Plan of Organization and Activity of a Democratic Movement," *CI* 2, no. 5 (Winter 1950), 3.

24. Diderot quoted in Josef Weber (as Wilhelm Lunen), "Appeal for an English Edition of Diderot's 'Jack the Fatalist,'" *CI* 4, no. 15 (July–Aug. 1953), 180. Murray used the Diderot quotation in "A Philosophical Naturalism," in *The Philosophy of Social Ecology* (Montreal: Black Rose Books, 1995), 28.

25. Murray Bob, handwritten drafts, n.d., courtesy Reni Bob.

26. Murray Bob on Murray Bookchin (ca. 1959), handwritten draft.

27. Dave Eisen to Andrew, Feb. 5, 1959, courtesy Dave Eisen.

28. "Correspondence," *CI*, no. 21 (Jan.–Feb. 1955), 80. The book was *Lebensgefährliche Lebensmittel* (Krailling bei München: Hanns Georg Müller Verlag, 1953).

29. Bookchin (unsigned), "We Cannot Let Russian Armor Crush the People of Hungary!" (Nov. 3, 1956), in *CI*, no. 30 (July–Aug. 1957), 444–46.

30. Bookchin, "The 1960s," interview by Doug Morris, in Bookchin, *Anarchism, Marxism, and the Future of the Left* (San Francisco: A.K. Press, 1999), 60–61.

31. Annette Jacobson to author, June 15, 2008.

32. Jack Grossman, "C.I.—Whence and Whither?" 1958, courtesy Jack Grossman.

33. Ibid.

34. Josef Weber (as Ernst Zander), "The Campaign against Remilitarization in Germany: A Letter and a Reply," *CI* 6, no. 23 (May–June 1955), 195; Grossman, "C.I.—Whence and Whither?"

35. Weber (as Zander), "Unfinished Letter," 92.

36. Grossman, "C.I.—Whence and Whither?"

37. Josef Weber (as Wilhelm Lunen), "The Problem of Social Consciousness in Our Time," *CI* 8, no. 31 (Oct.–Nov. 1957), 478–79, 516, 527, 502; Josef Weber (as Wilhelm Lunen), "Appeal

for an English Edition of Diderot's 'Jack the Fatalist,'" *CI* 4, no. 15 (July-Aug. 1953), 160, 190; Josef Weber (as Wilhelm Lunen), "The Ring of the Nibelung, or The Art and Freedom of Interpretation at the Middle of the 20th Century" (written May 1944), *CI* 5, no. 19 (Aug.-Sept. 1954), 180, 166, 162.

38. Weber (as Lunen), "Problem of Social Consciousness," 479, 506, 526. Weber seems to have been fond of making up laws.

39. Ibid., 516-17.

40. Weber to Dave Eisen, Jan. 3, 1959, copies in MBPTL and author's collection.

41. Grossman to author, Sept. 5, 2008. Schwartz wrote under the pseudonym Giacomo Troiano.

42. Steve Selzer to Dear Friends, Nov. 12, 1959, courtesy Jack Grossman; Weber (as Zander), "Unfinished Letter," 102.

43. Jack Grossman, "C.I.—Whence and Whither?"; Weber (as Zander), "Unfinished Letter," 100-103.

44. Weber to Eisen, Apr. 3, 1959, courtesy Dave Eisen.

45. Grossman, "C.I.—Whence and Whither?"

46. Bookchin (as Robert Keller), "Reply to Troiano," unpublished ms., copies in MBPTL and author's collection; Grossman, "C.I.—Whence and Whither?"

47. As recalled in Bookchin, "Journal for an End of a Century," vol. 5, "October 18–November 13, 1992," handwritten ms., 18-21, 29; and in Bookchin to Jacob Suhl, Feb. 20, 1993, MBPTL and author's collection.

48. Josef Weber (as Ernst Zander) to John Clarkson (written Oct. 19, 1958), published in "Sex and Society," *CI* 11, no. 42 (Sept.-Oct. 1961), 96.

49. Murray Bob (as Alan Dutscher), fragment, Nov. 25, 1959, courtesy Reni Bob.

50. Dave Eisen to Weber, Jan. 4, 1959, courtesy Dave Eisen.

51. Weber (as Ernst Zander), letter to Dave Eisen (as Paul Ecker) (written Apr. 3, 1959), published in "Sex and Society," *CI* 11, no. 42 (Sept.-Oct. 1961).

52. Weber to Eisen, Apr. 1959, courtesy Dave Eisen.

5

Eco-anarchist

IN 1960, AFTER five thousand people had been displaced from their homes, construction crews completed the East Tremont section of the Cross Bronx Expressway, and cars soon surged along the route day and night, belching exhaust fumes into nearby windows, creating unbearable noise.[1]

The following year Rose Bookchin died. An immature, narcissistic woman, she had never understood how to nurture her brilliant, high-strung son. But he had been decent and dutiful to her, giving her daily insulin injections for two decades. When her life ended, in a city hospital far from the rutted dirt roads of Bessarabia, he cried. So many significant losses were coming all at once now: his old neighborhood, his mother, his father figure, his marriage. He had an affair with a neighbor woman, looking for solace, but found none. A sympathetic friend reported that he was "drunk almost all of the time." At the movies, he would yell at the characters on screen.[2]

He continued to go to CI meetings—his friends there, still his surrogate family, trying to cobble together a new footing post-Weber. Meanwhile, the issue he knew best—chemicals in food—was turning out to be momentous, in an episode that gained national attention. On November 9, 1959, the US Food and Drug Administration (FDA) banned aminotriazole, a weed-killer used on cranberry crops; it had been found to cause thyroid cancer in rats. Just weeks before the Thanksgiving holiday, cranberry products disappeared from grocery shelves. When American families sat down at their dinner tables that year, the lack of traditional cranberry sauce forced them to contemplate the problem of chemicals in food.

The issue gave Bookchin a path forward—he could write his way out of his despair. *CI* had forbidden remuneration for writers, but perhaps now he could make a living as a science journalist. He would not even have to compromise his radical politics, because the political implications of chemicals-in-food, to his mind, were very radical indeed.

He spent much of 1960–62 writing up his important *CI* work for a book aimed at a general readership. Agrochemicals threaten the viability of the soil and hence the food supply; when they come into contact with living tissue, they, like nuclear fallout and radiation, threaten "early debilitation and a shortened life span," from heart disease and cancer. Gigantic cities further pollute and cause psychological stress. And what is driving these "unrelenting" physical challenges, fouling the air we breathe, the water we drink, and the food we eat? The answer, he explained, is our economic system, which puts short-term economic gain ahead of human biological welfare.[3]

Bookchin thought most people, being sensible and rational, would not and could not ignore these all-too-real threats to their health. Economic exploitation had not turned the proletariat revolutionary, but ordinary citizens, faced with the prospect of an early death, would surely rise up in outrage against the system that generated them, break up large-scale industrial agriculture, disperse the megalopolis, and embrace the only sensible alternative—eco-decentralism, a synthesis of humanity and nature, town and country. "The most compelling laws of biology," as Bookchin put it, would propel fundamental social change.[4]

It so happened that a neighbor of Bookchin's, Burton Lasky, had aspirations to become a literary agent. He accepted the manuscript as his first endeavor and showed it to Angus Cameron, an editor at Alfred A. Knopf, with whom he had connections. Cameron liked the manuscript and decided to publish it, giving it the title *Our Synthetic Environment*. Murray dedicated the book to his mother Rose.[5]

In the wake of Weber's *Götterdämmerung*, Bookchin redefined himself as an anarchist.[6] Perhaps the older man's betrayal, departure, and death freed him to adopt an ideology that had intrigued him for some time; or perhaps adopting the new ideology helped him move forward after the catastrophe; or perhaps both. In any case, he soon realized that his eco-decentralist ideas had an unmistakable affinity with anarchism, the ideology that asserts that the state (or what Americans call government) is irremediably oppressive and that people can and should free themselves from the centralized bureaucracy that renders them psychologically dependent and politically impotent and at the same time serves as an arm of the capitalist system that menaces their physical well-being.

Bookchin's interest in anarchism had not come from studying the writings of its nineteenth-century theorists Proudhon, Bakunin, and Kropotkin. His ideas bore some resemblance to those of Kropotkin's *Fields, Factories and Workshops*, but he did not come across that book "until long after I had worked out my own ideas."[7] Rather, he had absorbed some of Kropotkin's ideas secondhand, through Mumford.[8]

The writings of a little-known twentieth-century anarchist seem to have confirmed his choice of anarchism. Born in 1886, Erwin Anton Gutkind had been an architect and city planner in Weimar Berlin. In the 1930s he fled Hitler's Germany for

London. At the war's end, he returned to Berlin as a member of the Allied control commission that governed the British Zone, charged with helping to reconstruct the city. But he found the operation too bureaucratic, so he quit.

Gutkind shared Bookchin's abhorrence for the present-day megalopolis. His two books *Community and Environment* (1954) and *The Twilight of Cities* (1962) are arguments for decentralism, for small-scale communities that balance urban and rural.[9] Like Bookchin, he wanted to disperse the large city and redistribute new communities over a broad area. Small in scale and dense in structure, these communities, imbued with "mutual aid and cooperation," would rejuvenate humanity with "an inspiring diversity and a new élan vital."[10] Unlike Mumford, Gutkind was an explicit anarchist, considering "communities in a stateless world" to be "the highest ideal which we can discern at present." He gave his ideas a name: "Social Ecology," a term that "stresses the indivisibility of man's interaction with his environment."[11]

Equally crucial to Bookchin, anarchism as an ideology seemed suited to advance the ethical revolt against capitalism, instrumentalism, and commodification that he envisioned. In contrast to Marx's "scientific" socialism, which had expected social and economic forces to carry the world toward the good society, anarchism counted on individuals to hold the existing system (the *what is*) accountable to an ethical standard, find it wanting, consciously reject it, and choose to fashion an alternative society that would fulfill that standard. Moreover anarchism upheld, as a matter of principle, a unity of means and ends: only ethical means could produce a good society, and anarchist institutions and actions must be consistent with and indeed prefigure that society. That must have been attractive to Bookchin, who had come to despise Marxism-Leninism's acceptance and even endorsement of authoritarian, morally repugnant means to hasten the operative social forces toward their inevitable end, the socialist society.

In 1960 anarchists were scarce in the United States. Among them were Paul Goodman, an advocate of self-reliant, self-supporting communities; and Julian Beck and Judith Malina, who in 1947 had founded the Living Theatre, an anarchist-pacifist troupe that performed avant-garde plays.[12] A few old-timers huddled together in New York's Libertarian League and reminisced about the anarchist revolution in 1930s Spain and the valorous but doomed fight for it. But for the most part, anarchism seemed to Bookchin to be a historical relic. In fact, the historian George Woodcock had recently pronounced it all but dead: "Clearly, as a movement, anarchism has failed"; after experiencing many defeats, it had dwindled "almost to nothing."[13]

Perhaps there was life in the old ghost yet, Bookchin thought. Perhaps anarchism could be revived and renovated, in such a way as to become relevant to the second half of the twentieth century.

Contemporary Issues, post-Weber, rallied well. The magazine's early 1960s issues published credible and substantial articles on the civil rights movement, antiapartheid protests, and African anticolonialism, as well as nuclear weapons testing.[14] For the sake of group harmony, Bookchin agreed to shorten "The Limits of the City," and it was finally published in abbreviated form in 1960.[15]

Political protest, which had been so conspicuously scarce for a decade, was finally reappearing in the United States and even catching fire. In February 1960 a sit-in at the Greensboro, North Carolina, Woolworth's lunch counter launched a new stage in the movement against racial segregation. That August in Connecticut, members of the Committee for Non-Violent Action (CNVA) tried to block the launch of nuclear-weapons-equipped Polaris submarines, by rowing or swimming out to them and physically sitting on top of them, refusing to move. And in November 1961, after reports that strontium 90 was present in human and cows' milk, some fifty thousand women in sixty American cities marched to protest nuclear radiation. Mostly middle class—and pushing baby carriages to make their point—their placards read, "End the Arms Race, not the Human Race."[16]

The new political protests used nonviolent means—and they were ethical, proceeding from outrage at injustice. An ethical revolt was also the basis of Students for a Democratic Society (SDS), founded in June 1962. Its manifesto, the Port Huron Statement, rejected the oppressive systems in American life—racism, corporations, the Cold War, the nuclear arms race, the power elite, and the military-industrial complex—in favor of "a new ethical politics" with broad moral and social goals. Lamenting "the felt powerlessness of ordinary people," the Port Huron Statement insisted that ordinary citizens must be able to participate in making decisions in all institutions—a concept that it called "participatory democracy."[17]

The Port Huron authors might as well have been reading *Contemporary Issues*.[18] But as SDS grew into, in effect, a movement for democracy of content, *CI* remained depressingly small. At meetings of its aging radicals, Bookchin argued that *CI* should reach out to these interesting young people and try to bring them into its fold. In January 1962 he proposed that the group dedicate every second meeting to introductory discussions. But Jack Suhl, a German friend of Weber's, objected, saying he found such discussions to be superficial and unrewarding. *CI* was, he said, an émigré organization. Bookchin replied that to achieve its goals as a vital, engaged political group, it must expand its ranks. Voices were raised; one member, Harold Wurf, threatened to punch Bookchin. Others pulled Wurf away before he could get violent. The meeting broke up.[19]

In the spring of 1962, after forty-one years in the Bronx, Bookchin moved into Manhattan, to a small apartment in the West Village.[20] His agent, Lasky, taught him how to create book indexes, which finally gave him a tolerable livelihood, as a freelance indexer.[21]

Our Synthetic Environment was published on June 11, 1962, a memorable date. That very week the *New Yorker* devoted its entire June 16 issue to a critique of pesticides, written by a marine biologist named Rachel Carson. Her lengthy piece continued over the magazine's next two issues. In September it was published between covers as *Silent Spring*, which attracted widespread attention. It called attention to some of the same abuses that Bookchin had been writing about—but it had the effect of stealing *Our Synthetic Environment*'s thunder. That coincidence of timing, Burt Lasky told me, "was always a source of great unhappiness to Murray."[22]

Carson's writing style was evocative and compelling. "She did a wonderful job with *Silent Spring*," Bookchin acknowledged, praising her "stylistic magic."[23] The warning bell that her book rang brought the issue of pesticides to public attention, and over the screams of manufacturers, scientific research on the hazards of DDT quickly got underway, leading to its ban in 1972. Carson's book sparked the modern environmental movement. But her treatment of the social and economic structures that promoted pesticide use was comparatively muted. As historian Yaakov Garb observes, *Silent Spring* "brought its readers to the threshold of difficult questions, . . . but Carson's avoidance of politics, abetted by her conceptions of nature, helped lead them away again."[24]

By contrast, *Our Synthetic Environment* discussed an array of environmental contaminants and ills—including but not limited to pesticides—and indicted the social and economic interests that underlay them. William Vogt (author of *Road to Survival*) reviewed the two books jointly, writing that Bookchin "ranges far more widely than Miss Carson and discusses not only herbicides and insecticides, but also nutrition, chemical fertilizers . . . soil structure." He lamented that Bookchin's book was given "the silent treatment" and said "it is to be hoped that he may now ride into the limelight on [Carson's] coattails." For the situation that both books addressed was urgent: "Will we move fast enough, now, to escape possible immolation?"[25]

And in 1964 T. G. Franklin wrote in the British periodical *Mother Earth* that *Our Synthetic Environment* is "one of the most important books issued since the war and I thoroughly recommend it to all who are interested in the way we live." The noted microbiologist René Dubos praised both *Our Synthetic Environment* and *Silent Spring* for alerting the public "to the dangers inherent in the thousands of new chemicals that technological civilization brings into our daily life."[26]

But the book failed to inspire the ethical revolt against industrial agriculture and megacities that Bookchin had hoped for. Years later the environmental writer Stephanie Mills would look back on the upstaging of *Our Synthetic Environment* in perplexity: "Perhaps Bookchin's wider range of concern put people off, or perhaps his quiet prescription of social revolution as the remedy for the problems he detailed was too much for people to swallow in '62." Culture critic Theodore Roszak was equally bemused: "Why did Bookchin's superior work receive so little supportive notice? The reason, I think was the staggering breadth and ethical challenge of Bookchin's analysis. Nobody, as of 1962, cared to believe the problem was so vast. Even the environmentalists preferred the liberal but narrowly focused Carson to the radical Bookchin." As Yaakov Garb concluded in 1995, "Bookchin's pill was clearly too big, bitter, and unfamiliar for most Americans to swallow at that time."[27]

After a few months in the West Village, Bookchin moved to the very neighborhood that his mother and grandmother, a generation earlier, had been eager to leave.[28]

Rents were cheap on the Lower East Side: fifty dollars a month got you a two- or three-room apartment, with a bathtub in the kitchen. Rock-bottom rents attracted bohemians, and in the early 1960s young Americans who rebelled against the conventional goals of the affluent society were drifting in. Here they could live on crumbs and dedicate themselves to exuberant self-expression, then go hear Sonny Rollins play his saxophone, or Bob Dylan perform in a café.

That fall Dave Eisen was rethinking *Contemporary Issues*: he proposed revamping the magazine into a less stodgy, large-format sixteen-page monthly, so that it could find its place in the new protest scene and establish itself "as a gathering point for social opposition."[29] He volunteered to take over the editorship, if Bookchin would agree to write for it, which he did. But as a result of *Our Synthetic Environment*, invitations to lecture were arriving in Bookchin's mailbox, making it hard for him to fulfill his commitment to his old friend.[30]

That December, New York's energy utility, Con Edison, applied for a permit to construct a thousand-megawatt nuclear reactor in Ravenswood, a Queens neighborhood on the East River. It would be the largest nuclear reactor ever built anywhere, and the first ever to be located in the heart of a large city.[31]

The Ravenswood proposal gave rise to yet another first: for the first time in American history, a local population organized itself to oppose a commercial nuclear power plant. On February 19, at a public meeting in Queens, Con Ed's representatives told neighborhood residents that the reactor posed little danger to their health. The utility's board chairman, Harland C. Forbes, said he was confident that the plant could be built in their neighborhood—"or in Times Square for that matter"—without presenting any danger to the community. But when a biology professor stood and objected that there was no safe dose of atomic radiation, the residents applauded unanimously. In April former Atomic Energy Commission chairman David Lilienthal announced that he agreed with the opposition: he "would not dream of living in the Borough of Queens," he said, "if there were a large atomic power plant in that region." A state senator pointed out that "the mind of man has not yet invented an accident proof piece of mechanical equipment."[32]

Shocked by the outrageous proposal and its even more outrageous siting, Bookchin published two reports, under the name Citizens Committee for Radiation Information. Nuclear power plants were inherently unsafe, he wrote, because no one knew how to dispose of radioactive wastes safely. Constructing a reactor in New York City would set a precedent "for locating power reactors in the major population centers."[33] New Yorkers, he insisted, must not allow that to happen.

The Ravenswood project seems to have led Bookchin, already a critic of urban gigantism, to think about the connection between cities' size and the various types of energy that they produced, consumed, and in general depended on. Large cities, with their towering apartment and office buildings, seemed to require commensurately large energy systems, ones that were moreover centralized and concentrated. In his next book, *Crisis in Our Cities*, he argued that today's gigantic cities depended for their existence on the availability of "huge packages of fuel" generated by "immense power plants." Nuclear power could provide energy in the needed quantities, but it was patently unsafe, given radioactive wastes.

Megacities had at hand another source of concentrated energy as well: fossil fuels—"mountains of coal and veritable oceans of petroleum." In fact, since nuclear power was too dangerous, we were left with fossil fuels. And since the end of the war, fossil fuels were what had been allowing giant cities to "reach immense proportions and merge into sprawling urban belts."[34]

But fossil fuels, Bookchin pointed out, are seriously compromised, perhaps even more so that nuclear power. For fossil fuels were contributing to a certain planet-wide disaster-in-the-making that everyone needed to become aware of. Over the previous century, he observed in the climactic chapter of his book, human beings had spewed 260 billion tons of carbon dioxide into the atmosphere.

> This blanket of carbon dioxide tends to raise the atmosphere's temperature by intercepting heat waves going from the earth into outer space. . . . Theoretically, after several centuries of fossil-fuel combustion, the increased heat of the atmosphere could even melt the polar ice caps of the earth and lead to the inundation of the continents with sea water. Remote as such a deluge may seem today, it is symbolic of the long-range catastrophic effects of our irrational civilization on the balance of nature.[35]

These sentences, written in 1964, could have been written today, except that the looming catastrophic effects are no longer either remote or symbolic.

Bookchin continued with relentless logic: fossil fuels, since they generate the greenhouse effect, have no future as a safe source of energy. And since giant cities depend on them, giant cities have no future either. So "if an industrial civilization is to survive, man will have to develop entirely new sources of energy." Those sources will be solar, wind, and tidal. From "the heat of the sun, the fury of the winds, the surge of the tides," humankind "could draw inexhaustible quantities of energy without impairing the environment." Certain "revolutionary lines of technological innovation," like solar reflectors, parabolic collectors, and more, "hold the promise of a lasting balance between man and the natural world."[36]

But the sun, the wind, and the tides, he realized, cannot supply "the large blocks of energy needed to sustain densely concentrated populations and highly centralized industries."[37] Much as fossil fuels are suitable for giant cities, renewable fuels are suitable only for smaller communities, with their small energy grids.

Eco-decentralism, then, was no mere utopian fantasy. The rescaling of town and country and rural-urban integration had a technological imperative: "To use solar, wind, and tidal power effectively, the megalopolis must be decentralized."[38] That is, because of the greenhouse effect—or climate change, as we call it today—eco-decentralism would be necessary for the survival of civilization.

Fortunately, decentralizing society into human-scale communities would have positive effects on our civilization. It would improve our health, because we could then eliminate chemicals from food production. Integrating town and country would bring the sun, the wind, the earth, back into people's daily lives, overcoming our alienation from nature. And thanks to new technologies of computerization, miniaturization, automation, and cybernetics, we could continue to enjoy "the amenities of an industrialized civilization." "We stand," he wrote, "on the brink of a new urban revolution."[39]

Crisis in Our Cities too was ignored, attracting little more attention than *Our Synthetic Environment*. A typical reviewer rejected its decentralist solution as unachievable.[40] The world was far from ready to hear about global warming, let alone a solution that required a profound transformation in the organization of society.

But Bookchin knew he was on to something, and as he studied scientific ecology in the works of Charles Elton, his philosophically oriented mind discerned striking parallels between natural processes and social processes.[41]

Both nature and society, he reflected, are processes of evolution, moving from the simple to the ever more complex. Nature has evolved from the inorganic to the organic, and from simple organisms to more complex ones; human social organization, in turn, has evolved from the band and the tribe to the nation-state, the international market, and numerous types of community and association in between. Both types of evolution achieve "a greater degree of specialization, complexity, and interrelatedness."[42]

Both kinds of evolution, in order to move forward, depend on the existence of a degree of spontaneity on the part of their components. In natural evolution, "variety emerges spontaneously," through complex ecological situations, random mutations, and the interactions of organisms with their environment. And in social evolution, it is people, who have the freedom to exercise their will, who move history forward. The existence of social freedom allows them to spontaneously "find their authentic order." Indeed, "spontaneity in social life converges with spontaneity in nature" to form the foundation of a society that would be both free and ecological.[43]

In both natural and social evolution, uniformity is a weakness. In agriculture, monocultures are prone to disease and insect infestation. Similarly, a centralized, authoritarian, conformist society is prone to breakdown under stress. (Witness the Aztecs, easily overpowered by the conquistadors.) Diversity, by contrast, promotes stability, strength, and resilience in both realms: diverse crops in agriculture, diverse life-forms in an ecosystem, diverse people in a free society.

Remarkably, anarchism and ecology seemed to parallel each other in striking ways. For one thing, spontaneity turned out to be not only a social concept but a specifically anarchist concept, as anarchism has an "ideal of spontaneous organization," and "both the ecologist and the anarchist place a strong emphasis on spontaneity." Both, too, avowed the importance of diversity. In natural evolution, diversity and differentiation produce complexity, but the same is true of anarchism, for as the anarchist Sir Herbert Read observed: "Progress is measured by the degree of differentiation within a society." Increasingly, in Bookchin's mind, ecology and anarchism were converging.[44]

Furthermore, in both natural and social evolution, diversity creates "wholeness and balance." In ecology, diverse species and elements create a healthy, rounded whole." Similarly, in an anarchist society, diverse individuals and communities "seek to achieve wholeness" in the sense of "complete, rounded" people living harmoniously with their natural surroundings. In ancient Athens and in the small cities of Renaissance Italy, agriculture harmonized with urban life, "synthesizing both into a rounded human, cultural, and social development."[45] Free people in an ecological, anarchist society could revive and approximate that wholeness and roundedness.

In sum, the principles of ecology and the principles of anarchism coincided in a compelling way. Spontaneity, differentiation, complexity, diversity, balance, and wholeness were all "as important in producing healthy human communities as they are in producing stable plant-animal communities." Their application would "promote human health because they produce a stable ecosystem."[46]

Equally fascinating for Bookchin, "unity in diversity"—another name for such wholeness—was also a theme in Hegelian philosophy. That was "an intellectual convergence that I do not regard as accidental," he would later observe, since "the language of ecology and dialectical philosophy"—concerned with processes of developmental change—"overlap to a remarkable degree."[47]

By contrast, Bookchin argued, modern society is sending the developmental processes of natural evolution into reverse: by polluting the air and water, by destroying topsoil and forests, by poisoning food with chemicals, and by promoting the hypertrophic growth of megacities, it was stripping nature of its complexity, "disassembling the biotic pyramid," replacing a complex environment with a simpler one that will be able to sustain only simpler life-forms. If current trends continue, then "the preconditions for advanced life will be irreparably damaged and the earth will prove to be incapable of supporting a viable, healthy human species."[48]

The convergence of anarchism, ecology, and dialectical philosophy gave rise to "Ecology and Revolutionary Thought," the first manifesto of radical political ecology, which Bookchin wrote in 1964. Only a revolution, one that creates an anarchist society, can create the deep social transformation that will be necessary to avert the ecological crisis. Far from being impractical, or "a remote ideal," an anarchist society "has become a precondition for the practice of ecological principles."[49] Rosa Luxemburg had said that civilization's choice was between "socialism or barbarism"; Bookchin restated it as a choice between anarchism and extinction.

As he traveled to lecture more frequently, he missed Friday-night *CI* meetings and, despite his promise to Dave Eisen, submitted nothing for publication. Eisen was perturbed. "You have given everything else precedence over your responsibility to the magazine," he scolded.[50] Finally, Eisen decided he did not want to have to wheedle and coax: in July 1963 he resigned both from the editorship and from the *CI* group. Bookchin followed, resigning in August.[51] *CI* would expire altogether a few years later.

To former *CI*-ers who were looking for a new political home, the most promising group around was CORE, the Congress of Racial Equality, which used Quaker- and Gandhi-inspired nonviolent resistance to fight southern racial segregation. CORE had helped organize the Freedom Rides, where blacks and whites partnered to ride together on interstate buses—and were beaten for it by white supremacists in Alabama. Now CORE was planning Freedom Summer, to conduct a voter registration drive in Mississippi. A member of New York's Downtown CORE chapter, Mickey Schwerner, had already gone to the South to lay the groundwork. In early 1964 Bookchin joined Downtown CORE.

Meetings took place at a storefront on East Thirteenth Street, where Bookchin met a twenty-year-old organizer named Allan Hoffman, with curly jet-black hair. At twenty, Allan was less than half Murray's age, but they struck up an acquaintance. Each recognized in the other something of himself: political passion and intellectual curiosity.[52]

Born in Brooklyn in 1944, Allan as a teenager had soaked up existentialism, poetry, and jazz in Greenwich Village, spending long evenings "drinking cheap wine and talking about the meaning of art, the existence of God, and who we were." He had joined the Committee for Non-Violent Action, a pacifist group, for its rejection of "all manifestations of violence." And he fell in with the anarchist-pacifist Living Theatre, performing in their space at Fourteenth Street and Sixth Avenue. The Living's cofounder Judith Malina praised him as "beautiful and far-out," committing "daring and pure acts of civil disobedience in the best Gandhian sense."[53] Now he was a field organizer for CORE.

After meetings Murray and Allan would head over to Stanley's Bar, on St. Mark's Place, in whose laid-back, dimly lit ambience they'd talk free-form politics.[54] Murray and Allan surely raised a glass of vodka together on January 6, 1964, when Con Ed withdrew its application to build the Ravenswood reactor. For the first time, a citizen movement to oppose the construction of a nuclear reactor had triumphed over the huge utility. It was an astounding victory, with strong implications for the power of popular, democratic local government.[55]

Murray and Allan did not see eye to eye on everything. For one thing, Allan was a pacifist. In the fall of 1961, in Times Square, he had been protesting nuclear weapons testing, when police mounted on horses had charged. Allan had sat down in front of them and stopped them, nonviolently, "with pure Gandhian zap," as Malina put it.[56] By contrast, back in the 1930s, when Murray and his comrades had been protesting, they had actively provoked mounted police to charge and then fought them with two-by-fours. Neither would have been impressed by the other's tactic.

The two men shared a philosophical bent, but while Murray was fascinated by the convergence of anarchism, ecology, and dialectical philosophy, Allan was drawn to Eastern mysticism—Taoism and the *I Ching*. Despite Murray's distaste for spirituality of any kind, he soon realized that Allan was talented, in fact "rather brilliant."

He found intellectual interchange with Allan a "sheer pleasure," and he was eager to teach him about radical history and theory, while Allan taught him about Albert Camus and existential revolt. "We complemented each other to an astonishing extent," Bookchin would later write. By the summer of 1964 they were close friends— "We loved each other dearly."[57]

In the spring of 1964, as the Johnson administration was sending troops to Vietnam, a radical group in New York decided to organize a protest march. M2M (for May 2nd Movement, the date of the march) was actually a front for a Marxist-Leninist vanguard party called Progressive Labor (PL).[58]

In December 1961 the American Communist Party had expelled a cadre of ultra-leftists for being too militant—they identified with the China of Mao Zedong. A few months later they had formed PL, which soon had six hundred members. They were champing at the bit to provoke the police into charging at them, so they could spark the revolution and win esteem as revolutionary leaders in the eyes of the working class.[59]

Under the name M2M, PL called a meeting to organize New York City's first anti–Vietnam War march. Bookchin attended—and was startled to see young Maoists declaiming in Bolshevik jargon and strutting imperiously. If there was one thing Bookchin knew, and knew to avoid, it was Marxist-Leninist politics. He scolded the organizers for their adherence to an outmoded ideology that had authoritarian implications. You shouldn't be organizing in terms of the proletariat, he berated them. You should be organizing in terms of anarchism and ecology. Bookchin's outburst didn't alter PL's plans, but it inspired some attendees "to begin reading about the origins and practice of contemporary anarchism."[60]

On April 22, 1964, in Queens, the New York World's Fair was scheduled to open to extravagant festivities. Its various pavilions would celebrate American capitalism and technological prowess. GE's exhibit, for example, would showcase electric power generation with a simulated thermonuclear fusion reaction.

New York's civil rights and peace groups loathed the message of the fair: social progress in America was a fraud as long as black citizens in the North and South were subject to discrimination and police brutality and poverty. So CORE organized a nonviolent protest for opening day. On April 22, at around nine in the morning, Murray and probably Allan arrived at the fair's entrance, along with James Farmer, Bayard Rustin, Michael Harrington, Paul Goodman, and 750 other protesters. They fanned out to the various displays. Some blocked the doors at the New York City Pavilion; some climbed the Unisphere, a large structure donated by US Steel; some stood atop the Florida Pavilion's giant orange; and some, at the Louisiana Pavilion, demonstrated the use of cattle prods, wielded on black prisoners in jails in New Orleans and Baton Rouge. Bookchin was among those arrested for their nonviolent civil disobedience and was detained for a week at Hart Island, a World War II–era prisoner camp.[61]

But the nonviolent phase of the civil rights movement was coming to an end. A few weeks later in Mississippi, CORE member Mickey Schwerner, along with two other civil rights workers, disappeared, brutally murdered by white supremacists. In New York on July 16, an off-duty policeman shot and killed a black fifteen-year-old. Two days later CORE rallied in Harlem to demand that the policeman be arrested for murder. The chairman of Downtown CORE announced, "I belong to a nonviolent organization, but I'm not nonviolent. When a cop shoots me, I will shoot him back." His fellow protesters agreed, clashing with police at the station house. "Over the next several hours, thousands rioted in the surrounding blocks—banging on cars, throwing bottles, smashing windows. The police pursued, firing revolvers into the air."[62]

At a rally in Harlem, Bayard Rustin pleaded with angry protesters to remain nonviolent, but they booed loudly, having no further use "for preachments on nonviolence." An urban rebellion erupted, and flames burned in Harlem for four days. One Harlem PL-er told a street-corner crowd: "We're going to have to kill a lot of cops, a lot of these judges, and we'll have to go up against their army."[63] Five people were killed, and several hundred were injured.

Malcolm X announced that "the day of turning the other cheek to those brute beasts is over." Waning too was the black-white alliance—white CORE members felt increasingly unwelcome. That fall Murray and Allan quit; two years later CORE would expel its white members.[64]

Since the end of World War II, mechanization and automation in the industrial workplace had been eliminating assembly-line jobs. In March 1964 several high-profile leftists issued a statement called *The Triple Revolution*, to warn that the technological revolution would soon render much of the manual workforce unemployed. In order to save jobs, they urged, government should place restraints on the development and use of technology.[65]

Bookchin and his old CI friends were perplexed—they considered technology's elimination of toil to be, not a problem, but a liberation. In the fall of 1964, to respond to *The Triple Revolution*, Frances Witlin, a friend and fellow ex-Trotskyist, created the Alliance for Jobs or Income Now, which sponsored a three-month lecture series called "Automation and Social Change." Bookchin was one of the lecturers, and in his presentation outlined what he was beginning to call his "postscarcity thesis."

For millennia, people had competed for scarce resources, and those few individuals who got control of the largest share ended up dominating everyone else. This tyranny of scarcity had always stymied revolutionary movements for freedom, since even after a revolution toppled a despot, most people were still burdened by material want and had to return to work, allowing new despots to arise. In the past, socialism had tried to put a good face on the problem by extolling toil as ennobling. But that just had the effect of equating socialism with industrial labor and full employment.[66]

But now machines could perform repetitive tasks, freeing people from that drudgery. The technology that had emerged since the end of the war had created a high level of material abundance, enough to provide for the basic needs of all. Most so-called work was, after all, useless. "Roughly seventy percent of the American labor force does absolutely no productive work," Bookchin observed—they just shift papers around. And most of the goods produced were superfluous, "such pure garbage that people would voluntarily stop consuming [them] in a rational society."[67] Let technology perform whatever toil was necessary, and let us eliminate make-work and reduce working hours—and a decent means of life could be available to all.[68]

The fact that everyone could now potentially have material security had enormous implications. People would no longer need to scramble to get a living, so competition for scarce resources could be consigned to the past. In this "postscarcity" world, even social and political domination could pass into history.[69]

In the winter of 1964–65 at Stanley's, as Murray and Allan talked about postscarcity and philosophy, other patrons clustered around, eager to listen in. At closing time, Stanley would call out, "Time gentlemen, time," and turn up the lights. It was "almost traumatic for patrons to have to leave," one denizen recalled. "Stanley almost had to press them."[70] So Bookchin invited the listeners who wanted more to come up to his apartment.[71] The conversations became regular meetings, and when the regulars kept coming and brought their friends, they formed a group.

Most of them were in their twenties, young bohemians seeking "meaningful alternatives to the self-destructive life around us."[72] They were hungry for a libertarian education, and Bookchin was happy to provide it, turning the group into a study group. His and Allan's shared interest in philosophy and theory set the tone. (Joyce Gardner called it the "metaphysical discussion group.")[73] Theory was necessary to the revolutionary project, Bookchin taught them, because it could "rescue the rationality of the individual revolutionary from the irrationality of the existing society." But theory alone was insufficient: revolutionaries must also have "the forces of history" on their side. And while waiting for those forces to emerge, they must not only analyze the ills of the existing society but plan the institutions and culture for the new society.[74]

He taught them about the potentialities latent in technological progress that made utopia no longer a vague ideal but a practical program for liberation, about decentralized eco-communities and solar energy, and about assembly democracy. He taught them about the French Revolution, the Paris Commune, and the democratic soviets in the early phases of the Russian Revolution. Martin Buber's *Paths in Utopia*, which they read aloud, gave them an introduction to nineteenth-century cooperativism.

They read *The Mass Psychology of Fascism* by Allan's favorite author, Wilhelm Reich, which fascinated them. Reich had argued that Hitler's rise to power had been made possible by sexual repression; the way to prevent a new fascism from emerging was therefore through sexual liberation. The book had been banned by a US district court a few years earlier, so the group published a pirate edition themselves. "We thought it was a revolutionary act," Gardner recalled. "We had printers who were into 'liberating' paper and printing stuff for free."[75]

Bookchin recommended the writings of Herbert Marcuse, from whose *Reason and Revolution* he had learned so much in his CI years. They read his *Eros and Civilization* (1955), which they applauded for its vision of a sexually nonrepressive civilization.

But *One-Dimensional Man*, which came to them hot off the press in 1964, was a disappointment, portraying the present society as a vast technocracy in which social control and repression were ubiquitous and monolithic. All oppositional behavior and thought, argued Marcuse, were integrated into this "one dimensional" social universe: even conceiving an alternative was all but impossible.[76] Bookchin, chagrined by the pessimistic direction Marcuse was taking, regarded the book as so bleak as to betray the revolutionary project.

Overall, by educating these young people in radical politics and theory and history, Bookchin hoped to generate a new Enlightenment that would spawn a revolutionary opposition, the way the French Enlightenment had fed into the Revolution of 1789.[77] He urged the group to identify itself as anarchist. Allan objected, citing "the violent, personal-terrorist aspect" of anarchist history. (Nineteenth-century anarchists had assassinated heads of state.) But in the end they had decided, probably at Bookchin's urging, that *anarchism* was "the only word which could potentially embrace our enthusiastic affirmation of life, together with the negation of all coercive systems."[78] Besides, the label was not co-optable. In the end, they agreed and took the name New York Federation of Anarchists.

Bookchin wrote the federation's manifesto, "The Legacy of Domination," which also summarized his own current thinking. Domination is based on scarcity, but scarcity no longer exists, so domination in all its forms—military, statist, economic, cultural, psychological—is historically redundant, lacking any further justification. It is merely a hollow "legacy of the past, a product of historical inertia." As such, we may easily overturn it. We may justifiably abolish all systems of domination and replace them with institutions appropriate to a balanced relationship with nature, especially "assemblies of free individuals who live in decentralized communities" and who "determine their social destiny in a direct, face-to-face democracy."[79]

Those assemblies would take over decision-making on social life, including postscarcity economic life. In a long, detailed article called "Towards a Liberatory Technology," he showed how new technologies could facilitate the transition to a decentralized, self-managed society and constitute its industrial skeleton. Today's giant centralized factories could be eliminated, replaced by small-scale plants producing for local needs. Within those small factories, automated machinery could replace most onerous human physical labor; through cybernation, such machines could correct their own errors, replacing human judgment. At the same time miniaturized electronic components—including the new digital computer—could coordinate most routine industrial operations and even perform much of the repetitive, tedious mental labor characteristic of industrial production. Various kinds of machines could be multipurposed to manufacture a variety of products, including other machines. Decentralized industry, like decentralized agriculture, would be tailored to regional economies, powered by solar and wind and tidal energy.

All these innovations could potentially replace "the realm of necessity" with the "realm of freedom" (terms that Bookchin borrowed from Marx) and could become a "technology for life" contributing to wholeness and balance, rather than a technology for mindless consumption and military destruction. Bookchin was optimistic about the prospects for such change because "a basic sense of decency, sympathy, and mutual aid lies at the core of human behavior."[80]

Allan marveled at how the "strange & inexorable laws of dialectical thought" had moved revolutionary theory from Marxism "onto a new level of speculation which becomes suddenly Anarchist." But how would the next revolution prevent a counterrevolution and a new revolutionary tyranny? he wondered. "We have already seen . . . too many revolutions which create greater evils than the ones they destroy."[81] The solution, he thought, was to change people at the psychological level, to alter individual character structure. Today, domination warps people's minds and bodies, but a revolutionary movement must try to generate "a new psychological order in which the fundamental drives (Eros) are liberated & allowed free play." It must remake "the totality"—not just the society but the individual person, sexuality included.

Bookchin, who had been grappling with the problem of revolutionary authoritarianism for a quarter-century without considering the psychological dimension, concurred. From the YCL commissars to the SWP Trotskyists to Herr Weber, the leading revolutionaries he had known had had dictatorial personalities. That they had all failed was fortunate, for "had society fallen into our hands, it would have been a world-historical disaster." He knew he had to get rid of his own inner commissar and "remake the revolutionary": "to the degree that it is humanly possible, [the revolutionary] must reflect in his [or her] own life what he is fighting for in social life."[82]

Agreeing with Martin Buber that the fate of the human race hung on the "rebirth of the commune," the New York *fédérés* made themselves into an urban commune, renting a loft at Bowery and Second Street.[83] They fitted it out with a library and set up a communal kitchen and dining area. They refurbished two discarded printing presses and set up a radical print shop, where they published a journal called *Good Soup*. The first issue contained Bookchin's "The Legacy of Domination" and Allan's rejoinder on totality, as well as poems by Joyce Gardner and Judith Malina.[84]

Early in 1964 Lower East Side poets found themselves without a place to read their work aloud. (The city had recently shut down the readings at Le Metro café, saying it didn't have a cabaret license.) Poet David Henderson, who lived upstairs from the New York Federation's loft, suggested to Allan that the anarchists host the poets' readings.[85]

They agreed, and thereafter Lower East Side poets—including the likes of Ishmael Reed and Allen Ginsberg—read their work weekly at the loft; Allan named the whole project the Bowery Poets Coop.[86] Bookchin didn't care for the poetry he heard there, finding it "harsh and shrill. People strode up and down the aisles, screaming at the audience. 'Up against the wall motherfucker. White honky.' Who cares? . . . What was interesting was that a cultural dimension had to be added, woven into a political outlook." After the readings, the *fédérés* would pass the hat to pay the rent. They let the poets use their printing presses to publish their work. The Poets Coop continued for a year or so, until police and landlord harassment, and lack of funds, broke it up. In the summer of 1966, the poets found a new home at St. Mark's Church.[87]

In 1965 the *fédérés* opened the Torch Bookstore on East Ninth Street, the first anarchist bookstore in New York since World War I. Used books and magazines lay scattered about, and historic posters adorned the walls. It was organized—or not—along libertarian lines. "No one really runs the Torch," wrote the *East Village Other*. "There is no cash register. The rent is paid by voluntary contribution." It had no regular hours—"it is open only when someone wants to open it."[88]

The New York Federation's urban commune was small and its efforts fragile, but its message nonetheless had a pronounced influence on other bohemians in the Lower East Side. At antiwar demonstrations, the *fédérés* marched under a banner that read "Community, anarchy, ecology." People understood the first two words but asked, *What's ecology?* No one had ever heard of it. Meanwhile the group circulated Bookchin's "Towards a Liberatory Technology" and "Ecology and Revolutionary Thought." The postscarcity thesis caught on, enough that, for example, in 1966 Abbie Hoffman could remark that he was thinking in terms of a "postabundant society" and affirm that "due to our technological capabilities, we can really have an anarchist utopian future in this country." The SDS newspaper *New Left Notes* observed approvingly that "the concept of post-scarcity" means "the liberation of men and the creation of a socialist society."[89] Postscarcity soon became the de facto economic theory of New York's youth movement.

In February 1966 the *fédérés* demonstrated publicly as anarchists—carrying black anarchist flags for the first time since Emma Goldman had done so during World War I. They were protesting the arrest, in the USSR, of two Russian novelists for criticizing the system. Sally Kempton, who covered the protest for the *Village Voice*, called it "historic."[90]

In 1965, forty-four-year-old Bookchin found himself associating with people half his age. Looking for generational peers, he started attending meetings of the Libertarian League, although its ten or so members turned out to be elderly, like Sam Dolgoff, who was sixty-three.[91] League members spent a lot of time memorializing the Spanish Revolution of 1936–37 and the Confederación Nacional del Trabajo (CNT), the anarcho-syndicalist trade union that had carried out the workplace collectivizations in Catalonia.

One league member, Russell Blackwell, was intriguing: he had actually fought in Spain. Now sixty-one, Blackwell had stowed aboard a freighter in 1936 and arrived in Barcelona in 1936 to gather information for his Trotskyist comrades back home. Speaking fluent Spanish, he had mixed easily with the anti-Stalinist leftists of the Partido Obrero de Unificación Marxista (POUM, in whose militia George Orwell had fought), as well as the anarcho-syndicalists of the CNT and the anarchist-communists of the Federación Anarquista Ibérica (FAI).[92] He had joined the militant anarchist group Friends of Durruti.

In May 1937, in that momentous Barcelona street battle in which the Stalinists crushed the anarchist revolution, Blackwell had fought alongside the anarchists and was wounded. After vanquishing the uprising, the Stalinists had arrested him on charges of being an American spy and, of course, "an agent of Trotsky." He was imprisoned, interrogated, and beaten. His friends back home formed a defense committee, and thanks to their agitation, he was released.[93]

When I knew Murray, decades later, he spoke of no one except his grandmother with more reverence than Russell Blackwell. It had been a privilege to be his *compañero*, he said. (*Compañero* was what Spanish anarchists called each other, as opposed to the Stalinist *comrade*.) Blackwell gave him intimate details of the Spanish anarchist revolution. He told him about their remarkable organizational unit, the *grupo de afinidad*, a small group of *compañeros* who functioned closely together and could not be penetrated by outsiders.

Bookchin admired the Spanish Revolution immensely, but no matter how glorious it had been, he thought the league's anarchists were living too much in the past. He told them about his project to renovate anarchism for the present era, through ecology and postscarcity. The younger generation, he said, the New Leftists and the New Bohemians, had an anarchistic outlook, and his New York Federation was explicitly anarchist. Out in San Francisco, a group of anarchistic actors called the Diggers were performing plays for free in Golden Gate Park and serving free food. They had founded a free store that gave away secondhand clothes and set up free housing.[94] These institutions were part of a new and growing money-free economy, Bookchin told the old-timers. It was part of the new society, a new technology, and potentially a new utopia.

But Dolgoff remained skeptical—he had a low opinion of the young bohemians. Whenever they came to league meetings, they talked mainly about sex and drugs: "They were disruptive and did little constructive work. The problem became how to remove or expel them." Bookchin could not persuade him otherwise and resigned himself to the aging anarchists' hostile incomprehension of their youthful counterparts.[95]

Only Russell Blackwell lent Bookchin a sympathetic ear. But one evening he took his young *compañero* aside and gave him an unexpected warning: *Don't use the word* anarchist *for your political label*, he said; *if you do, you'll attract every nut for miles around*. Bookchin listened but decided not to heed the advice: that would not be a problem, he felt, since he was sure he could renovate anarchism into something relevant for the present—and enlist enough spontaneous, diverse, ethical young people, whom he could educate in radical politics, to create a movement for a free, rational society.

One of the scruffy young bohemians who turned up at league meetings was Ben Morea, a dark-haired twenty-four-year-old, short in stature but intense in demeanor, a swaggering street kid.[96] A former teenage junkie, he had kicked the habit and taken up abstract painting. Morea's canvases, Bookchin recalled, "consisted of vast panels of black. Swirling nebulae. Completely black."[97] He had learned about anarchism from the Living Theatre and now turned up at Libertarian League meetings, where he too was fascinated by Russell Blackwell.

At Bookchin's invitation, Morea attended a New York Federation meeting as well. He listened for a while, then scowled that they were all petty bourgeois white honkies and stormed out with a stream of curses. He came to the next meeting, but the same thing happened. He found Murray "too intellectual," while Allan was just "a passive hippy." No love was lost: Ben's tantrums did not impress Murray, while nonviolent Allan found him destructive and nihilistic.[98]

But the Lower East Side of 1965 had room for them all. Near the *fédérés*' loft lay Tompkins Square Park, and on spring and summer afternoons, Bookchin would meet his friends there. Lounging on a sunny bench, Allan, with his long, curly black beard, would tilt his head back, lower his eyelids, and soliloquize, as a barefoot Joyce nestled at his side. In the evening they'd go to the loft to dine on vegetables and brown rice (or in Murray's case, tuna sandwiches). And on sweltering August nights, they'd sleep in the park or stay up all night and watch the sun rise over the river.[99]

As ever, Murray and Allan carried on their dialogue, Murray holding forth on revolutionary history and politics, Allan on the transformation of the individual psyche, and sexual liberation, and totality. Their respective writings reveal that they listened to each other closely. Murray echoed Allan when he wrote: "The established order tends to be totalistic—it stakes out its sovereignty" over the self's "innermost recesses"; so liberation, too, "can be totalistic, implicating every facet of life and experience."[100] And Allan's writings are shot through with allusions to postscarcity and cybernetics and the Paris Commune. Murray thought their dialogue had "a tremendous potential ... for creating a rich cross-cultural, visionary, utopian, even communistic movement." Between them they sought to meld "the potentially utopistic visions of the counterculture with the socialist visions of the New Left and with anarchism."[101]

Bookchin would always cherish "the beauty, the experimentalism, the openness, the generosity" of that summer of 1965, when the *fédérés*' communal life made it seem that "everything of a loving nature was possible." In later years, in later movements, he would look for that mutual trust and shared idealism in the people he worked with. For without it, "nothing will have been gained or will be worth gaining."[102]

NOTES

1. Robert A. Caro, *The Power Broker: Robert Moses and the Fall of New York* (New York: Alfred A. Knopf, 1974), 889ff.

2. Sandy Shortlidge cited in Robert Palmer papers, courtesy Bob Erler. The drinking bout was specific to this crisis; Murray was no alcoholic. This episode is not to be confused with a 1965 incident involving *A Thousand Clowns*.

3. Bookchin (as Lewis Herber), *Our Synthetic Environment* (New York: Alfred A. Knopf, 1962), 28, hereafter *OSE*.

4. Ibid., 26. On decentralism, see 237ff.

5. Burton Lasky, interview by author; Bookchin, *OSE*, v. At this time Bookchin was living, however tenuously, with Beatrice and their two small children, Debbie (born in 1955) and Joseph (born in 1961), at 3470 Cannon Place in the Bronx.

6. MBVB, part 30.

7. Bookchin, "Postwar Period," interview by Doug Morris, in Bookchin, *Anarchism, Marxism and the Future of the Left* (San Francisco: A.K. Press, 1999), 58.

8. See, for example, Lewis Mumford, "The Natural History of Urbanization," in William L. Thomas Jr., ed., *Man's Role in Changing the Face of the Earth* (Chicago: University of Chicago Press, 1956), 395–96. Wayne Hayes told me that Murray regarded this article as particularly important. In 1974 Bookchin would note that he benefited "immensely . . . from Lewis Mumford's studies on urban development." *OSE*, xviii.

9. E. A. Gutkind, *Community and Environment: A Discourse on Social Ecology* (New York: Philosophical Library, 1954); and *The Twilight of Cities* (New York: Macmillan, 1962).

10. Gutkind, *Twilight of Cities*, 183; *Community and Environment*, 76. "Mutual aid" is the title of a book by the anarchist Kropotkin.

11. Gutkind, *Community and Environment*, 81, 47, 50. The subtitle of this book, notably, is *A Discourse on Social Ecology*. Gutkind and Lewis Mumford both taught at the University of Pennsylvania in the 1950s; Mumford used the term *social ecology* in his introduction to the reprint edition of *Technics and Civilization* (1932; reprinted New York: Harcourt Brace & World, 1963); the book presented "technical development within the setting of a more general social ecology" (n.p.). In the 1970s Bookchin would make the name *social ecology* famous and give Gutkind credit for originating it. See chapter 8 of this book.

12. Judith Malina, *The Enormous Despair* (New York: Random House, 1972), 27.

13. George Woodcock, *Anarchism: A History of Libertarian Ideas and Movements* (Cleveland: World, 1962), 468.

14. See issues no. 34–45 passim, published from 1958 to 1962.

15. Bookchin (as Lewis Herber), "Limits of the City," *CI* 10, no. 39 (Aug.–Sept. 1960). On the circumstances of publication, see Murray Bob (as Dutscher) to friends, Nov. 25, 1959, courtesy Reni Bob.

16. Mina Grossman, "Focus on Independent Committees," Mar. 15, 1966, courtesy Jack Grossman. The group, which helped generate the movement to ban nuclear weapons testing, was called Women Strike for Peace.

17. "Port Huron Statement of the Students for a Democratic Society" (1962), online at http://www.h-net.org/~hst306/documents/huron.html.

18. The ideas were similar, but the Port Huron Statement was not, to my knowledge, influenced by *CI*. The inspirations were, rather, C. Wright Mills, Albert Camus, and Paul Goodman.

19. Dave Eisen files, Augusta, Me.

20. The new apartment was at 152 Seventh Avenue South. Bookchin to Tom (handwritten notification of address change), August 14, 1962, courtesy Juan Diego Pérez Cebada. His divorce from Beatrice would be finalized in October 1963. They would remain friends for the remainder of his life.

21. He would index Will and Ariel Durant's *Rousseau and Revolution* (in 1967) and the first complete English translation of Max Weber's *Economy and Society* (in 1968).

22. Lasky interview.

23. Bookchin, "Postwar Period," 54.

24. Yaakov Garb, "Rachel Carson's *Silent Spring*," *Dissent* (Fall 1995), 540–45.

25. William Vogt, "On Man the Destroyer," *Natural History: The Journal of the American Museum of Natural History* [New York] 62 (Jan. 1963), 3–5. See also "Cassandra in the Cornfields," *Economist* [London] (Feb. 23, 1963), 711. While Carson's book upstaged Murray's in the United States, *Our Synthetic Environment* received much more attention elsewhere in the English-speaking world, as Juan Diego Pérez Cebada has shown. The book was reviewed much more widely in the United Kingdom, and reviews appeared even in Australia, South Africa, and Pakistan. See Cebada, "An Editorial Flop Revisited: The Initial Impact of *Our Synthetic Environment* (M. Bookchin, 1962)," publication forthcoming. I'm grateful to Cebada for sharing his manuscript with me.

26. T. G. Franklin, "Evolution in Reverse?" *Mother Earth* [journal of the Soil Association, UK] (July 1963); René Dubos, *Man Adapting* (New Haven, CT: Yale University Press, 1965), 196, 415. Brian Morris observes that Lewis Mumford, René Dubos, and Bookchin were all part of an intellectual current that he calls ecological humanism; see his *Pioneers of Ecological Humanism* (Bristol: Book Guild, 2012).

27. Stephanie Mills, "Peter Kropotkin, Murray Bookchin, Ivan Illich," *CoEvolution Quarterly*, Summer 1975, 67; Theodore Roszak, "The Obsessive Drive to Dominate the Environment," *San Francisco Chronicle*, May 16, 1982; Garb, "Rachel Carson's *Silent Spring*," 540.

28. As of August 1963 he was living at 217 Avenue A, near Thirteenth Street. Bookchin to Dave Eisen, August 23, 1963, courtesy Dave Eisen.

29. Dave Eisen to Friends, Nov. 12, 1962, courtesy Dave Eisen.

30. See, for example, Bookchin (as Lewis Herber), "Dangerous Environment of Man," *Consumer Bulletin*, Aug. 1962; and Bookchin (as Lewis Herber), "The Transformation of Our Environment," WBAI, New York, October 1962.

31. Richard P. Hunt, "Atomic Question for the City," *New York Times Magazine*, Oct. 6, 1963, 46ff.

32. George T. Mazuzan, "Very Risky Business: A Power Reactor for New York City," *Technology and Culture* 27, no. 2 (Apr. 1986), 270–74, 283; Murray Illson, "City Hall Witnesses Split on Bills to Bar Reactor in Queens," *New York Times*, June 15, 1963.

33. Bookchin (as Lewis Herber), *The Ravenswood Reactor: A Preliminary Report to the Public* (New York: Citizens Committee for Radiation Information, 1963), in MBPTL and copy in author's collection.

34. Bookchin (as Lewis Herber), *Crisis in Our Cities* (Englewood Cliffs, NJ: Prentice-Hall, 1965), 186, 194–95.

35. Ibid., 187.

36. Ibid., 186, 188–89.

37. Bookchin, "Towards a Liberatory Technology" (1965), in *Post-scarcity Anarchism* (Berkeley, CA: Ramparts Press, 1971), 128–29. *Post-scarcity Anarchism* is hereafter *PSA*.

38. Bookchin (as Herber), *Crisis in Our Cities*, 194.

39. Bookchin, "Towards a Liberatory Technology," 102, 107; *Crisis in Our Cities*, 194–95.

40. Edward T. Chase, "The Effluent Society," *Book Week*, May 2, 1965.

41. His source for scientific ecology was Charles Elton, *Ecology of Invasions by Animals and Plants* (London: Methuen, 1958). He credits Elton in Bookchin, *OSE*, xvii.

42. Bookchin, *OSE*, 59.

43. Bookchin, "Ecology and Revolutionary Thought" (1964), in *PSA*, 76; Bookchin, "Introduction" (1970), in *PSA*, 21.

44. Bookchin, "Towards a Liberatory Technology," 91; Bookchin, "Ecology and Revolutionary Thought," 77. Murray quoted this sentence from Read's essay "Philosophy of Anarchism" (1940) in "Ecology and Revolutionary Thought," 77.

45. Bookchin, "Ecology and Revolutionary Thought," 77; "Towards a Liberatory Technology," 138; *OSE*, 242.

46. Bookchin, "Ecology and Revolutionary Thought," 77–78; *OSE*, 209.

47. Bookchin, "Toward an Ecological Society" (1973), in *Toward an Ecological Society* (Montreal: Black Rose Books, 1980), 59; Bookchin, "Spontaneity and Organization" (1971), in *TES*, 272. *Toward an Ecological Society* is hereafter *TES*.

48. Bookchin, *OSE*, 60.

49. Bookchin, "Ecology and Revolutionary Thought," was reprinted frequently but most accessibly today in *PSA*.

50. Eisen to Bookchin, Jan. 20, 1963, courtesy Dave Eisen.

51. Dave Eisen files.

52. Marty Jezer, introduction to Allan Hoffman, "Earth—Life—Reclamation," *WIN Magazine* 6, no. 19 (Nov. 15, 1970), 4.

53. Ibid.; Allan Hoffman (as Totalist), "Eighteen Rounds of Total Revolution," *Anarchos*, no. 2 (Spring 1968), 13–14; Malina, *Enormous Despair*, 176–77.

54. Bookchin, "When Everything Was Possible" (originally written in 1973), in *Mesechabe: The Journal of Surregionalism* [New Orleans], nos. 9–10 (Winter 1991). On Stanley's as a political forum, see Peniel E. Joseph, *Waiting 'Til the Midnight Hour: A Narrative History of Black Power in America* (New York: Holt, 2007), 41–42.

55. Thomas Buckley, "City to Weigh Peril of Nuclear Plant Sought for Queens," *New York Times*, May 10, 1963, 1; "Atomic Plant in Queens," *New York Times*, Aug. 26, 1963, 26; and Richard P. Hunt, "Atomic Question for the City," *New York Times Magazine*, Oct. 6, 1963, 46ff. See also Mazuzan, "Very Risky Business," 263.

56. Malina, *Enormous Despair*, 176–77.

57. Bookchin, "When Everything Was Possible," 2.

58. Newfield, *Prophetic Minority*.

59. Ibid.

60. Peter Berg, "Some Encounters with Murray Bookchin," http://www.planetdrum.org/bookchin.htm. The antiwar march took place on May 2, 1964. A thousand people marched from Washington Square to UN Plaza. Police arrested about fifty.

61. Junius Griffin, "Stall-in Leaders Defy Plea to Bar Tie-up Tomorrow," *New York Times*, Apr. 21, 1964; Fred Powledge, "Demonstrations Atop Unisphere and Giant Orange Considered," *New York Times*, Apr. 21, 1964; and Homer Bigart, "Fair Opens, Rights Stall-in Fails; Protesters Drown Out Johnson; 300 Arrested in Demonstrations," *New York Times*, Apr. 23, 1964. See also Seth Cagin and Philip Dray, *We Are Not Afraid: The Story of Goodman, Schwerner, and Chaney and the Civil Rights Campaign for Mississippi* (New York: Scribner, 1988); and Bookchin, "The 1960s," interview by Doug Morris, in Bookchin, *Anarchism, Marxism, and the Future of the Left* (San Francisco: A.K. Press, 1999), 65.

62. John D'Emilio, *Lost Prophet: Bayard Rustin and the Quest for Peace and Justice in America* (New York: Free Press, 2003), 382; August Meier and Elliott Rudwick, *CORE: A Study in the Civil Rights Movement, 1942–1968* (1973; Urbana: University of Illinois Press, 1975), 301–2.

63. D'Emilio, *Lost Prophet*, 383; Newfield, *Prophetic Minority*, 161.

64. D'Emilio, *Lost Prophet*; Bookchin, "1960s," 75–76.

65. The document was signed by, among others, Michael Harrington, Irving Howe, Linus Pauling, Bayard Rustin, and its principal author, Robert Theobald. See http://bit.ly/r92PGJ.

66. Bookchin, "Towards a Liberatory Technology," 90–93.

67. Bookchin, "Discussion on 'Listen, Marxist!'" (1970), in *PSA*, 236.

68. To my knowledge, no record survives of what Bookchin said at his JOIN lectures; I am inferring from articles he wrote around the same time.

69. Bookchin, "The Legacy of Domination," *Good Soup*, no. 1 (ca. 1965).

70. Judith Watson, "Memoirs: Living in New York City's East Village (1963)," East-Village.com, accessed Jan. 19, 2008.

71. The apartment was at 718 East Ninth Street, near Avenue C.

72. Joyce Gardner and Allan Hoffman, "Introduction," *Good Soup*, no. 1 (ca. 1965), n.p.

73. Joyce Gardner, *Cold Mountain Farm: An Attempt at Community* (n.p.; ca. 1968).

74. Bookchin, "Legacy of Domination."

75. Joy Gardner interview, Apr. 5, 2008. *The Mass Psychology of Fascism* and other works by Reich, as well as his "orgone accumulators," were banned on March 19, 1954, by Judge John D. Clifford in U.S. District Court in Portland, Maine, case no. 1056.

76. Herbert Marcuse, *One-Dimensional Man* (Boston: Beacon, 1964), 253.

77. Bookchin, "Legacy of Domination."

78. Gardner and Hoffman, "Introduction."

79. Bookchin, "Legacy of Domination."

80. Bookchin, "Towards a Liberatory Technology," 130, 138.

81. Allan Hoffman, "New Dimensions in Revolutionary Theory" (written Nov. 1965), *Good Soup*, no. 1 (ca. 1965), n.p.

82. Bookchin, "Youth Culture: An Anarcho-Communist View," in *Hip Culture: Six Essays on Its Revolutionary Potential* (New York: Times Change Press, 1970), 47; Bookchin, "Introduction" (1970), in *PSA*, 16; "Legacy of Domination," n.p.

83. Martin Buber, *Paths in Utopia* (1949; Syracuse, NY: Syracuse University Press, 1996), 136. See Gardner interview; Gardner and Hoffman, "Introduction"; and Bookchin, "When Everything Was Possible." Members of the New York Federation of Anarchists didn't call themselves *fédérés* to my knowledge, but I'm doing so here, as Murray in later years would use that word to refer to the French revolutionaries of 1789; it literally means "people who belong to a federation or confederation."

84. *Good Soup*, published by the New York Federation of Anarchists, no. 1 (ca. 1965).

85. See Daniel Kane, *All Poets Welcome: The Lower East Side Poetry Scene in the 1960s* (Berkeley: University of California Press, 2003), 20ff, 230–31 n.109.

86. Gardner interview; Paul Prensky, interview by author, Mar. 31, 2008.

87. MBVB, part 32; Gardner interview; Gardner and Hoffman, "Introduction"; Kane, *All Poets Welcome*, 55–56.

88. The storefront was at 641 East Ninth Street. Prensky interview; Bookchin, "When Everything Was Possible," 3; Don Newlove, "The Torch, but Yet a Candle," *East Village Other*, ca. 1966. On June 22, 1966, Murray spoke there on "The Relevance of Anarchism Today": see advertisement in *Village Voice*, June 23, 1966, in Bob Palmer notes, courtesy Bob Erler.

89. Abbie Hoffman, interview, in Allen Katzman, ed., *Our Time: An Anthology of Interviews from the East Village Other* (New York: Dial, 1969), 290, 296; Robert Gottlieb, Gerry Tenney, and David Gilbert, "Toward a Theory of Social Change," *New Left Notes*, May 22, 1967, 5.

90. MBVB, part 30; Sally Kempton, "Anarchists Picket Soviets, Proselytize Park Avenue," *Village Voice*, Feb. 24, 1966, 3.

91. The Libertarian League met at 813 Broadway, between Eleventh and Twelfth Streets.

92. Sidney Lens, *Unrepentant Radical* (Boston: Beacon Press, 1980), 83–84.

93. Ibid. Blackwell was supported by John Dewey, Carlo Tresca, Norman Thomas, James T. Farrell, Suzanne LaFollette, Sidney Hook, A. J. Muste, Dwight Macdonald, Bertram Wolfe, James Cannon, and Max Eastman, among many others. "Russ Blackwell, Longtime Fighter for Libertarianism," *Industrial Worker*, Oct. 1969. See also his obituary in *New York Times*, Aug. 22, 1969.

94. Michael William Doyle, "Staging the Revolution: Guerrilla Theater as a Countercultural Practice, 1965–1968," in Peter Braunstein and Michael William Doyle, eds., *Imagine Nation: The American Counterculture of the 1960s and '70s* (New York: Routledge, 2002), 71–98.

95. "Sam Dolgoff," in Paul Avrich, ed., *Anarchist Voices: An Oral History of Anarchism in America* (San Francisco: A.K. Press, 2005), 427.

96. Osha Neumann, *Up Against the Wall Motherf**ker* (New York: Seven Stories Press, 2008), 56.

97. Quoted in Anthony Haden-Guest, *True Colors: The Real Life of the Art World* (New York: Atlantic Monthly Press, 1996), 30–31.

98. Ben Morea, interview by author, May 1, 2008; "Jumbled Notes: A Critical Hidden History of King Mob" (2008), http://bit.ly/pcpoS8; Bookchin, "When Everything Was Possible," 4.

99. Bookchin, "When Everything Was Possible," 7.

100. Bookchin, "Desire and Need," (1967), in *PSA*, 283, 286.

101. Bookchin, "1960s," 71, 74.

102. Bookchin, "When Everything Was Possible," 6.

6

Counterculture Elder

AS AMERICAN MILITARY involvement in Vietnam escalated, the New Left shifted from demanding participatory democracy to protesting the war. On April 17, 1965, SDS held an antiwar demonstration in Washington, D.C.—the showing of twenty-five thousand was unexpectedly large. Against the advice of their Old Left elders, SDS-ers had decided to allow Progressive Labor, the Maoist group, to participate. Why be so exclusive? they wondered. What could go wrong?

Meanwhile the whole youth revolt was becoming "totalistic" in Allan's sense—favoring cultural and psychological as well as political liberation. Rock and roll and a new generational self-awareness were inspiring not simply bohemians but multitudes of young people to shake off social conventions. Starting in California in November 1965, they gathered in large venues under strobe lights, drank acid-spiked punch, and danced with abandon. LSD, these hippies believed, could tap into the interior spaces of personality and unlock their natural, spontaneous selves—and those selves were peaceful and communal. If enough people underwent the psychedelic experience, it could solve the problems of humankind, leading to revolutionary social and political change.

In 1966 and 1967, hippies streamed into the Lower East Side, supplanting the earlier bohemians, bringing with them a new kind of alternative culture, living together in communes. The younger New York *fédérés* too donned hippie garb and consumed LSD enthusiastically, considering it a revolutionary tool for the liberation of the psyche. They "incorporated sexual radicalism with all our other radicalism," Gardner told me. "Many members of the commune experimented with sexual freedom."[1]

Bookchin didn't wear hippie clothes—he stuck to his usual plaid or polyester shirts and olive drab field jacket. He kept his hair short and his face shaved. And he stayed aloof from the sexual experimentation. "I don't recall Murray being lovers with anybody," recalled Gardner. Sexually, he preferred emotional intimacy to promiscuity. "My sexual life was guided by my feelings," he later told an interviewer.[2]

But even as he stayed aloof from counterculture practices, he admired the "beautiful innocence" in his friends' glowing faces and trusting behavior,[3] seeing in it the potential for a profound social transformation. Rebellious youth had disavowed money and commodities and dropped out of a destructive culture; they had embraced the values of cooperation and communality. It was an ethical revolt, and their values could possibly fulfill the promise of the new technology, guiding it toward liberatory uses rather than wasteful or destructive ends. An ethically grounded youth movement, he thought, might achieve the role that Marx had assigned to the proletariat, as the agent of broad social revolution.[4] Young people might lack a socially revolutionary consciousness, but they could learn, and he would cultivate it in his young friends.

What irritated him was the drugs. At political meetings, joints would come out, and then next thing he knew, people wouldn't be talking anymore—"they were looking at their feet."[5] The New York Federation never did publish a second issue of *Good Soup* because the *fédéré* who was supposed to handle it was too stoned.[6] Nor did he agree that psychedelic drugs could lead to revolutionary social transformation: they were "devices for mere sensation, for formless states of mind."[7] Personally, he said, "I like clarity of thinking," so "I would not take LSD."[8]

By 1965 hippies were flooding into SDS, the largest New Left and antiwar organization, bringing with them a preference for sex, drugs, and rock 'n' roll over politics and demonstrations. At SDS's June 1965 national convention at Kewadin, Michigan, the many sensual distractions made focused discussion difficult. Those who tried to present position papers were ignored; those who tried to provide leadership were shrugged off as elitist. In a Dionysian haze, the participants voted to allow Communists to join SDS.[9]

No sooner had they done so than Progressive Labor burst in, having waited a long time to enter the fast-growing student antiwar organization. Clean-cut and disciplined, these Maoists had studied the Chinese and Cuban revolutions and now were prepared to assume the role of the New Left's vanguard. While other SDS-ers merely demanded American withdrawal from Vietnam, PL-ers actively supported the struggle of the Vietcong against the imperialist aggressor; now they would turn SDS's orientation from antiwar to pro-Vietcong. They diligently formed clubs on campuses to attract student supporters. "PL grew because it projected an image of fearless militance," noted one observer.[10]

Traditional Marxism seemed to be infecting an old hero of Bookchin's as well. In 1966 Herbert Marcuse spoke at the NYU Law School, where Murray was stunned to hear the author of *Reason and Revolution* and *Eros and Civilization* mouthing a crudely orthodox Marxist class analysis. According to this new incarnation of Marcuse, the peasant-based anticolonial revolts constituted a globalized proletariat that was even now rising up against the globalized bourgeoisie—the wealthy industrialized nations. Bookchin met him here and perhaps chided him for returning to Marxist formulas just when a rebellious youth culture with fervent dreams of a peaceful, egalitarian, postscarcity society was booming all around them.[11]

In March 1967 the two met again after Marcuse spoke at the School of Visual Arts, telling his hearers that the "totalitarian" nature of the United States "easily absorbs all non-conformist activities" and anesthetizes dissent. The only hope for changing this society lay in an art that assumes "a position of protest, denial and refusal." Once again, Bookchin tried to speak past his pessimism and call his attention to the new possibilities for a liberatory, ecological, and decentralized society.[12]

Bookchin had remained on good terms with the anarchist painter Ben Morea, despite their disagreements. Sometime in 1966 Morea formed a small group called Black Mask, as if to embody Marcuse's SVA call for an art of protest, denial, and refusal. An art that changed society, Black Mask held, had to exist outside museums and be integrated into everyday life.[13] Hence its members produced poster-size broadsheets that featured extravagant drawings with captions like "The proletarian revolution is the sexual revolution," and "The revolution is sexuality trampling civilization." Black Maskers plastered them onto walls around the Lower East Side, where they became a familiar sight.[14]

Increasingly, some of the New York *fédérés* dreamed of leaving New York and moving to the countryside. In 1966 they paid a visit to the School of Living at Heathcote, an intentional community in rural Maryland. Founded in the 1930s by Ralph Borsodi and now led by Mildred Loomis, the School of Living taught cooperative lifeways, communal land ownership, and self-sufficiency. The New Yorkers learned that they too could form a rural commune, where they could cultivate the soil together, produce their own food, and live self-sufficiently without much money. The visit was "a transformational revelatory experience," its atmosphere "akin to that of a church camp or revival or music festival." The *fédérés* decided "to go ahead and do it."[15]

They chose a rundown 450-acre farm in upstate New York, near Hobart, and named it Cold Mountain Farm. A group of the *fédérés* would go there and farm communally, organically, and send fresh vegetables back to the rest of the group in New York. Some harbored visions of free love.[16] Murray expected Allan to stay behind in the city with him, but in the spring of 1967 he decided to join Joyce and headed for the countryside. His departure pained Murray and brought "a coldness into our relationship that lasted for some time."[17]

After they left, Bookchin mulled over the compulsion that had driven half his group to betake themselves to the countryside.

The film *Marat/Sade* was then showing in New York, and as he listened to the inmates of the Charenton insane asylum singing about copulation amid the bloodshed of revolutionary France, he pondered the question of expanding the revolutionary project to include sexual desire. The problem of Need (the economic agony of the masses) would have to be solved by social revolution, while Desire (individual pleasure and sensuality) could be answered by private satisfaction. But Desire and Need were not in contradiction, because both are necessary for full human liberation. People in revolutionary situations historically demanded both "the filled belly and the heightened sensibility." For the hippie culture of the Lower East Side, the problem of Need had been solved by the technology-generated affluence of postwar America, and if the neighborhood now floated on a sea of sex and drugs, so had revolutionary Paris of 1793, which "floated on a sea of alcohol—for months everyone in the Belleville district was magnificently drunk."[18]

Still, the counterculture's new apotheosis of the individual self, he warned in his essay "Desire and Need," must not be taken to such an extreme that people give up on social action. For without social action, the self "contracts to banalities and trivia" and takes a placid "journey inward."[19]

Just at that moment, in March 1967, two members of the New York Federation who had declined to go to Cold Mountain Farm decided to join the Situationists, a group of artists and poets based in Paris. Since the 1950s, the Situationists had been criticizing the postwar cityscape, along lines similar to Bookchin's, calling it dehumanizing and anonymous and banal and gigantic. But where Bookchin offered a program—eco-decentralism—to contest and replace it, the Situationists' solutions were more fleeting: taking boulevard strolls without a destination; bringing art into the fabric of everyday life; and creating transient "situations" and "irruptive acts," such as building occupations, street demonstrations, guerrilla theater, and graffiti. Such *détournements* would transform everyday life momentarily, if not in any lasting way.[20]

The Situationists had recently formed an international (SI), and Tony Verlaan, an official representative of the SI, had arrived from Paris to form the American section; it was he who recruited the two *fédérés*. Between Cold Mountain Farm and Situationism, Bookchin felt his group evaporating. Rather than mourn another loss, he decided to go to Europe to investigate the political scene there. He'd been told that in Europe, too, disaffected children of affluence were rejecting consumerism and yearning for a revolutionary transformation of everyday life. Perhaps he could bring them together into an international anarchist movement. So in mid-August 1967 he boarded a plane for Paris.[21]

Upon his arrival, he spent a day with his friend Judith Malina of the Living Theatre, explaining that he wanted to "build up an anarchist program" and consolidate anarchist tendencies on both continents. She found him "full of grace and openness."[22] He visited members of the French Anarchist Federation, a small group of old-time anarcho-syndicalists, as well as Noir et Rouge, a group of younger anarchists. But old and young, in that summer of 1967, the French anarchists all told him the same thing: as far as radical politics was concerned, France—*C'est mort!* And indeed, compared to the growing social uproar over the Vietnam War back home, Paris did seem eerily quiet.

In Amsterdam, Bookchin went looking for the Provos, another youth movement, whose reputation for "happenings," theatrical gestures, and performance art had crossed the Atlantic. A philosophy student named Roel Van Duijn had founded the Provos back in 1965, based on the idea that artists, dropouts, street kids, juvenile delinquents, and beatniks could create a viable alternative to capitalist society. Van Duijn wanted to "awaken their latent instincts for subversion" and channel them "into revolutionary consciousness" and "anarchist action."[23] The Provos called for banning automobiles from the inner city and setting up a system of white-painted bicycles, to be shared, gratis, for urban transport.[24] Many members of the Dutch public sympathized with them—so much so that the Provos became superficially popular, even among the bourgeoisie. The founders considered such widespread acceptance intolerable and in the spring of 1967 declared the "Death of Provo." Bookchin arrived a few months later to find Amsterdam, too, quiet.

Returning to Paris, Bookchin visited (or was granted an audience with) the principal Situationists: Guy Debord, Raoul Vaneigem, and Mustapha Khayati. They, too, were depressed about the placid French political situation. Murray tried to give them hope. He told them about anarchism, but they weren't interested.[25] He told them about the counterculture and the huge impact it was having in the States, but the Frenchmen dismissed it as "politically regressive and petty bourgeois." Finally, he told them about the sleeper issue, ecology—people weren't yet complaining much about deforestation, or air and water pollution, or chemicals in agriculture, or the destruction of topsoil, but they would, sooner or later. But that only led to more derision—the Situationists mocked Bookchin as "Smokey the Bear." He concluded with a shrug that they were not serious politically—they were "basically literary."[26]

Bookchin had a second reason for coming to Europe. The Spanish anarchist revolution of 1936–37, with its confederations and collectives, had been a high point in radical history, but no good study of the Spanish anarchist movement existed. He had decided to write that book himself. Surviving veterans of the 1936 Spanish Revolution—CNT-FAI men—were still alive, living as expatriates in France. He would track them down and interview them, starting at a meeting of the CNT in Paris.

Sure enough, there he met veterans of '36 who told him about "the fiery anarchist proletariat of Saragossa, the authentic black-flag center of the movement." He asked many of them about the *grupos de afinidad* that Russell Blackwell had mentioned, and they explained these close-knit, nonhierarchical groups for political action. After hearing their tales, he resolved that the "magnificent" Spanish working class "must never be forgotten."[27]

He came across Gaston Leval, a historian, who answered his questions about the revolution's industrial and agricultural collectives. He met Pablo Ruiz, veteran of a militia column of fifteen thousand Barcelona workers known as the Durruti Column. As they marched northwest to the Aragon front to fight Franco, they had spread social revolution through the Catalan countryside. At one point, someone introduced Bookchin to Cipriano Mera, commander of the anarchist troops in Madrid. Murray had thought he was dead, but here he was, standing before him in a long blue army overcoat. They sat down together in a Parisian bistro, where Mera detailed for him the organizational structure of the anarchist militias, using salt and pepper shakers to illustrate their maneuvers.[28]

Pablo Ruiz, the Durruti Column veteran, told Bookchin how stunned he had been by the CNT leadership's betrayals of 1936. Late that year the Popular Front had invited the anarcho-syndicalist union to join the government, and to the horror of the anarchist rank and file, the leaders had accepted. That meant they agreed to collaborate not merely with the hated nation-state—that would have been bad enough—but with their arch-enemies the Spanish Stalinists, who were part of the Popular Front coalition. Aghast, Ruiz had helped found a militantly antistatist *grupo* called the Friends of Durruti. The Friends at their peak had had only about 250 members, Ruiz explained, but their periodical, *El Amigo del Pueblo*, had been widely read. By adhering to antistatist principle at a time when the CNT leaders were abandoning it, Bookchin thought admiringly, they had upheld the honor of Spanish anarchism.[29]

Ruiz told Murray about the momentous May 1937 uprising of the Barcelona proletariat, in which he had personally participated. It had been mind-boggling: government troops, including Stalinists, aided by the treacherous CNT leadership, had reduced anarcho-syndicalist Barcelona to a "shambles."[30]

Suppose the uprising had prevailed, Bookchin asked Ruiz, *and the anarchists had regained control of Barcelona. What would you have done?*

Ruiz exclaimed, *We would have marched to Valencia*—the temporary republican capital—*and kicked those bastards out!*

The answer was superb in its nerviness. Ruiz and the other militants, Bookchin felt, knew the meaning of principled commitment. And even after three decades, their political passion remained unabated. Their viewpoint on the Spanish Revolution—including their hatred of the CNT leadership's collaboration—would become his position as well, and that of the book he was planning.

Traveling south, Bookchin met José Peirats in Toulouse, a tailor who was also the author of an important three-volume history of the anarcho-syndicalist confederation. Peirats welcomed Bookchin warmly: he too was self-educated, having dropped out of school at twelve. During their conversations, Peirats unscrambled for Murray the complicated organizational structure of the CNT and the FAI.[31]

Bookchin then crossed the border into Spain itself, where Franco, the general who had destroyed the greatest proletarian revolution in history, still held power. Walking the Ramblas in Barcelona, Bookchin mentally reconstructed the May 1937 uprising—he knew the positions of the various militias and the Stalinist forces, even down to individuals. When he got up close to building walls, he could still see the bullet holes.

But Spain was no longer the same country. The Ramblas was filled with "as many American-style attaché cases as lunch boxes." In Andalusia, once an anarchist heartland, the houses had television antennas.[32] Residents seemed to have forgotten the anarchist history that he was researching, or to have suppressed their memories of it.

When he stepped into a bookshop and asked the elderly proprietor for material on the *anarquistas*, the man's face turned pale, his eyes anxious. But after a pause, he held up a finger, indicating that Bookchin should wait, and went into a back room. After a few minutes, he returned with some material retrieved from a hiding place, then handed it over. The fear that Murray saw in people's eyes caused him to reexamine what he was doing, and he decided that "it would be imprudent to continue the research I had planned."[33] Now that he was a published author, the Spanish authorities could easily figure out who he was, and he didn't want to endanger anyone in Spain by association with him. So he cut his trip short and returned home in mid-November.[34]

Back in New York, he discovered, lo and behold, that Allan Hoffman had returned from the countryside. Murray must have been delighted to see his protégé. But what happened to Cold Mountain Farm? he asked.

The farmhouse turned out to be more isolated than we expected, Allan explained, and it was more primitive—it had no electricity. The would-be members of the rural commune were inexperienced farmers—they didn't plant their crops until the weather warmed up, by which time it was too late in the season. They had bought a tractor, but it broke down.

At first the group members had lived together. Some practiced casual nudity, working outdoors all day. But the neighboring farmers got upset and turned against the hippies. Some commune members drifted away, new people arrived—and the unstable population made it impossible to organize household tasks. A hepatitis epidemic struck. Joyce and Allan broke up. Finally the local health inspector demanded that they install electricity, refrigeration, and indoor plumbing, including a toilet—but they couldn't afford it. Stranded in a hostile environment, they declared the experiment over and left.[35]

Murray was glad to be reunited with Allan, and the two of them resumed their close friendship. Allan was "starved for some kind of theoretical stimulation," Murray noticed, so he showed him the Situationist articles he'd picked up in Europe, including "Totality for Kids," a translation of an essay by Raoul Vaneigem. Allan loved the article: in fact, it inspired him to write one, combining as always his own ideas with ideas he got from Bookchin—and now with Vaneigem's as well. "The goal of revolution," Allan wrote, "is the liberation of the entirety of daily life," for "in our time the revolution will be total or it will not be!" He called the piece "Eighteen Rounds of Total Revolution"—an unusual title for a pacifist—and signed it "the Totalist."[36]

Just then the Situationist author himself, Vaneigem, arrived in New York. Allan naturally asked to meet him. Vaneigem agreed, and he read "Eighteen Rounds," but he wasn't impressed—Allan had misunderstood him, he complained.[37] And after meeting Allan, he dismissed him as a "mystic" and an "acidhead."[38]

Then Ben Morea sought an audience with Vaneigem, understandably, since Black Mask's broadsheets were a Situationist-style *detournement* in their own right. But Vaneigem refused to meet him. Baffled, Morea sat down and wrote a letter to Guy Debord, demanding an explanation. The letter he got in response explained that he had been rejected because of his friendly relations with the "mystic" Allan Hoffman.

Meanwhile Bookchin had personally met with Vaneigem, defended Morea to him, and urged him to reconsider.[39] His appeal did not go over well, not at all. On December 21, 1967, the Situationist leadership in Paris thundered a denunciation of Bookchin: "Confusionist cretin, your suspicious efforts to act as mediator in New York in favor of pathetic Morea and his mystical associate have finished you off. You are only spit in the horrible communitarian soup. . . . Never more hope to meet a situationist (if you see one, it will be a false one)."[40] In another letter they denounced Morea, and in yet another they expelled the SI's whole British section, for the crime of solidarizing with Bookchin and Morea.[41]

Somehow Bookchin managed to survive denunciation by this strange clique. In the next years the Situationists would become ever more notorious for expelling people. Their dogmatic severity contrasted mightily with the saucy high spirits of, say, the Dutch Provos. "The full measure of the degeneration that occurred between 1965 and 1968," Bookchin observed, "can be understood by placing these two tendencies in juxtaposition to each other."[42]

Allan was a man in crisis. The failure of Cold Mountain Farm and his breakup with Joyce distressed him deeply. Then one day in late 1967 a friend, Walter Caughey, was stabbed to death by burglars who'd entered his apartment. The killers got away clean—the police never found them. Bookchin thought Caughey's unpunished murder was what propelled Allan into a new way of life.[43]

He gave up on his long-standing commitment to nonviolence, concluding that it was merely "a re-statement of christian ideals in an east Indian (gandhian) guise" and amounted to merely "begging the established order to rectify itself."[44] He swung to the other extreme—and became an advocate of revolutionary violence, adopting the motto "Armed Love." He took to writing "rabid poetry about vengeance," recalled Judith Malina. Even though Allan was "in his personal manner extremely gentle," sometimes his "passion is roused and he jumps up and shouts—'You gotta kill them.'"[45]

Other young radicals were making a similar transition, at the same moment. Black activists were abandoning nonviolent resistance, in the face of brutal, seemingly intractable racism. And New Leftists found that all their peaceful antiwar protests had accomplished exactly nothing: the United States was still bombing North Vietnam, still destroying villages, still dropping napalm. Nonviolent resistance and bearing moral witness seemed ineffectual. So when Frantz Fanon's 1961 anticolonial tract *The Wretched of the Earth* began passing from hand to hand, young radicals eagerly absorbed its lesson that using violence against a dehumanizing oppressor was justifiable as part of attaining manhood. Riots broke out in Watts in August 1965, in Newark and Detroit in 1967, and in many other American cities; young radicals reframed them as instances of revolutionary guerrilla warfare, waged by a colonized people seeking liberation.

National liberation struggles were firing the imaginations of activists, black and white. The examples of Mao Zedong and Fidel Castro and Che Guevara taught them a new revolutionary zeal and beckoned them to create "two, three, many Vietnams."

Above all they admired Ho Chi Minh, for his heroic resistance to imperialist aggression. Some even saluted him as a comrade, shouting at protests "Ho, Ho, Ho Chi Minh / The NLF is gonna win." When Clark Kissinger (a onetime national secretary of SDS) was asked at his draft board hearing, "Are you willing to fight in Vietnam?" he responded, "Sure I am—just not on your side."[46]

An international revolutionary armed struggle seemed to be underway: Herbert Marcuse, lecturing at NYU Law School, told his listeners that the Third World constituted the real proletariat, and the developed world constituted the bourgeoisie.[47] And within that international struggle, the American wing was the student and black movements. SDS must become "a revolutionary organization," argued national secretary Greg Calvert, and its leadership must function as a "steering committee," a vanguard, leading radical whites, urban blacks, and industrial workers to make a socialist revolution in the United States.[48]

From Mao's Little Red Book they learned to "dare to struggle, dare to win." From the film *The Battle of Algiers* (released in the United States in September 1967) they learned how to wage urban guerrilla warfare. And from Regis Debray's writings they learned to create *focos*, or small, fast-moving guerrilla groups that would fight in the streets and provoke the government into responding with excessive force, which would generate sympathy among the general population, who would then join the revolutionary struggle.

From these pieces, antiwar radicals assembled a new strategy that they called "mobile tactics." They put it to work in October 1967 in Stop the Draft Week, an action to shut down the induction center in Oakland, California. Donning motorcycle helmets, they ran into the streets and overturned garbage cans, ripped down fences, stopped traffic, and damaged property. Like Third World urban guerrillas, they used walkie-talkies and coordinated their actions from "command posts." They succeeded in shutting the center down for a few hours—and thereby paused the war machine, however briefly.

With mobile tactics, they realized, they need not settle merely for protesting the war or bearing nonviolent witness—they could actively disrupt the war machine. They shifted the antiwar movement's goal "from protest to resistance." Another Stop the Draft Week was planned, for New York in early December. Thousands of demonstrators converged at the Whitehall Induction Center, near Wall Street. From there would-be guerrillas ran up Broadway and smashed windows and knocked things over, aiming to provoke the police into a confrontation. It worked—on Friday the cops clubbed them in Union Square.[49]

Bookchin participated in Stop the Draft Week but as a peaceful protester. He admired the young people's revolutionary zeal (perhaps he saw his younger self in them), but he decided that mobile tactics were mere adventurism, a fight-the-cops diversion that succeeded mainly in disrupting traffic. Even more dismayingly, he thought, SDS leaders were leading the antiwar and black movements astray, by giving up their vision of participatory democracy in favor of Marxism. The Third World revolutionaries they admired—Mao, Castro, Che—were authoritarians, centralizers, irrelevant to the egalitarian aspirations of American political culture. Bookchin saw it as his task to try to persuade SDS to return to its original participatory democracy program.

He organized some old friends into a collective to publish a periodical, *Anarchos*, which would "advance nonauthoritarian approaches to revolutionary theory and practice."[50] The first issue contained three of his articles, under his old *CI* pseudonyms Robert Keller and Lewis Herber, as well as Murray Bookchin. It went to press at the end of December 1967.[51]

The fact is, he explained in "Revolution in America," that in the United States "there is no 'revolutionary situation' at this time" and "no immediate prospect of a revolutionary challenge to the established order." That being the case, the call to urban warfare in the streets is "a demagogic exercise in adventuristic sloganeering." On the other hand, "the potential for a future revolution is greater" here "than in any other industrialized country."[52] Among youth and blacks, opposition to the war and to racism was growing into a profound disrespect for established institutions and "a hatred of political manipulation." Ordinary Americans were "emerging from the inertness of the 1950s" and becoming irritated by the war. They had a long-standing cultural suspicion of government and authority. By the millions, they were questioning *what is* (the actuality of domination and capitalism) and envisioning *what could be* (the potentiality of freedom), the kind of questioning that Bookchin since the 1950s had thought necessary for a revolutionary movement.[53]

Sometime soon, mainstream America might become interested in participatory democracy. But just at this moment, the New Left was giving up on democracy and embracing Leninist propaganda. Instead of calling for greater democracy, they were emulating Red Guards in their inflexible authoritarianism, their insufferable cadres, and their acrid scorn for the populist impulses of ordinary people. This was no way to achieve freedom.

The fact is, Bookchin explained, the American movements are not part of some common international revolutionary struggle. The Third World revolutions that the New Left admires are taking place in societies that have yet to overcome the problem of scarcity. China, Vietnam, and Cuba are struggling to industrialize, a task the United States and western Europe achieved long ago. Moreover, Mao and Ho and Castro are not just promoting industrial development, they are forcing it, using dictatorial means. The states they created in the name of socialism aren't really socialist—they are police states, draped in red flags and adorned with portraits of Marx, Engels, and Lenin. Their little red books, invoking socialism, offer only the fraudulent maxims of tyranny.[54]

Really, Bookchin said, the best way for American radical youth to help Third World peoples would be to make revolutionary change here at home. "Don't cop out by hiding under Ho's and Mao's skirts," he implored, "when your real job is to overthrow domestic capitalism!" The New Left must cease speaking German (that is, Marxism) and Russian (that is, Leninism). It must cease propounding the absurd notions that the whole Third World is the proletariat, and that college students are workers, and that Third World dictators represent liberation.[55]

Rather, they must talk to Americans in their own terms, appealing to their indigenous discontents and aspirations. Americans were demanding "the self-realization of all human potentialities in a fully rounded, balanced, totalistic way of life." Meanwhile, modern technology was making it possible for them to achieve a civilization of "unprecedented freedom." That technological potential was overripe: "Like hanging fruit whose seeds have matured fully, the structure may fall at the lightest blow." Americans had utopian dreams; radicals must give them a utopian vision, based on their issues. Given their deep suspicion of government, Americans were much more likely to respond to the language of anarchism than to Marxism and to join a movement that tried to end hierarchy and domination.[56]

If and when the revolution comes to the United States, he said, it will not be created by a Marxist-Leninist vanguard. It will emerge spontaneously, from below, by a "molecular movement of 'the masses.'" It will seek to abolish domination by the centralized, bureaucratic state, and authoritarianism of all kinds including racial and sexual, in order to emancipate the spontaneous individual, the creative personality, and the diverse eco-community. It will try, that is, to dissolve "hierarchy as such." At most, revolutionaries may act as "catalysts" but never as commissars.[57]

The "moment of confrontation" will in fact involve "no confrontation at all"— because at that point defenders of the old order will have vanished, leaving its institutions available for the revolutionaries simply to seize. If the United States and Europe were to undergo such a libertarian revolution, it would spread abroad, to the Soviet Union and to the rest of Asia. Then the planet as a whole could become eco-anarchist. To achieve a libertarian revolution in America would be "the highest act of internationalism and solidarity with oppressed people abroad."[58]

After the moment of liberation, societies will need new, democratic institutions— "forms of freedom," as he called them in an article by that name. Ancient Athens long ago achieved a popular, face-to-face democracy. And revolutionary movements of the past also formed democratic assemblies. In 1789, as the French ancien régime was on the brink of collapse, the city of Paris was divided into sections, where residents met in ad hoc assemblies to draw up lists of grievances. Once they performed that task, they were scheduled to disband, but the *citoyens* refused. In the next years, these citizens' assemblies became "genuine forms of self-management" in the heart of revolutionary Paris, "the very soul of the Great Revolution." Young revolutionaries today could create similarly self-managing institutions, embodying the heart of the ethics-laden youth revolt. And because they are of, by, and for the emancipated people, they could become "the universal solvent," dissolving bureaucracy, war-making state, and hierarchy generally, as well as industrial agriculture and the gigantic city.[59]

Bookchin's foundational 1960s trilogy—on ecology ("Ecology and Revolutionary Thought"), technology ("Towards a Liberatory Technology"), and democracy ("The Forms of Freedom")—was published in the first issues of *Anarchos*. The response was like nothing he had seen before. The mere two or three thousand copies that were printed circulated widely, passed from hand to hand. Letters poured in from young people forming anarchist groups around the country. The times were "buoyant," Bookchin recalled decades later—"What a time to live!"[60]

To Murray's disappointment, Allan didn't join the *Anarchos* collective. In early 1968 he decided that the time for theorizing and writing and publishing was over. He would become an urban guerrilla and try to make "the street into the arena of social change." He transformed himself, as Bookchin described it, "from the rustic to the tough urban 'street person.'"[61] And in the streets, he suddenly found Ben Morea—whom he'd once detested—and Black Mask, with its provocative broadsheets, to be fascinating. He joined the other artistically minded radicals who were also making their way into Morea's orbit.

In February 1968, the city's sanitation workers went on strike. Wealthy New Yorkers hired private haulers to remove their refuse, but on the Lower East Side mounds of garbage piled up on the sidewalks and spilled into the streets. "Rats were having a field day," recalled Osha Neumann. One evening, disgusted by the city's inaction, Morea and his friends stuffed some stinking, rotting refuse into large plastic bags, boarded the uptown subway, and got off on the Upper West Side, where the glittering new palace of culture, Lincoln Center, welcomed elite patrons dressed in their finest. Morea and friends approached the illuminated fountain and emptied some of the trash into it, then dumped the rest on the marble steps. The leaflet they distributed read, "We propose a cultural exchange: Garbage for garbage." And it was signed, "Up Against the Wall Motherfucker," a line from a poem by Amiri Baraka. That signature became the new group's name.[62]

UAWMF was Black Mask on an expanded scale, a "street gang with analysis." Its florid broadsheets featured skeletons, skulls, and smoking guns and bore captions like "We are the ultimate Horror Show" and "Armed Love striking terror into the vacant hearts of the plastic Mother & pig-faced Father."[63] One poster reproduced a picture of Geronimo holding a .30-caliber lever-action rifle, over the caption, "We're looking for people who like to draw." The Motherfuckers wore leather jackets and clicked open folding knives with one hand, recalled Neumann. While the counterculture talked of peace and love, "we talked of rage—its reality in us and the dangerous rage of society against us."[64]

Anarchos and the Motherfuckers maintained friendly relations. Bookchin viewed their violence as basically rhetorical, believing "that their main job was to 'blow people's minds.'" But Allan the poet was now positively attracted to these artistic street rebels, whose totalistic vision fused "revolutionary & economic consciousness w/liberation of the unconscious (the psychological project) & artistic reconstruction of life." Murray warned Allan that the Motherfuckers mostly practiced street theater: Morea thought he could incite people to a revolution with artistic antics, but street theater "rarely makes [people] think, and it can get out of control and undermine serious organizations." Above all, it could make you lose sight of politics, which had to be the top priority. "However personalized, individuated, or dadaesque may be the attack upon prevailing institutions, a liberatory revolution always poses the question of what social forms will replace existing ones"—that is, what concrete institutions, what forms of freedom. Art cannot answer that question. Serious revolutionary thought must "speak directly to the problems and forms of social management."[65]

But by this time Allan was already within Morea's circle, sympathizing with UAWMF on a gut, generational level. "The thing that pushed Allan toward us," Morea told me, "was that we were part of the youth rebellion. We used drugs rampantly. We saw LSD as a revolutionary tool. We thought the mind had to be changed as well as the environment." When Allan finally joined the Motherfuckers, Murray was stunned.[66] He had given Allan a unique education in revolutionary history and theory—and Allan was setting it aside in favor of street theater.

In the spring of 1968, Columbia University announced its plans to build a large gymnasium in Morningside Park, which was used mainly by blacks in neighboring Harlem. On April 23 student radicals, in outraged protest, marched into Hamilton Hall, occupied it, and renamed it Malcolm X Hall. Others took over Low Library and occupied the university president Grayson Kirk's office. "It is the opening shot in a war of liberation," SDS leader Mark Rudd wrote to Kirk. "Up against the wall, motherfucker, this is a stick-up."[67] A thousand students "liberated" three more buildings, barricaded themselves inside, created communes, and held discussions on politics, society, ideology, and the role of the university. Bookchin admired the students for shedding "the internalized structure of authority, the long-cultivated body of conditioned reflexes, the pattern of submission sustained by guilt that tie one to the system."[68] He saw them as courageous and beautiful. On April 25 some conservative student athletes tried to blockade a building to keep food from getting through; Bookchin participated in a defense squad to prevent them.

On April 30, at President Kirk's insistence, a thousand city police arrived and began to arrest students, who surrendered peacefully (although some yelled "Up against the wall, motherfucker"). Students and faculty stood before the occupied buildings, attempting to block the arrests nonviolently. The cops brought out billy clubs and brass knuckles and rioted, injuring more than one hundred. Bleeding students were dragged to paddy wagons.[69] Thereafter students went on strike; the administration shut the university down.

The student occupation and the police repression that it provoked could have come out of Debray's and Fanon's playbooks for urban guerrillas. It gave rise to a new revolutionary strategy. Radical students would occupy more university buildings and build strong barricades. Police would be unable to overcome the barricades, and the universities would have to shut down, as Columbia had. The government would send troops to restore order. Then the rebellion would spread to the cities, where black militants would join the students, and the Black Panthers would mount guerrilla rebellions. The crisis would be so massive that the police could not vanquish it. Having no alternative, the government would finally have to end the war in Vietnam. This was the strategy in pursuit of which Tom Hayden, echoing Che Guevara, called for "two, three, many Columbias." A university's only function now, announced SDS leader Mark Rudd, was "the creation and expansion of a revolutionary movement."[70]

Bookchin, aghast, recognized that the whole plan was foolhardy. Ordinary Americans were repulsed by posters of Mao and Ho and Che and Fidel, and they didn't like cops being provoked. Revolutionary rhetoric and quotations from the Little Red Book would do nothing but drive them away.

Even as Bookchin was writing such words, students in France were doing more or less what he was advocating.

In the early spring of 1968, students at Nanterre University occupied school buildings, protesting the university's bureaucratic structure and its archaic curriculum. On May 2 the administration shut the school down. The next day, in Paris, students rallied outside the Sorbonne to protest the Nanterre closing. In the late afternoon, black vans pulled up, carrying the paramilitary riot police called CRS. The students attacked the vans with fists and rocks, overturning cars and creating makeshift barricades. The CRS responded with tear gas and truncheons.

The battle continued into the following week. On Friday, May 10, students ripped paving stones from the streets and piled them up to make several dozen new barricades. After midnight the CRS attacked with more tear gas and smoke bombs. Neighborhood people living above the barricades helped the students by pouring water from their windows to douse the gas; they dropped flowerpots onto the police below. At daybreak the police fired gas grenades and dispersed the students.

A national outpouring of sympathy for the student rebels followed—they had courageously taken on the much-hated CRS. On May 13 one million people thronged the streets of Paris in support. The government yielded, reopening the Sorbonne and releasing four students from jail. Students then occupied the Sorbonne, hoisted the red flag, and declared it a "people's university." They turned it into a twenty-four-hour debating forum, insurrectionary headquarters, and revolutionary commune.

Throughout the Left Bank, students plastered walls with posters bearing slogans like "Power to the imagination"; "It is forbidden to forbid"; "Be realistic—demand the impossible!"; and "We take our desires to be reality because we believe in the reality of our desires." Even the prime minister, Georges Pompidou, understood that "our civilization is being questioned—not the government, not the institutions, not even France, but the materialistic and soulless modern society."[71]

Then industrial workers went on strike, again in sympathy for the students. On May 14 in Nantes, workers occupied the Sud-Aviation plant. Farmers came from the countryside on tractors and joined them; for a week a revolutionary soviet ran Nantes. The general strike spread to Paris, shutting down banks, post offices, insurance firms, and department stores. The strikers' grievances, remarkably, did not involve wages and working conditions: they complained, rather, about the "alienated atmosphere" in factories and workplaces and the bureaucratized society. "It was not only a workers' strike," Bookchin would observe—"it was a *people's* strike that cut across almost all class lines."[72]

The media reported on the strike sympathetically. It swept to other major French cities and to the manufacturing, mining, utilities, and transport sectors. Workers occupied factories. Stores, teachers, civil servants, and physicians walked off their jobs. Reaching 9 million workers by May 22, it was the largest general strike in French history. Bookchin was champing at the bit to get there, but because the transport workers were on strike, no planes were landing at Orly Airport.

On May 27 the government offered large increases in wages and fringe benefits and half pay for strike time. Rank-and-file workers rejected it, and the general strike continued. On May 30 President Charles de Gaulle dissolved the National Assembly and announced new parliamentary elections. His party organized large counter-demonstrations; the armed forces conducted ostentatious troop movements. The Communists, colluding with de Gaulle to end the revolt, tricked Metro workers into resuming operations. The strike wave subsided, and the revolutionary mood waned. On June 16 police cleared the Sorbonne. Two days later the Renault plant surrendered. A week after that, de Gaulle's party triumphed in the parliamentary elections.

Finally, on July 13, Bookchin arrived in Paris. The streets had been repaved, and the graffiti was painted over. CRS loitered in full battle gear, some carrying submachine guns. On the eve of Bastille Day, he marched with a crowd of Africans to the Rue de Madeleine, singing "The Internationale." The CRS chased them into the Rue Soufflot; he escaped near the Panthéon. Street fighting flared up sporadically during that night.[73]

Bookchin spent the next few weeks interviewing leaders of the student uprising, to find out what had happened, and how, and why.[74]

In hitherto quiescent Paris (*C'est mort!*), the May-June uprising had surprised the whole Left, be it Marxist, anarchist, or Situationist: it had emerged spontaneously, without leadership.[75] It had had majority support, based on no single class.

Structurally it had been, in a word, anarchic. Students had exercised "self-management," in general assemblies, action committees, and strike committees. The Sorbonne assembly had been nonhierarchical and open to all, on the principle that "the direct entry of the people into the social arena is the very *essence* of revolution." The students had made decisions by finding "the sense of the assembly."[76]

The strikers' grievances were less about material issues than about the character of modern society. Young people saw bleak lives stretching out ahead of them, empty of meaning, geared toward routine work and the consumption of useless, shabby goods; their elders despaired that their lives fulfilled precisely that expectation. But the opportunity to rebel called forth their passions and aroused their senses: "A fever for life gripped millions. . . . Tongues were loosened, ears and eyes acquired a new acuity. . . . Many factory floors were turned into dance floors." Repressed people demanded the liberation of experience. The tenor of the revolt was creative, even artistic, as reflected in the clever graffiti Bookchin saw.[77]

As for the Marxist-Leninist groups, of which Paris had several, the naive observer might expect them to have scrambled to support the uprising, but in fact they had played a retarding or even counterrevolutionary role. "Virtually every one of these 'vanguard' groups," Bookchin noted, "disdained the student uprising up to May 7." The Trotskyists and the Maoists had criticized the student revolt as peripheral; the Trotskyists had condemned its street actions as "adventuristic." Once the general strike erupted, the Maoists had opposed demands for workers' control and occupation of factories. "Not a single Bolshevik-type party in France raised the demand for self-management," he reported. "The demand was raised only by the anarchists and the Situationists."[78]

The Communists, who controlled the powerful industrial trade union CGT, joined the revolt only to try to retard it, Bookchin maintained, using trickery and deceit. In the Sorbonne assembly their machinations had interrupted the democratic proceedings and thereby "demoralized the entire body." In workplaces, instead of promoting the general strike, they had tried to dampen the enthusiasm and even sent some factory workers home. They had attempted to keep the students and workers apart. They told Metro workers at some subway stations, untruthfully, that others were going back to work, to induce them to give up. Bookchin called their actions "shameless" and "treacherous."[79]

He wrote up his findings in a series of letters, enough to fill an entire special issue of *Anarchos*. But the collective didn't have enough money to publish such an edition, so *New Left Notes* and the *Rat* published several of them, and the new underground press around the country picked them up.[80]

When he returned to New York on August 1,[81] Bookchin was shocked to find that the May-June *événements* had hardly registered on the American New Left. But in retrospect, the reason was clear: the French general strike had taken place in a Western, industrialized country, and its outlook had not coincided with Marxist ideology.

And Marxist ideology was now de rigueur among the New Left's leadership. On June 9–15, at an SDS national conference in East Lansing, Michigan, pictures of Lenin and Trotsky had hung on the walls. In the plenary, two sets of Marxist-Leninists glowered at each other from across the room in mutually seething hatred. One was Progressive Labor, its clean-cut Maoists present in force, disciplined and unwaveringly self-assured. The other was a group of Marxists who preferred Cuba to China, organized by the Bernardine Dohrn of the SDS National Office to counteract PL. Both groups wore red armbands and raised clenched fists.

PL promoted its ideas aggressively at the conference, pushing its line in every plenary and workshop. Whenever a regular SDS militant made a proposal, PL introduced a ratcheted-up counterproposal; "they waved their little red books at every juncture." They resisted all suggested compromises and counterarguments. Unprepared intellectually to rebut PL or offer a coherent alternative, Dohrn's group simply competed in revolutionary braggadocio: "Motions from the floor outdid one another in their claims to be revolutionary."

Someone from Dohrn's group attempted to expel PL on the grounds that it was an "external cadre"; PL dug in its heels and screamed that that was "red-baiting." Dohrn's people on the floor chanted "PL out!," but the PL delegates sat immobile. They successfully obstructed every move to sideline them.[82]

In July 1968 the well-known anarchist Paul Goodman, whose writings had helped to inspire the Port Huron Statement, published an article in the *New York Times Magazine* berating the SDS leadership for embracing Marxism and for manipulating the youth revolt's "lively energy and moral fervor."[83] At the time, Bookchin was still trying to persuade the New Left to abandon Marxism in favor of anarchism. He felt Goodman's prominent article made that task more difficult, so he chastised Goodman publicly, in the *Rat*, in a piece signed "Incontrollado."[84]

But Bookchin's efforts to promote anarchism were not bearing much fruit. In November 1968, writing in *New Left Notes*, Huey Newton, cofounder of the Black Panthers, explained why he rejected it: anarchists were disorganized and unruly. "You cannot oppose a system such as this is," he said, "without organization that's even more disciplined and dedicated than the structure you're opposing." Newton called instead for "a real, disciplined vanguard movement."[85] Newton's piece cut to the heart of the matter. Bookchin wrote a response, arguing that, contrary to popular belief, anarchists were no strangers to organization; most notably the Spanish anarchists had maintained their trade unions for decades. The relevant issue wasn't organization as such but what kind: authoritarian or libertarian; top down or bottom up.[86]

The Spanish anarchists, however, were all but unknown in the United States, especially to the young people around him. So Bookchin spent the rest of 1968 working on his book on "the largest organized Anarchist movement to appear in our century."[87] It would be a history of that movement but also an argument for anarchism as an ideology and an illustration of an alternative, enlightened, nonhierarchical way for the New Left and the counterculture to organize themselves.

The Spanish anarchists' trade unions had been organized—indeed, intricately, elaborately. "In individual shops, factories, and agricultural committees," workers in assemblies had "elected from their midst the committees that preside over the affairs of the *secciones de oficio* and the *federaciones locales*; these were federated into regional . . . committees of nearly every geographic area of Spain."[88] The local federations mandated their delegates to CNT congresses, a practice that kept them accountable to the base and kept the movement democratic.

The Spanish anarchists' sophisticated training in organization bore fruit in July 1936, when they made their revolution to "gain full, direct, face-to-face control over their everyday lives, to manage society . . . as thoroughly liberated individuals."[89] They collectivized factories and farms; they established councils and committees to function as local self-government; they set up armed patrols to serve as police; and they formed militias to defend the revolution against the generals. This was organization indeed, melding structures of the traditional village with those of modern industry.

At the same time, the Spanish anarchists resembled the 1960s counterculture in their lifestyle changes, their spontaneity, and their fondness for initiative from below—and in their antihierarchical outlook. They, too, wanted to develop integral (in the sense of authentic and well-rounded) personalities.

The Spanish anarchists' unit of revolutionary organization (as opposed to institutions of governance) was the affinity group, a small action group whose bonds were both political and personal. (Murray had learned about affinity groups from Russell

Blackwell and from the Spanish anarchists he interviewed in France in 1967.) Affinity group members were friends and comrades who shared "common revolutionary ideas and practice."

As a way of organizing, Bookchin thought, the affinity group should be of particular interest to the New Left—it was the anarchist answer to the Marxist-Leninist vanguard. Marxist vanguards attempted to constitute a movement's command center, coercing uniformity, but affinity groups acted as "catalysts" within "the spontaneous revolutionary movement of the people." While vanguards were hierarchical, affinity groups were local and could confederate regionally for joint action. While vanguards commanded and coerced, affinity groups achieved coordination through education and the formulation of common policies. While vanguards sought to seize power, affinity groups sought to dissolve it.[90]

In August 1968, the New Left's urban guerrillas and the Chicago police waged their climactic battle at the Democratic National Convention. Radical youth turned over garbage cans and pounded on cars. They threw bottles and rocks, bricks, eggs, chunks of concrete, and urine-filled balloons. Police responded by clubbing indiscriminately, beating protesters unconscious, and rioting: they cracked skulls and broke knees for twenty minutes. If the New Left's strategy had been to provoke, the use of force by the police was excessive beyond anyone's wildest predictions.

But as Bookchin predicted, it did not arouse sympathy in mainstream Americans; in fact, it appalled them. Even many who opposed the Vietnam War thought the Chicago cops had been too restrained, had used insufficient force against the protesters. Ordinary Americans, it turned out, hated the protesters more than they hated either the atrocious war or the thuggish cops. In fact, antiwar protesters were by now "the most despised political group in the country."[91]

But the SDS leaders in the National Office were not listening—they ratcheted up their own Marxist zeal to counter that of PL. In December they proposed that SDS should no longer be a student movement—it should become an organization for working-class youth called Revolutionary Youth Movement (RYM). Their vision was all but indistinguishable from PL's program for a worker-student alliance. They differed only on the issue of black nationalism: PL opposed all nationalism as reactionary and bourgeois, while RYM said that in the United States the Black Panthers were the revolutionary vanguard and that SDS must develop disciplined cadre to support them.[92]

The RYM and PL would wage their own climactic battle for the soul of SDS at its Ninth Annual Convention, to start on June 18, 1969, at the Chicago Coliseum. In advance, Bookchin wrote a long essay called "Listen, Marxist!" in an effort to break the Marxist-Leninist spell that had such a powerful hold on both factions. "All the old crap of the thirties is coming back again," he began with confidence born of experience. Marxism, with it fantasies of proletarian dictatorship, belonged to the past, he continued, and good riddance. Its cadres and vanguards had done nothing more than seek power for themselves, then maintain it at any cost. During the revolutionary ferment of 1917 in Russia, workers had organized themselves into relatively democratic factory councils, called soviets, which the Bolsheviks had pretended to favor and thereby won the support of the workers. That support had been their stepping-stones to power. But once in charge of Russia, they had suppressed dissent and transformed the soviets into instruments for top-down rule.

Marxist-Leninist organizations behaved the same way everywhere, Bookchin argued: they sought above all to build their own organization and to gain power, and they would pursue both goals at the expense of all other considerations, including social revolution itself. In Paris a few months earlier, the Bolshevik groups had been "prepared to destroy the Sorbonne student assembly in order to increase their influence and membership." Today in SDS, PL was trying to do the same thing. He appealed to SDS to look not to the past but to the present, when the revolutionary agent is not the worker but "the great majority of society," especially the youth. SDS should resist Marxification, he pleaded, and return to its original anarchistic call for participatory democracy.[93]

He also wrote a program for an alternative SDS, called "The Radical Decentralist Project," that called on SDS chapters to reconstruct themselves as affinity groups—and to focus on issues of ecology and community. It urged university students to create liberated spaces "in the fashion of the Sorbonne" and to "help the Third World . . . by changing the First."[94] A printer friend ran off thousands of copies, and the *Anarchos* group loaded them into a truck and drove off to Chicago.

At the Coliseum, a huge, barren hall constructed of steel girders and cement blocks, all the SDS politicos were present: the National Office's RYM group; the Black Panthers; and socialists of all different stripes. PL bused in its people from around the country—organized into squads, they constituted about one-third of those present. And mingling among them all were hundreds of plainclothes Chicago police, with cameras, and assorted other government officials.[95]

The *Anarchos* group set up a literature table with copies of "Listen, Marxist!" and "The Radical Decentralist Project." By the end of the first day, all copies were gone.

Clashes between RYM and PL erupted right away. When a speaker said something that PL found objectionable, its cadres tossed Little Red Books at him or her. PL would chant, "Mao, Mao, Mao Tse-tung," while RYM answered with "Ho, Ho, Ho Chi-Minh."

Then RYM played its trump card: its alliance with the Black Panthers. Rufus Walls, minister of information from the Illinois Black Panther Party, took the mike and declared that the Panthers were the true vanguard because while the white Left was sitting comfortably in armchairs, the Panthers were out in the streets shedding blood. Walls denounced PL, in what RYM hoped would be a fatal blow.

He continued by saying that the Panthers supported women's liberation—and believed in "pussy power." The room gasped; then the PL-ers started chanting, "Fight male chauvinism!" RYM was mortified but recovered enough to send female Panther leader Jewel Cook out to the mike to try to salvage the situation. She delivered an ultimatum to PL: its members had to change their politics and support the Panthers, or else get their privileged white behinds out of SDS.

PL didn't bat an eye. Refusing to be bullied, it switched its chant to "Smash red-baiting!" and "Read Mao!" Against them, RYM and its supporters cried "Power to the people!" PL came back with "Power to the workers!"

Amid the screaming, National Office and RYM member Bernardine Dohrn rushed onto the stage and took the mike. PL, she said, by virtue of its failure to support the Panthers, was objectively racist. It was up to SDS "to decide whether the racist PL was fit political company." Then she initiated a split: anyone interested in helping decide should join her into the next room.[96]

PL chanted "Sit down, sit down, no split, no split."

The RYM supporters chanted back "Join us, join us."

Six hundred delegates followed Dohrn out of the room, where they voted to expel PL from SDS. Dohrn came back into the main hall and delivered their verdict: PL was no longer part of SDS.

PL was stunned into momentary silence. Then its members began to shout, "Shame!" Dohrn's contingent cried, "Power to the people," and, "Ho Ho Ho Chi Minh," as they went out into the night.

By now Bookchin had found enough sympathizers, alienated by the posturing, to form an anarchist caucus. About 250 met in the old IWW hall nearby and voted to endorse "The Radical Decentralist Project" as their program.

The next morning PL opened its meeting in the main hall, while the RYM groups opened a second convention in a smaller auditorium nearby. Jeff Jones, of RYM, stood before the packed room and announced, *I hereby declare the opening of the real SDS convention. Who wants the floor?* Bookchin leaped to his feet, the first to speak. SDS, he said, had to find a way to win the support of the American people. It shouldn't just denounce "Amerika." It had to offer a broad social vision that was ecological and decentralist and anarchist. The people listening cheered: *Right on, right on!*

Then someone else stood up to talk and said the opposite of what Bookchin had been saying, and the audience shouted *Right on, right on!* He realized he couldn't achieve anything more here. He and *Anarchos* would have to organize an alternative movement. They headed for the exits, ready to plan a follow-up meeting.

Back in New York, *Anarchos* invited all the caucus members to the meeting on September 6, 1969, to be held at a campground in west-central Wisconsin called Black River. Once again Bookchin and friends traveled halfway across the continent, this time to develop the radical decentralist alternative. About eighty people showed up, gathering under a big outdoor canvas tent. Bookchin wondered, when he saw the tent, if that venue would really be conducive to the serious work at hand.[97]

He wondered the same thing when the local organizers insisted that the participants sit in a circle—so as to be nonhierarchical—and speak one after the other, going around the circle. What was needed, Bookchin thought, was a back-and-forth discussion of issues and courses of action. But the artificial, circular structure made that impossible. People said whatever was on their mind, without response or consequence. The meeting deteriorated into "irrelevant words, rhetoric, and exhaustion."

The underground press, the Black River conferees were told, was waiting for them to issue a statement. Movement newspapers all over the country would publish it, and millions of readers would see it. Bookchin told the conferees that the *Anarchos* policy statement was available, but the conferees said they preferred to write their own. A single individual writing a statement would be hierarchical, they said. Rather, the statement must be a collective effort to which each member contributed.

One of the conference organizers had a pet theory that eroticism should have a central place in political proceedings and was hitting on participants to join in some revelry happening elsewhere. The nonhierarchical Black River gathering, needless to say, produced no statement. Bookchin and his *compañeros* who had been serious about establishing a libertarian alternative to the now-Bolshevik SDS went away empty-handed and disappointed.[98]

As for RYM, after Chicago it changed its name to Weather Underground and in 1970 declared war on the federal government. It embarked on a career of bombing ROTC buildings on college campuses, draft board headquarters, army induction facilities, research laboratories, and corporate headquarters. These bombings, which continued for several years, attracted not the slightest sympathy from mainstream Americans.

A few months later, in May 1970, the United States invaded Cambodia, and students at universities around the country protested. At Kent State University in Ohio, the National Guard killed four unarmed student demonstrators. Outraged, 4 million American students went on strike, shutting down more than nine hundred colleges, universities, and high schools. It was the largest student strike ever in the United States. But mainstream public opinion sympathized not with the students but with the National Guard—150,000 people paraded for flag and country down New York's Broadway.[99] Again, the majority of Americans who by now opposed the war did so in spite of New Left, not because of it.

On the Lower East Side, the utopian hippie counterculture was being displaced by a criminal element. LSD was giving way to heroin. Drug dealers preyed on teenage runaways. Bookchin's apartment was burglarized several times.[100]

The Motherfuckers deteriorated into outright criminality. One day an apartment superintendent, a Puerto Rican, said something disrespectful about the scruffiness of a few passing Motherfuckers. The next day some UAWMF street kids came by and stabbed the super to death. "We had shouted 'Off the pig!' but the first person to die was a Puerto Rican superintendent," Neumann (who was not involved in the confrontation with the superintendent) noted ruefully. Soon after, the Motherfuckers hightailed it out of New York on a bus, shoplifting their way to New Mexico. Neuman followed later.[101]

Out west along the California and Oregon coast, a loose network of communes was being formed, known as Armed Love. Its spiritual center was Black Bear Ranch, in northern California. That was where Allan Hoffman landed. The setting seems to have calmed him. The future lies with "those who return to the land & to communal forms of living together," ready to give it another try. He decided to "abandon ideologies" and explore Reichian psychology.[102]

On September 21, 1970, Allan and some of his friends drove their Dodge Power Wagon into Arcata to get supplies. On the return trip, along a twisting mountain road, Allan was lying in the open back. An eighteen-wheel lumber truck rear-ended the Dodge, and the impact threw Allan into the air. He landed on his head. Rushed to a hospital, he died a week later.[103]

His body was returned to New York, and Bookchin delivered the eulogy at the funeral on Long Island. Janis Joplin and Jimi Hendrix had recently self-destructed; the Hell's Angels had killed someone at a Stones concert at Altamont. But for Murray, it was Allan's death that marked the end of the 1960s.

Still, throughout the 1970s he would defend the counterculture, or at least its potential. Yes, its pleas for "a vision of love, community, material simplicity, and an uninhibited openness and directness in human relationships" had surely been naive. But the counterculture had added new ethical and psychological dimensions to traditional socialist theory. Its revolt had been driven by moral outrage; it had linked the psychological and the aesthetic with the social and political. Those dimensions, along with its communal spirit, its trusting open-heartedness, and its sensuality, remained necessary to any future utopian vision. The counterculture hadn't died; it "remains incomplete as a project," awaiting "a richer, more perceptive, and more conscious development."[104]

NOTES

1. Joy (formerly Joyce) Gardner, interview by author, Apr. 5, 2008.

2. MBVB, part 37.

3. Bookchin, "When Everything Was Possible" (1973), *Mesechabe: The Journal of Surregionalism* [New Orleans], nos. 9–10 (Winter 1991), 2.

4. Bookchin, "Post-scarcity Anarchism" (1970), in *PSA*, 40; Bookchin, "Toward an Ecological Solution," *Ramparts* (May 1970), 61.

5. MBVB, part 37.

6. Bookchin, "The 1960s," interview by Doug Morris, in Bookchin, *Anarchism Marxism, and the Future of the Left* (San Francisco: A.K. Press, 1999), 74.

7. Bookchin, "Desire and Need," *Black Flag* (1967), 6. This passage does not appear in the version of the article published in *PSA*.

8. MBVB, part 37.

9. Terry H. Anderson, *The Movement and the Sixties: Protest in America from Greensboro to Wounded Knee* (New York: Oxford University Press, 1995), 149.

10. Jack Newfield, *A Prophetic Minority* (New York: New American Library, 1966), 158.

11. Bookchin thought his urgings bore fruit in *Essay on Liberation* (Boston: Beacon Press, 1969), where Marcuse wrote lines that echoed Bookchin's essay "Towards a Liberatory Technology": "Freedom indeed depends largely on technical progress"; if science and technology were "reconstructed in accord with a new sensibility—the demands of the life instincts," they could become "vehicles of freedom" for "a human universe without exploitation and toil" (19). Bookchin to Laurence Veysey, November 10, 1973, unpublished letter, MBPTL and author's collection.

12. Marcuse's SVA talk was published as "Art in the One-Dimensional Society," *Arts Magazine* 41, no. 7 (May 1967), 27–28, 53–55. Bookchin later identified the various themes of his conversations and correspondence with Marcuse in "Response to Andrew Light's 'Rereading Bookchin and Marcuse' as Environmental Materialists," *Capitalism Nature Socialism* 4, no. 2 (June 1993), 189–91.

13. Ben Morea, interview by Iain McIntyre, 2006, online at http://libcom.org/history/against-wall-motherfucker-interview-ben-morea, accessed Oct. 8, 2011.

14. Bookchin, "1960s," 83; Bookchin, "When Everything Was Possible"; Ron Hahne and Ben Morea, *Black Mask & Up Against the Wall Motherfucker: The Incomplete Works of Ron Hahne, Ben Morea, and the Black Mask Group* (Oakland, CA: PM Press, 2011).

15. Gardner interview; Timothy Miller, *The 60s Communes: Hippies and Beyond* (Syracuse, NY: Syracuse University Press, 1999), 60; Joyce Gardner, *Cold Mountain Farm: An Attempt at Community* (n.p.; ca. 1968), 1.

16. Gardner, *Cold Mountain Farm*, 6–8.

17. Bookchin, "When Everything Was Possible," 5. He would move around a great deal during the next few years. As of June 6, 1966, he was living at 718 East Ninth Street (source: *United States v. W. Mark Felt and Edward S. Miller*, US District Court, Washington, D.C., Sept. 1980, notes, author's collection and MBPTL). In 1967 he lived—nonmaritally, in separate quarters—in Beatrice's apartment at 100 Bleecker Street (source: conversations with author). From August to December 1967 he traveled in Europe (source: passports, author's collection; Bookchin to Debbie, Bea, and Joey, Aug. 11, 1967, MBPTL). In May and June 1968, perhaps longer, he was in California (source: Bookchin to Dave Eisen, May 24 and June 13, 1968, courtesy Dave Eisen). In July 1968 he (and Beatrice and the two children) went to Paris for his research (source: conversations with author). Afterward they returned to 100 Bleecker Street. In December 1968 he moved to California because many interruptions made it impossible to work in the Bleecker Street apartment (source: Bookchin to Eisen, Dec. 20, 1968). At some point he returned to Bleecker Street. In September 1969 he moved to 235 Second Avenue (Felt-Miller trial transcript and FBI file). There he lived for an unknown period of time with a girlfriend; she moved out on April 9, 1971 (Robert Palmer papers).

18. Bookchin, "Desire and Need" (1967), *PSA*, 277.

19. Ibid., 279.

20. Robert Chasse and Bruce Elwell, "A Field Study in the Dwindling Force of Cognition, Where It Is Expected Least: A Critique of the Situationist International as a Revolutionary Organization," Feb.–Mar. 1970, online at http://bit.ly/oBGPoi, 4.

21. Bookchin to Debbie, Bea, and Joey, Aug. 11, 1967, MBPTL.

22. Judith Malina, *The Enormous Despair* (New York: Random House, 1972), 24–25.

23. For a detailed account of Provo, see Teun Voeten, "Dutch Provos," *High Times*, Jan. 1990, 32–36ff.

24. It didn't work back then—people stole the bikes. But Provo's White Bicycle program was the ancestor of bike sharing, common in many cities today, the problem of theft solved by digital swipe cards.

25. Bookchin, "1960s," 86.

26. Bookchin, "Reflections: An Overview of the Roots of Social Ecology," *Harbinger*, Spring 2003, 6; Bookchin, "1960s," 86.

27. Bookchin, introduction to *The Spanish Anarchists: The Heroic Years, 1868–1936*, 2nd ed. (San Francisco: A.K. Press, 1998), 6.

28. Murray recollected his meetings with Ruiz, Leval, Mera, and Peirats in "A Journal for an End to a Century," vol. 3, "November 5, 1991–January 9, 1992," handwritten ms., 3–15.

29. Ibid. See also Bookchin, "Reflections on Spanish Anarchism," *Our Generation* 10, no. 1 (Spring 1974), 28.

30. Bookchin, "The Forms of Freedom" (1968), in *PSA*, 154.

31. Bookchin, *Spanish Anarchists*, 9.

32. Bookchin, "Reflections on Spanish Anarchism," 29. See also Bookchin, "Intelligentsia and the New Intellectuals," *Alternative Forum*, Fall 1991.

33. Bookchin, *Spanish Anarchists*, 11.

34. U.S. passport, copy in author's collection.

35. Gardner, *Cold Mountain Farm*.

36. Bookchin, "When Everything Was Possible," 5; Allan Hoffman (as Totalist), "Eighteen Rounds of Total Revolution," *Anarchos*, no. 2 (Spring 1968), 11, 18.

37. One possible reason for Vaneigem's disapproval of Allan's article may have been that it criticized the Situationists: "Despite its claim, the IS does not have the total theory & represents merely the theory of the spectacle ... it is not the theory of its [the spectacle's] transcendence & the birth of what will come to be." Situationists did not take kindly to criticism. Hoffman (as Totalist), "Eighteen Rounds," 12.

38. *Jumbled Notes: A Critical Hidden History of King Mob* (2008), http://bit.ly/pcpoS8.

39. Chasse and Elwell, "Field Study." See also Guy Debord, "Circular to All the Sections, December 21, 1967," http://bit.ly/vSL9LB.

40. Published in *Guy Debord Correspondance, Vol "o": Septembre 1951—Juillet 1957: Complete des "lettres retrouvées" et d l'index général des noms cites* (Librairie Arthème Fayard, October 2010); translated from the French, at http://bit.ly/oUlOBV. Murray considered the letter ridiculous and chortled whenever he mentioned it to me.

41. Charles Radcliffe to author, July 10, 2010.

42. Bookchin, introduction to the second edition of *Post-scarcity Anarchism* (Montreal: Black Rose Books, 1985), 38n.

43. Bookchin, "When Everything Was Possible"; Ben Morea, interview by author, May 1, 2008.

44. Hoffman (as Totalist), "Eighteen Rounds," 13–14.

45. Malina, *Enormous Despair*, 177.

46. Tom Wells, The *War Within: America's Battle over Vietnam* (Berkeley: University of California Press, 1994), 172.

47. Bookchin, "1960s," 78.

48. Greg Calvert, in *New Left Notes*, Dec. 1967, quoted in Kirkpatrick Sale, *SDS: The Rise and Development of Students for a Democratic Society* (New York: Random House, 1973), 393.

49. Wells, *War Within*, 217; Sally Kempton, "A Week of Confrontation: The Rough Side of New York," *Village Voice*, Dec. 14, 1967; Thomas Maier, *Dr. Spock: An American Life* (New York: Basic Books, 1993); Vincent J. Cannato, *The Ungovernable City: John Lindsay and His Struggle to Save New York* (New York: Basic Books, 2001), 151–52.

50. Publication statement, *Anarchos*, no. 1 (Feb. 1968). Most *Anarchos* members were part of an urban commune called East Side Anarchists, centered at Avenue B and Third Street, later at Second Avenue and Tenth Street.

51. Bookchin, "1960s," 84–85. It published about two thousand copies of *Anarchos*, no. 1.

52. Bookchin, "Revolution in America," *Anarchos*, no. 1 (Dec. 1967), 3, 10.

53. Bookchin, "Post-scarcity Anarchism" (1968), in *PSA*, 48; Bookchin, "Revolution in America," 10; Bookchin, "Introduction" (1970), in *PSA*, 12.

54. See Radical Decentralist Project, "Toward a Post-scarcity Society: The American Perspective and S.D.S.," Resolution #1, Platform of the Radical Decentralist Project at the last SDS Conference (May 1969), published in Howard J. Ehrlich et al., eds., *Reinventing Anarchy: What Are Anarchists Thinking These Days?* (Boston: Routledge & Kegan Paul, 1979), 125; Bookchin, "A Discussion on 'Listen, Marxist!'" (1970), in *PSA*, 235.

55. Bookchin, "Discussion," 235.

56. Bookchin, "Revolution in America," 7; publication statement, *Anarchos*, no. 1 (1968), 2; Bookchin, "Introduction," 24.

57. Bookchin, "Forms of Freedom," 167; Bookchin, "Revolution in America," 3, 8.

58. Bookchin, "Revolution in America," 12; Bookchin, "Introduction," 26.

59. Bookchin, "Forms of Freedom," 160–61, 169.

60. Simon Sebag-Montefiore, "The Amiable Anarchist," *Stop Press with Varsity*, Nov. 22, 1985.

61. Hoffman (as Totalist), "Eighteen Rounds," 21; Bookchin, "When Everything Was Possible," 5.

62. Osha Neumann, *Up Against the Wall Motherf**ker* (New York: Seven Stories Press, 2008), 57. The 1968 Third World Newsreel film *Garbage* documented the action; on Newsreel, see Leo Braudy, "Newsreel: A Report," *Film Quarterly* 21, no. 2 (Winter 1968–69).

63. John McMillian, "Garbage Guerrilla: The Mystery Man behind the East Village Art Gang with the Unprintable Name," in Clayton Patterson, ed., *Resistance: A Radical Political and Social History of the Lower East Side* (New York: Seven Stories Press, 2007), 532–42.

64. Neumann, *Up Against the Wall*, 57.

65. Hoffman (as Totalist), "Eighteen Rounds," 20; Bookchin, "1960s," 84; Bookchin, "Forms of Freedom," 143.

66. Morea interview by author.

67. Quoted in Anderson, *Movement and Sixties*, 195.

68. Bookchin, "Post-scarcity Anarchism" (1971), in *PSA*, 51.

69. Todd Gitlin, *The Sixties: Years of Hope, Days of Rage* (New York: Bantam, 1987), 307–8.

70. Hayden quoted ibid., 308; Rudd quoted in Anderson, *Movement and Sixties*, 200.

71. Quoted in Jules Witcover, *The Year the Dream Died* (New York: Warner Books, 1997), 215.

72. Bookchin, "May-June Events in France: 2" (1968), in *PSA*, 262.

73. Bookchin described this episode in detail in "Whither Anarchism" (1998), in *Anarchism, Marxism*, 238–40. See John L. Hess, "De Gaulle Insists on Public Order," *New York Times*, July 14, 1968, 1.

74. His description and analysis may be found in "The May-June Events in France," parts 1 and 2 (1968), in *PSA*.

75. Bookchin, "May-June Events in France," part 1, 249–50.

76. Ibid., 251, 253, 254.

77. Ibid., 251.

78. Bookchin, "Listen, Marxist!" (1969), in *PSA*, 194–95, 216; Bookchin, "May-June Events 2," in PSA, 269.

79. Bookchin, "Listen, Marxist!" 195.

80. One of Murray's letters from France was published in *New Left Notes*, July 29, 1968; *The Rat*, July 1968, contains an interview with Daniel Cohn-Bendit "furnished by *Anarchos* Magazine."

81. Bookchin passport, copy, author's collection.

82. Sale, *SDS*, 455, 461.

83. Paul Goodman, "The Black Flag of Anarchism," *New York Times Magazine*, July 14, 1968, 10ff; reprinted in Taylor Stoehr, ed., *Drawing the Line: The Political Essays of Paul Goodman* (New York: Free Life Editions, 1977).

84. Bookchin, "An Open Letter to Paul Goodman," *Rat*, no. 1 (Aug. 9–22, 1968), 2ff. A few years later, in 1971, Bookchin recognized that Paul Goodman had been right and publicly apologized for his 1968 attack. Letter to the editor, *WIN Magazine* 7, no. 14 (Sept. 15, 1971), 32. In 1972 he met Goodman at a conference in New Hampshire and apologized personally, acknowledging Goodman's perspicacity. The conference, titled "Adequate Action for a Human Future," was held Jan. 23–25, 1972, at World Fellowship Camp in Conway, N.H. It was "for all who want to bring a better balance between agriculture and industry, country and city, decentralization and centralization. Focus on homesteading and alternative power." The panel members, in addition to Bookchin and Goodman, included Ralph Borsodi and Scott and Helen Nearing.

85. Huey Newton, "In Defense of Self-Defense," *New Left Notes*, Dec. 23, 1968.

86. Bookchin, "Anarchy and Organization: A Letter to the Left," *New Left Notes*, Jan. 15, 1969.

87. Bookchin, *Spanish Anarchists*, 4.

88. Ibid., 54–55.

89. Ibid., 8.

90. Bookchin, "A Note on Affinity Groups" (n.d.), in *PSA*, 221–22; "Post-scarcity Anarchism" (1970), in *PSA*, 47; "Listen, Marxist!" 215.

91. Wells, *War Within*, 284.

92. Ibid., 304.

93. Bookchin, "Listen, Marxist!" 173, 191, 206.

94. Radical Decentralist Project, "Toward a Post-scarcity Society: The American Perspective and S.D.S.," in Ehrlich et al., *Reinventing Anarchy*, 123.

95. Wells, *War Within*, 304.

96. Ibid.

97. Dick Young, "The Black River Movement: Manifesto for an Anarchist Revolution," in *Black River: A Journal of Correspondence*, no. 2 (Oct. 22, 1969), 9, MBPTL and author's collection.

98. Bookchin, "1960s," 104–5.

99. Bruce Schulman, *The Seventies: The Great Shift in American Culture, Society, and Politics* (New York: Da Capo Press, 2002), 39.

100. His apartment was at 235 Second Avenue. *United States v. W. Mark Felt and Edward S. Miller*, US District Court, Washington, DC, Sept. 1980, transcript, 1760.

101. Neumann, *Up Against the Wall*, 63–65.

102. Allan Hoffman, "Earth—Life—Reclamation," *WIN Magazine* 6, no. 19 (Nov. 15, 1970).

103. Joy Gardner to author.

104. Bookchin, "Introduction" (1979), in *TES*, 23; Bookchin, "Statement of Purpose," *Comment*, Jan. 14, 1979.

7

Man of the Moment

ANTIWAR ACTIVISTS DESPAIRED: for six years they had rallied and marched on an unprecedented scale; they had been arrested and tear-gassed and beaten with clubs; and they had mounted an enormous strike. Yet the war in Indochina continued. Meanwhile, a broad, seething, culture-wide repugnance was setting in among young Americans, who now rejected not only the war, and not only the war machine that prosecuted it, but the very society that supported it. They rejected its racism, its synthetics and consumerism and advertising, its chemical-laden food, its sexual repression, its hierarchical institutions, and its prefabricated housing, careers, and lives.

They were learning to reject its sexism as well. By this time, radical feminists were denouncing the sex caste system, marriage, and the patriarchal family. Challenging this basic feature of social organization, they were, to Bookchin's mind, among the most vocal critics of hierarchy, giving their movement a "revolutionary core." He supported them personally as well as politically: the radical feminist movement "has made me acutely aware of [patriarchal] traits in myself," such as using the word *man* to mean *people*, "and like many men I've tried earnestly to rid myself of them." It was part of his ongoing effort to try "to produce [a self] free of the values spawned by hierarchy and domination."[1]

Then, in January 1969, something happened that catapulted the counterculture's rejection of hierarchical society to a whole new level. Just offshore from the white sand beaches of Santa Barbara, California, hundreds of oil wells were sucking up crude from below the ocean floor. On January 28 a blowout spewed more than 3 million gallons into those waters; within days the tides spread a thick, gooey layer over the beaches. Birds soaked in sludge, unable to raise their feathers, died in misery on the mucky shoreline.

Two months later astronaut Rusty Schweickart, journeying to the moon on the *Apollo* 9 mission, took photos of Earth from space. The images he transmitted showed the green and blue ball glistening wondrously, but its ability to sustain life seemed inexpressibly fragile.

"When the hell are we finally going to create a movement that looks to the future instead of to the past?" Bookchin demanded in June, as Ohio's Cuyahoga River burst into flames, having become an industrial sewer.[2] The answer was: now. The world was finally catching up with what he had been saying since 1952. History, having ignored him for decades, came back into his life with the emergence of the environmental movement.

His basic articles had been circulating around the counterculture and the international New Left, copied, republished, and anthologized.[3] Richard Merrill, a graduate student in biology at the University of California, Santa Barbara, was typical of many young readers. After the oil spill, Merrill wanted to get involved in political protest, but he was "not sure of where to start. I joined various student groups but found no serious purpose in any of them." When a friend passed him a copy of Bookchin's "Ecology and Revolutionary Thought," he read and reread it. "I was never the same again. Murray had made it clear that the real political problem was . . . centralized power itself. For me it was a catharsis and clarion call. . . . For society to survive in a truly ecological manner, it needed to decentralize power."[4]

In 1970 Ramparts Press decided to publish his 1960s essays under the title *Post-scarcity Anarchism*. Bookchin dedicated the collection to Josef Weber (who "formulated more than twenty years ago the outlines of the utopian project developed in this book") and to Allan Hoffman (who had given Bookchin "a broader sense of the totality sought by the counterculture and youth revolt").[5]

But as the book went to press, Bookchin's phone was ringing off the hook with speaking invitations. People even followed him into movie theaters and demanded that he talk to them. The old soapbox orator reemerged, now stirring audiences on circuits around the country on behalf of the earth and a humane social order. In 1969–70 he started teaching at Alternate University, a free school recently established in a loft on Fourteenth Street in New York—issuing no grades or credit, it taught unconventional subjects. The faculty were radicals, but "Murray was the star attraction," recalled Arthur Lothstein, who taught a class on youth culture.[6]

Simply cleaning up air and water pollution, Bookchin told the rapt students in his class on anarchism, was not going to solve the ecological crisis. Yes, we need to install smog-control devices and filtration systems on smokestacks; yes, we must reduce automobile traffic; yes, we must protect endangered species. But these measures are merely delaying actions—they can't arrest our society's headlong momentum toward destruction, because the ecological crisis cannot be solved within the framework of the capitalist system.[7]

But that very system, he told them, was even now losing its raison d'être. Under conditions of material scarcity, people compete for resources; brute necessity forces them to become egoists; the historic result was the capitalist economic system. But now, thanks to technological development, material scarcity has receded into the past—the means of life are now available to all. Under present conditions of material abundance, people need no longer compete for goods but are free to become generous and sociable—and to create a cooperative society. Freed from toil, they can do away with hierarchy (he was popularizing that word) and domination, advance to the realm of freedom, and overcome "the conflicts between humanity and nature, between mind and sensuousness, between reason and feeling."[8]

Indeed, the environmental crisis was giving many young radicals a new orientation. Their task was not simply to end a war or fight for pollution controls; it was to get to the root of the problem and build a new society from the ground up, one that would be peaceful and cooperative and nature-friendly. So in growing numbers they left the culture of destruction behind—the polluted, stressful cities—and set out "up the country," journeying to rural Oregon, New Mexico, California, Vancouver, and New England. There they would disconnect from the consumer society and make what they needed for themselves. They would grow their own food, organically. They would live communally, with equal gender roles, and divide household chores fairly. They would form cooperatives—farms and shops, newspapers and radio stations and medical clinics.

In the early 1970s, young Americans formed about two thousand rural communes. Over the decade, they would create more than one thousand food and other consumer-goods coops, with more than a million members, and eight hundred housing cooperatives. Some 7.5 million people would belong to insurance coops. Cooperative-style healthcare programs had enrollments of 4.5 million. Seventeen hundred cooperative day-care centers dotted the land.[9]

Initially, this eco-counterculture was mistrustful of technology, but as young people moved to the land, they found that creating a society off the mainstream grid required those small-scale, cheap, easy, and environmentally benign technologies. Stewart Brand's *Whole Earth Catalogue* explained to them in practical terms how to make solar collectors and wind generators, but Bookchin's writings on technology also strongly influenced the eco-counterculture. "Towards a Liberatory Technology" had laid out the intellectual framework for renewable forms of energy that, as Bookchin put it, "the individual can understand, control, maintain, and even build."[10] He insisted that "eco-technics" (as he preferred to call alternative technology) were inextricably linked to a decentralized social context. Wind and solar and geothermal energy, for example, made little sense in giant cities but were entirely suited for smaller communities. So was energy generated from solid wastes. Fuel cells could supply energy for individual buildings. And electric cars, their range limited by the need to recharge their batteries, could provide clean transportation for trips of up to eighty miles.[11]

In 1970 thirty-three-year-old John Shuttleworth began publishing the magazine *Mother Earth News*, offering practical advice for the back-to-the-landers as well as interviews with cutting-edge thinkers. Right away, Shuttleworth said, readers from all over the country began sending him excited messages: "Do you know about Murray Bookchin?" "Have you read Bookchin?" Shuttleworth hesitated, because Bookchin was an anarchist, and "I had the usual idiotic idea that anarchists were people with bushy, black beards and Molotov cocktails." Then he actually met the man, at a conference in Connecticut in 1971. "My mind was boggled and my heart won," he said. "Murray Bookchin is one of the warmest, most thoughtful and sensible men I've ever met." And he had been "writing about ecology more years than some of today's staunchest environmentalists are old."[12]

In the spring of 1969, Bookchin had journeyed to California to study the political situation there. In Los Angeles, he found that noxious fumes and periodic smog alerts had children and the elderly running for cover just to breathe. The West Coast megalopolis was as unlivable as its eastern counterpart. In Santa Barbara, he visited Richard Merrill, who had become involved with an urban farming project, El Mirasol. "We talked non-stop," Merrill recalled. "I fed him information about basic ecological theory and he filled my head with an entirely new way of seeing the political process. It was Nirvana."[13]

Even in California, Bookchin found, some on the Left were slow to grasp the importance of the ecology issue, still considering it (as Guy Debord had done in 1967) petty-bourgeois.[14] In La Jolla, he dropped in on Herbert Marcuse, who was now teaching there. Surely by now, Bookchin thought, Marcuse would acknowledge ecology's revolutionary potential, but no: "He dismissed environmental problems as technologically resolvable by the bourgeoisie. Indeed, I found him to be singularly unresponsive to any discussion of ecology generally."[15] And in Berkeley, Bookchin tried to explain it to Tom Hayden, who was then living in a commune called the Red Family. He too, Murray later told me, criticized the issue as bourgeois.

But as he traveled up the West Coast, the many newly minted, militant ecology groups he encountered thrilled him. Among the savviest was Berkeley Ecology Action (BEA), which had not only developed a community recycling center (one of the country's first) but was using "festivity and celebration, guerrilla theater, direct action, sabotage, and disruption" to dramatize ecology issues. BEA members understood that mere pollution-control devices could not remedy the ecological crisis. They would don white coats and gas masks, go into traffic jams, and pass out leaflets that urged upon the stalled drivers the need for mass transit. Others staged a mock funeral procession for the internal combustion engine.[16] Here eco-anarchism was being born as an on-the-ground direct action movement.

Back in New York that fall, inspired by Ecology Action, Bookchin organized students in his Alternate U class as Ecology Action East (EAE).[17] He wrote its stirring manifesto, "The Power to Destroy, the Power to Create," affirming that the ecological crisis is an immense problem that requires an immense solution: fundamental social, political, technological, agricultural, urban, industrial, psychological, and cultural change. It requires reconciling "the great splits which divided human from human, humanity from nature, individual from society, town from country, mental from physical activity, reason from emotion, and generation from generation." And it requires a thoroughgoing social revolution to bring about "politically independent communities whose boundaries and populations will be defined by a new ecological consciousness." This is our agenda, he warned: "If we are to survive, we must begin to live."[18]

In the spring of 1970, Ecology Action East protested the annual Automobile Show at the New York Coliseum, under the slogan "Boycott the Deathmobile Show!" Its leaflet (likely drafted by Bookchin) pointed out that the automobile spews poisonous exhausts into the air and kills or injures thousands of people each year. By transforming cities into homes for cars rather than people, it pollutes our psyches and "steals our integrity as human beings." Instead of constructing ever more infrastructure for cars, we should put resources into public transportation. Above all, we should reconstruct our civilization to advance human needs rather than profit.[19]

In late 1969, Senator Gaylord Nelson called for a national environmental teach-in, to be called Earth Day, to channel student antiwar energy into the environmental question and thereby place the issue squarely on the nation's political agenda. Many organizers got to work, but at the University of Michigan at Ann Arbor, exams were scheduled for the selected day, April 22, 1970, so it chose to hold its extensive teach-in early, over five days in mid-March. Most high-profile Earth Day organizers wanted to keep the teach-ins politically moderate, but the Ann Arbor organizers "realized we were against the automobile, and that means being against road-building and oil. Right away, we were fighting the corporate structure. It wasn't a nice safe thing." The Ann Arbor teach-in drew fifty thousand attendees and ran from early morning to well after midnight. Bookchin spoke there, in top form, advising the students that "capitalism is inherently anti-ecological." Not only does it pit human beings against one another, it pits "the mass of humanity against the natural world." It "plunder[s] the earth." Action taken to prevent ecological disaster must therefore be "revolutionary action" against capitalism, "or it is nothing at all." The stark reality is that "either we will create an ecotopia based on ecological principles, or we will simply go under as a species."[20]

Decades before, Rosa Luxemburg had insisted that the alternative to socialism was barbarism, but now humanity faced alternatives that were even more extreme. How to word it? "Anarchism or annihilation" was one attempt at a new apocalyptic binary; "utopia or social extinction" was another; "ecotopia or ecological devastation" still another. But by the time of the Ann Arbor teach-in, he seems to have hit on the best of all updates of "socialism or barbarism." Back in 1968 he'd seen some graffiti on a wall in Paris: "Be realistic, do the impossible." Now he appended to it "because otherwise we will have the unthinkable." It would become his signature closing for his speeches.[21]

Recalled one attendee at Ann Arbor, "It was the more radical speakers, such as Murray Bookchin . . . who made the greatest impression."[22]

A few weeks later, on April 22, 20 million Americans participated in the enormous Earth Day teach-in, the largest of all 1960s-era mobilizations. All over the country, schoolchildren planted trees and picked up litter, speakers lectured, and students protested. Outside the Department of Interior in Washington, D.C., 2,500 radical demonstrators protested oil leases, chanting, "Give Earth a chance," while others poured oil onto the sidewalk.[23]

In New York, cars were banned along a stretch of Fifth Avenue, while Fourteenth Street became an "ecological carnival," bristling with exhibits on pollution. Con Edison, whose headquarters was at Union Square, draped lampposts with bunting and provided extra current for the occasion. Perhaps its executives hoped New Yorkers had forgotten its effort to build a nuclear power plant in Queens a few years earlier.[24]

Bookchin and EAE dismissed Fourteenth Street as a carnival of co-optation and went uptown to Columbus Circle, where the American Management Association was holding its annual packaging show at the New York Coliseum. There the merchants of garbage were showing off the latest in plastic, aluminum, and paper packaging, intended to dazzle consumers into buying products they didn't need. Out in front of the Coliseum, EAE staged a protest, calling for an end to deceptive and wasteful packaging, which ended up as clutter in people's lives and trash in the environment. Then EAE held a funeral procession for garbage (echoing the Berkeley funeral for the automobile), in which 150 people ceremoniously dumped more than a hundred pounds of used packaging into a coffin, for return to its progenitor, the packaging industry: "Ashes to ashes . . . Garbage to garbage." The action drew a crowd of two thousand and succeeded in shutting down the trade show for half an hour.[25]

In those days, EAE's neighborhood, around Fourteenth Street and Third Avenue, was low-slung and lined with small shops. In 1970 real estate developers wanted to build high-rise apartment buildings there and asked the city planning commission to alter the zoning accordingly. Local community groups opposed the plan, and so did Sandy Brownstein, an EAE member. At the group's weekly meeting at Alternate U, she persuaded the group to take a stand.

Perhaps thinking of People's Park in Berkeley, EAE decided to transform a vacant lot at Tenth Street and Third Avenue into "Squatters' Park," an action that they hoped would not only mobilize the community against the developers but spark a movement for "one, two, three . . . more squatters' parks." On the morning of June 13, EAE members showed up with shovels and wheelbarrows and started clearing rubbish. Within a few hours, several hundred people were helping out. Someone brought tools and constructed a sandbox and flowerbeds. Others contributed rubber tires and cartons for playground facilities. The work was joyous, reported EAE member Robert Palmer (writing as Grey Fox). "It was the sense of being able to take control of something—if only for a while," indeed "to make a utopia in the midst of the old society."

But on June 17 one of the developers erected a cyclone fence around the park. The volunteers held a community meeting and decided to cut the fence. That Saturday they chopped open a hole large enough for several people to walk through. Over the next week, community members tore down more sections of the fence, so that "fence cutting became a community pastime." A few months later, thanks to community pressure, the city's board of estimate rejected the developer's rezoning proposal. Grey Fox commented that Squatters' Park "as a community rallying point and symbol made the difference." Here was a new way to fight capitalist development, as Bookchin and many others at the time were realizing: community organizing.[26]

Almost four hundred miles to the north, the city of Montreal was in a state of political ferment since, as at Squatters' Park, citizens were protesting the destruction of neighborhoods. Hearing rumors about its feisty New Left and its passionate citizens', feminist, and tenants' movements, Bookchin visited the city to see for himself. He knocked on the door at a publishing office on Boulevard St.-Laurent. A tall man with a goatee admitted him and introduced himself as Dimitri Roussopoulos, editor of the New Left journal *Our Generation*.

Roussopoulos, in his mid-thirties, had been active in the Canadian Left at least since 1959, when he chaired a campaign to prevent Canada from acquiring nuclear weapons and edited its journal, *Sanity*. In 1965 he had organized a large sit-in in front of the American embassy, to protest the Vietnam War, and had founded an important Canadian student organization. In the summer of 1968 he had organized the first meeting of the international New Left, in Ljubljana.

How much of this he recounted to the stranger from New York is unknown, but when the conversation turned to Montreal, he surely told him that the city had a lot of problems. The municipal services—housing, transportation—were dismal. Above all, the longtime mayor, Jean Drapeau, was an oligarch. His government was not only unresponsive to citizen pressure—its workings were kept all but secret. The city council consisted of his handpicked operatives. Citizens who wanted to attend council meetings had to sit in a high gallery, where it was hard to hear; so did the press, and all were forbidden to record whatever they managed to hear. The authoritarian mayor was determined to keep citizens, with their irritating demands for transparency and democracy, out of city government.[27]

Meanwhile, Drapeau was orchestrating grandiose building projects and even a world's fair, Expo 67. To hide the city's neglected neighborhoods from fair visitors, he erected pastel-colored fences around them—upon which angry Montrealers scrawled a message: "Visitez les slums!"[28]

Recently, the mayor was making it easy for developers to buy up real estate. Just a few blocks from where Bookchin and Roussopoulos were talking lay a beautiful six-block urban village, Milton-Park. Its gray-limestone nineteenth-century townhouses, two and three stories tall on tree-lined streets, featured lovely outdoor staircases, perfect for neighborly conversations. Rents were reasonable, so these stately, once-elegant old buildings were home to working poor and students and immigrants and elderly people and artists. But recently a developer, Concordia Estates, had been buying up the land under the limestones in order to tear them down and build luxury high-rise condo towers. Roussopoulos and his wife, Lucia Kowaluk, were organizing to prevent it. Fortunately, Roussopoulos might have told Bookchin, the neighborhood had a strong sense of community, and by knocking on doors, they'd been able to build up an opposition.[29]

Then Roussopoulos asked his visitor about New York. "Do you know someone called Murray Bookchin? I've just come across a magazine called *Anarchos* with an interesting article by him."

> Bookchin said, "I'm Murray Bookchin."
> "The issue has another interesting article too, by Lewis Herber."
> Bookchin: "I'm Lewis Herber."
> "And this one, 'Revolution in America' by Robert Keller—"
> "Yes, I'm Keller too."[30]

Dimitri then invited Murray-Lewis-Robert to give a lecture at the University Settlement. He did so in the fall of 1969, to a packed house, on the topic of "revolutionary anarchism." So began a decades-long collaboration. *Our Generation* was an early adopter of many of Bookchin's ideas, editorializing in 1970: "The ecological crisis is fundamentally a social problem . . . cities must be decentralized into regional communities . . . the earth's resources must be utilized on a regional community basis," and "all social institutions of domination and exploitation must be dissolved."[31]

In 1969 Roussopoulos and his friends in Montreal, eager to develop and sophisticate the radical ideas of the 1960s, decided to establish a press, one that would publish works on anarchism, participatory democracy, and ecology. As they began to organize it, they tried to figure out what to call it, tossing around titles like *Journal of Canadian Studies* and *Canadian Sociology*.

Murray rolled his eyes at the straitlaced monikers, then told Roussopoulos, "I've got it—call it Black Rose Books!"

Roussopoulos was skeptical. What did black roses have to do with anything?

"Don't you know the legend, Dimitri?" Murray said. The black rose was a symbol of rebellion in central Europe, he explained. It doesn't exist in nature, but legend says that if we ever do somehow find a black rose in nature, we'll find freedom. And if we ever achieve freedom, we'll find the black rose in nature.[32]

Roussopoulos was delighted, and Black Rose Books was born, but they resolved never to publish a graven image of the black rose—it would remain imaginary, never a logo. Bookchin wrote the introduction to its inaugural offering (Ida Mett's *The Kronstadt Sailors*), and on the book's inside cover, Roussopoulos printed the black rose story. When Murray saw it, he confessed with embarrassment that he'd made it up. But by then no one cared—the name was perfect. The story did not, however, appear in the press's future publications.

Post-scarcity Anarchism, when it appeared in 1971, hit the New Left and the counterculture like a thunderclap. Its several essays dismissed the stereotype of anarchism as mere bomb-throwing and redefined it into a socially and ethically constructive alternative. It thereby offered a way for young radicals, disillusioned by Marxism-Leninism, to step over the debris of the imploded New Left and move forward. Onetime SDS leader Todd Gitlin praised it as "the closest thing I've seen to a vision both practical and transcendent. Anyone who wants to make revolution in this country should read it and reckon with it."[33]

The yearning for such a reworking—rethinking, even—of radical doctrine was genuine. In the spring of 1972, Brian Tokar, a physics and biology student at MIT, organized a protest against the mining of the Haiphong harbor. But Cambridge's local Marxist-Leninists, "spouting their ideology" and "claiming to be the legacy of the New Left," wrecked the protest. Tokar could not understand how such a thing could happen. An older friend gave him a copy of *Post-scarcity*, with "Listen, Marxist!" which detailed how Stalinists, despite their revolutionary rhetoric, were functionally counterrevolutionary. It came as "an incredible revelation."[34]

Similarly, in New York, the New Left had been "full of Guevara, Mao, Situationists, all jumbled in together," Bill Koehnlein told me, but Bookchin's essays "drew a line that distinguished anarchists from the rest of the soup." And Bookchin "wasn't just someone who latched onto anarchism and publicized it—he developed it."[35] In Baltimore, Howard Ehrlich, a budding sociologist, agreed that Bookchin, better than anyone else during those years, "articulated the fundamental differences among anarchists, liberals, and marxists."[36] *Post-scarcity* revived and renovated anarchism for a new generation. Decades later, the historian Peter Marshall would say that Murray was "the thinker who has most renewed anarchist thought and action since the Second World War."[37]

Bookchin stepped into the limelight as naturally as he had climbed onto that East Tremont soapbox as a young teenager. In the early 1970s, he spoke at colleges and universities all over the United States and Canada. He knew from his youthful training that history didn't come around to the radical side often, and now that it had, he had to try to channel its powerful forces toward revolutionary change; he must guide these angry, idealistic young people to form a movement, not to try to seize power, in Marxist fashion, but to demand the dissolution of power—and to transform society in a nonhierarchical, democratic, ecological way. After a lecture, he'd stick around a campus for days to talk to the students, to mentor them. He tried to compress a lifetime of radical lessons into a brief visit, answering questions, holding forth, and above all inspiring them to "be realistic, do the impossible, because otherwise we will have the unthinkable."[38]

Bookchin would have expected existing anarchists to welcome the fresh ideas and new thinking in *Post-scarcity Anarchism*, but curiously, certain anarchist reviewers reacted to the book with suspicion at best and open hostility at worst. Some mistrusted Bookchin due to his Marxist background and were irritated by what they saw as Marxist residues in his work. For one thing, he celebrated advances in technology so glowingly. Technology, complained one reviewer, "rests on the basic principle of centralized authority," which made it compatible with Marxism. Therefore the pro-technology Bookchin was "essentially a Marxist determinist waiting for St.

Karl's predictions to come true in their pre-ordained sequence." To which Murray replied that he had long ago rejected Marxism but made no apologies for his Marxist background and education in revolutionary theory and practice: "If anarchists learn nothing from Marx and Hegel, then they are destined to remain a theoretically impoverished lot."[39]

To be sure, Bookchin's affirmations of technology made him somewhat anomalous in the counterculture, which had rightly come to loathe armaments and nuclear weapons and chemical weapons like napalm.[40] But he understood that overall, machines had improved the human condition, and that a rational, postscarcity society could harness them for more constructive, humane ends rather than destructive ones and indeed must do so. Furthermore, new, more benign eco-technologies, like renewable energy, were emerging that could facilitate the decentralization of society along anarchist lines.

Other anarchist critics were annoyed that Bookchin had failed to pay due reverence to the canonical nineteenth-century anarchist writers like Bakunin and Kropotkin. Bookchin's ideas on town and country, one pointed out, could be found in Kropotkin's *Fields, Factories, and Workshops*.[41] But Bookchin hadn't read Kropotkin before developing his eco-decentralist ideas, he explained, and the Russian prince's writings had not affected his thinking.

In 1971 this puzzling issue was clarified when Bookchin finally met Lewis Mumford at a conference called "Technology and Man" at the University of Pennsylvania, where Bookchin spoke about anarchism. Between sessions, the two men sat and talked for hours. Bookchin would have told Mumford that his writings on cities had profoundly influenced him in the 1950s. Mumford explained that Kropotkin's *Fields, Factories, and Workshops* had inspired him enormously. Bookchin came to understand that through Mumford he had absorbed Kropotkin's call for "an ecological integration of town and countryside." Kropotkin, he would acknowledge, was "the real pioneer in the eco-anarchist tradition."[42]

But while existing anarchists groused, Bookchin championed their ideology intellectually, defending it against Marxists and liberals alike, and won over many converts. In the fall of 1971, to take one example, he attended a conference in Buffalo, organized by the journal *Telos*, to discuss the role of intellectuals in radical social movements. The attendees were mostly New Left–influenced scholars who were in the process of building up radical studies programs in university history and sociology departments. To Bookchin's chagrin, the presenters mostly spoke in Marxist terms, influenced by all the talk of student-as-worker from 1968. Herbert Marcuse, as one of the young academics' mentors, reinforced it when he explained that "all of us can be a part of the vanguard."[43] The French theorist André Gorz even told the conference that radical intellectuals should "leave the university ... and go work with the low-income industrial proletariat" to help develop "workers' councils in the factories."[44]

Bookchin, as one of the few actual proletarian intellectuals in attendance, must have shaken his head in dismay. What was the point of trying to redeem Marx? he asked the conference when his turn came. The really pressing imperative was not to organize the working class but to address the ecological crisis. An eco-counterculture was burgeoning outside the groves of academe, as young people sought to transcend mainstream American lifeways and replace individualism and hierarchy with a new cooperativism. They weren't looking for a Marxist-style vanguard to lead them—on the contrary, they were trying to free themselves of "the values spawned by hierarchy and domination"—values associated with Marxism. Instead, they were "seeding" the country with new organizational forms, "with affinity groups, ... communes and collectives—in cities, in the countryside, in schools and in factories." These nonhierarchical "catalysts" for transforming society, Bookchin reproached the *Telos* conference, were far more relevant and constructive than anything that Marxism had to offer.[45]

In January 1971 members of Ecology Action East, unable to prevail over developers on the Lower East Side, were pining to join the counterculture's exodus to rural America, to build communes and organic farms. The resourceful Sandy Brownstein made her way north, where she scouted out Vermont. Historically this rural state had been a haven for outliers—nonconformists, freethinkers, troublemakers, utopians. Now thousands of young people were choosing to move there. By 1970 several dozen communes dotted the state of Vermont, some left-radical, others apolitical, a culture that gave rise to food coops, collective schools, radio stations, and even a People's Bank. Perhaps Brownstein met some of the members of Free Vermont, a network of radical communes that considered it their revolutionary task to create a socialist society in the Green Mountains, organic farm by organic farm.[46]

If the state's political air was intoxicating, Brownstein discovered, so was its physical beauty. She visited Burlington, a human-scaled city nestled on the eastern shore of Lake Champlain, whose downtown—a few square blocks—featured bacon-and-egg diners, hardware stores, and gun shops (for the state's many hunters).[47] As she mingled with the hippies at Battery Park, perhaps she gazed westward at midday and saw the lake glittering in the sunlight; or perhaps she saw it in the evening and watched the sun dipping behind the purple-blue Adirondack ridgeline on the far shore.

She dashed back to the Lower East Side and told Bookchin and EAE about Vermont. What better locale for an anarchist collective? It was practically utopia already. They seem to have wasted no time—they arrived in Burlington that June. They renovated a building at 135 Church Street and transformed it into the Fresh Ground Coffee House, which they ran as a collective: it served not only java but fresh-baked bread and homemade soup to warm both body and soul during the bitter-cold winters.

Just as his political family was moving north, Bookchin was barnstorming around the country on his crowded speaking schedule. But when his apartment on Second Avenue was subjected to yet another of many break-ins, he'd had enough of the megalopolis and decided to join them in Vermont. He packed up some of his books on Spain and ecology and urban history to take with him. But how would he make a living in Vermont? he must have wondered as he packed. He wasn't sure. And the apartment, however prone to break-ins, was rent-controlled and therefore not to be given up lightly. So he decided to hold on to it and left most of his possessions there. Locking the multiple locks behind him, he pocketed the keys and headed north.[48]

Shortly before Thanksgiving, during a snowstorm, he arrived in Burlington. He carried his boxes up a flight of creaking wooden stairs to a second-floor flat on Main Street, around the corner from the Fresh Ground Coffee House.[49] Other members of the collective had set up housekeeping nearby, as did his ex-wife Beatrice, with their two children. He had lived for all of his fifty years in the New York megalopolis. Now he could finally yield to the relaxed tempo, small scale, and natural beauty of the *arcadie vermontaise*.

Some of the local communards were even trying to eliminate monogamy. A commune in Burlington, about fifteen people, removed all the doors in their house, laid the mattresses side by side, and changed sex partners regularly. Predictably, the members were soon consumed by jealousy and spent endless hours talking it all out. Someone from this commune who visited Murray's Main Street apartment exclaimed, "Oh, you have doors!" and criticized him for it. "You have your structure, and I have mine," he explained. "I need a door, I need a private life."[50]

Each day behind that door, he sat at a desk overlooking Main Street to write his book on the Spanish anarchists. He had also started to write a book about hierarchy, which he by now considered "the authentic 'social question.'"[51] Much as Marx had traced the historical rise of class society in *Capital*, Bookchin wanted to trace the rise of hierarchy in *The Ecology of Freedom*. Angus Cameron at Knopf was interested, and Bookchin applied for and received a grant from the Louis M. Rabinowitz Foundation to write it.[52] Since the rise of hierarchy predated the historical record, he had to study anthropology, so he gave himself a crash course, poring over writings on tribal cultures by Paul Radin (who had researched the Winnebago, in Wisconsin) and Dorothy Lee (who had studied the Wintu, in northern California), as well as overviews by the political anthropologist Elman Service and archaeologists V. Gordon Childe and Jacquetta Hawkes.

The earliest human societies, he grasped, had been small-scale, nonhierarchical, and egalitarian in nature. In Wintu society, Dorothy Lee had written, "equality exists in the very nature of things." Correspondingly, the Wintu language was free of hierarchical words and concepts. A mother didn't say, "I took the baby," which would have left no room for the child's will; she said, "I went with the baby."[53]

These early societies were not only egalitarian, they were mutualistic and matricentric. Women's work was as important as men's and complemented it. Tribespeople gathered or hunted food in groups and shared. Everyone had a right to what Radin called the "irreducible minimum" of subsistence goods—the means of life. People took what they needed and nothing more, operating on an unspoken principle of usufruct. (The Wintu language, said Lee, had no words for possession or ownership: a hunter didn't "own" bows and arrows, he "stood by" them.) In such societies, accumulated wealth tended to be suspect, so chiefs disaccumulated material goods, in order to create or solidify social bonds.[54]

Early societies lived in harmony, indeed in symbiosis, with the natural world. And since their social organization was egalitarian and cooperative, they perceived nature as egalitarian and cooperative as well. For the Wintu, Lee observed, "every aspect of nature, plants and rocks and animals . . . all have a cooperative share in the maintenance of the universal order." A culture's view of the flora and fauna around it, and its relations with them, Bookchin noticed, seemed to reflect its own social organization.[55]

Out of this "organic society," as he called it, hierarchy had gradually emerged, immanently. Elderly people, feeling vulnerable as their powers failed, claimed privileges over the young for self-protection, giving rise to gerontocracy. Shamans also experienced insecurity (magic being an unreliable means of healing), so for protection, they sought alliances with warriors. In return for protection, shamans validated warriors' authority by endowing it with a magico-political aura.

Priests then emerged, as privileged religious formulators. Male warriors became patriarchs who dominated women and young men; tribal chiefs became commanding rulers. Hierarchy—divisions of age and race and gender—became part of the unconscious apparatus of humanity. Instead of "going with" their children into the shade, mothers now took their children into the shade. Concepts of ownership developed, and words connoting inferiority and superiority entered vocabularies.[56]

The advent of agriculture and the domestication of animals produced surpluses, which created social classes—privileged minorities who could live off the labor of others, using them for production. Elites who wanted to increase their power no longer dispersed their wealth to others—they hoarded it for themselves. Collective societies became states, with armies, monarchs, and kings. For millions of years, people had lived without the state, but once it came into existence, it came to seem essential.

The rise of capitalism and the market economy dissolved mutualistic social ties. The quest for the maximum profitable deal replaced usufruct, and once-cooperative groups became mutually competing buyers and sellers. Cities became marketplaces. Once social hierarchies were in place, people projected a hierarchical mentality outward onto nature. As people and their labor power were increasingly objectified "as mere instruments of production," so nature was increasingly objectified as "natural resources."[57] As people's labor power was converted into a commodity, so were those natural resources, to be manufactured and merchandised for profit. And since commodities were considered scarce and thus to be fought over, capitalism pitted the mass of humanity against the natural world—even as it pitted humans against one another.

In 1972, as Bookchin was developing this narrative framework, a decades-old book by his old heroes Max Horkheimer and Theodor Adorno was finally translated into English.[58] *Dialectic of Enlightenment* (originally published in 1947) focused on a now-timely subject: the domination of nature. The two Frankfurt School authors had argued that the European Enlightenment fundamentally transformed the West's relationship to the natural world. It stripped precapitalist concepts of nature of their ethically prescriptive force; it rendered natural phenomena measurable, making it possible to calculate an object's potential utility; and it thus transformed the phenomena of the natural world into commodities. The Industrial Revolution harnessed new technologies to perform this exploitation with destructive rapacity, and capitalism created an expanded marketplace in which to sell them. The domination of nature was thus the fundamental ideology of industrial capitalism.

Bookchin absorbed Horkheimer and Adorno's idea that the values of the "bourgeois Enlightenment" were a project of rationalization and control. He agreed, too, that the Enlightenment's instrumental rationality—its warped form of reason—had rendered nature measurable and created apparatuses of rule through bureaucratization and rationalization.[59] Reason has been reduced to instrumentalism; technology has advanced at the expense of ethics. Incorporating these aspects of their work into his own, Bookchin hoped, would "place 'eco-anarchism' . . . on a theoretical and intellectual par with the best systematic works in radical social theory."[60] He developed a maxim that became basic to social ecology: "The domination of human by human preceded the domination of nature."[61] That sentence lends itself to several interpretations, but I think in its fullest meaning for Bookchin, the first part of the sentence refers to the anthropological and historical rise of hierarchy and nation-states; the second part refers to the domination of nature initially as outlined by Horkheimer and Adorno. While its meaning evolved, it became Bookchin's watchword for emphasizing the social origins of the ecological crisis—and the necessity of a social transformation to solve it.

As much as he admired the Frankfurt Schoolers, their extreme pessimism, at least since *One-Dimensional Man*, had been a problem for him. Despite the many insights he gained from *Dialectic of Enlightenment*, he found it equally bleak with pessimism, especially its argument that the Enlightenment had culminated in twentieth-century totalitarianism.

But where the Frankfurters had despaired, Bookchin—a buoyant optimist by temperament—used their analysis as the basis for a theory of social renewal. As fond as he was of narratives of decline, to explain unjust and unfree present conditions, he was equally enamored of narratives of ascent, to provide sources of hope for change. He saw both narratives as dialectically intertwined. "I share the Hegelian view," he once said, "that humanity had to be expelled from the Garden of Eden to attain the fullness of its humanness." For even a history of decline may ultimately lead to our creating, based on greater wisdom, "more fecund gardens than Eden itself." And he agreed with the anarchist Bakunin that the rise of hierarchy and domination had been "historically necessary" as part of an evolution toward a free, nonstate society; from Roman imperial times, for example, had arisen the valuable notion of a universal *humanitas*. The current process of social revolution would "annul" the state but retain the progressive idea of universal humanity, in an *Aufhebung* (transcendence) toward "a richly variegated completeness."[62] Paradoxically, the optimistic Bookchin could be energized and recharged by works of despair.

Thus as a counterpoint to his hierarchy narrative, he also developed a freedom narrative. Through history and across cultures, he explained in his new manuscript, people have dreamed of achieving freedom, "the land of Cokaygne," a utopia of peace and harmony, where the lion and lamb lie down alongside each other. The slave revolts of the ancient Mediterranean, the peasant upsurges of the European Middle Ages, the heretical Christians, the Levellers and Diggers of the English Civil War, the Enragés of the French Revolution, the Paris Communards of 1871, the Russian Socialist Revolutionaries of 1917, and the anarchists of the 1936 Spanish Revolution—all, to one degree or another, had sought a redistribution of wealth and an end to social hierarchies. These recurrent goals were, in essence, unconscious yearnings for the principles of organic society—usufruct, complementarity, and the irreducible minimum. In Bookchin's vision, in the eco-anarchist society they would be infused with the best advances of civilization, "extended from the kin group to humanity as a whole" in "a modern vision of freedom."[63]

The 1960s rebellions—of young people and blacks, women and gays, anarchists and ecologists, Native Americans and Latinos—were the most recent recurrence of the utopian impulse. But now their dreams finally stood a chance of realization. Hierarchical society had reached its limits. Its malaises—state, megalopolis, bureaucracy, centralized economy, patriarchy—were no longer "historically necessary" and, if anything, were historically redundant, even vestigial. Hierarchical society could now be invalidated by the creation of "a society in which every individual is seen as capable of participating directly in the formulation of social policy." It could be replaced by a society of small-scale, organic, renewable-energy-powered, self-governing communities. Against all things impersonal, anonymous, authoritarian, and commodified, Bookchin championed the cooperative, mutually responsible, libertarian, and ethical. Authoritarian approaches to reason, science, technology, and ethics would be overturned in favor of libertarian ones. And this transformation not only could but must happen, because the ecology crisis affected everyone, giving rise to the need for a general social revolution.[64] Such was the basic argument of the book that, written in fits and starts over the following years, would be called *The Ecology of Freedom*.

In Burlington, after a day of researching and writing, Bookchin would rise from his desk and head over to the Fresh Ground Coffee House. By this time, a stream of hippies, scholars, and activists were journeying to Vermont precisely to hear him hold forth at its wooden tables. In April 1974 a West German student activist, Karl-Ludwig Schibel, arrived to interview him—he was researching the American commune movement. As a sociology student at the University of Frankfurt, Schibel had studied with none other than Theodor Adorno himself. A freak April snowstorm snowed them all in, so they smoked and drank vodka and talked for several days.[65]

Perhaps Schibel told Bookchin about the new anarchist scene in Frankfurt—the city had a housing shortage, so students and youth were squatting communally in abandoned buildings. Some had hung black flags from the windows and fought back when the city tried to evict them.[66] Bookchin doubtless talked to Schibel about the communes he'd known since 1965, and about anarchism and ecology and the critique of Marxism. "For me this was a revelation," recalled Schibel. "The inconsistencies between my Marxist/Critical Theory training and my commitment to ecological issues fell into place."[67] In his book on communes, Schibel would call anarchism the commune movement's "organizational reality." He and his colleague Bernd Leineweber went on to translate some of Bookchin's essays into German. "At the time," he told me, Bookchin "was a *Geheimtipp* [go-to guy] among anarchists and urban planners."[68]

He dazzled his visitors with his erudition and his sense of possibility. As he discoursed on the Enragés of the French Revolution and the Communards of 1871, the coffeehouse seemed to transform itself into a café like those of revolutionary Paris. He charmed them, too, with his warmth and geniality, so remote from the demeanor of authoritarian Marxists.[69] But mostly he would have told his new *compañeros* about the remarkable Spanish anarchists of 1936–37. Under their leadership, ordinary Spaniards "took immense segments of the economy into their own hands, collectivized them, administered them, even abolished money and lived by communistic principles of work and distribution—all of this in the midst of a terrible civil war."[70] The anarchist Fredy Perlman was by now mocking Bookchin as an "institution freak," but Bookchin thought it obvious that anyone who seriously advocated a stateless society was obligated to explain to the world just what institutions would replace the state. The Spanish anarchists had demonstrated the institutions that anarchists could organize.

Around this time, Bookchin learned from his New York friends about the activities of Roel van Duijn, the Dutchman who had founded the short-lived Provo, which had innovated provocative stunts and anarchistic "happenings" in Amsterdam. Since Provo's demise in 1965, Van Duijn had read Bookchin's *Crisis in Our Cities* (in Dutch translation) and had gone on in 1969 to write *De boodschap van een wijze kabouter*, "a discussion of the political and philosophical work of Peter Kropotkin in connection with our current choice between catastrophe and a Kabouter city." The Kabouters, his new group, were doing community organizing, helping to create neighborhood associations, food cooperatives, communes, child-care centers, and farmers' markets.

Van Duijn wanted these organizations to have a voice in governing the city, so the Kabouters did something truly innovative, at least for anarchists: in June 1970 they ran for seats in the Amsterdam city council, on a platform of radical ecology and anarchism. Their plan, should they be elected, was to democratize city government, to restructure the municipal governing institutions to "form a council of the people in each section of the city." These councils would be accountable to a large citizens' assembly that met once a week, which would "solve local problems of housing, social problems, and also ecological problems." Ultimately, the Kabouters wanted to replace Amsterdam's city government with an assembly democracy. In the election, they gained five (out of forty-five) seats.[71]

Bookchin the institution freak was intrigued.[72] The Kabouters gave him the seed of an idea: that anarchists could enter the local political arena and call for the devolution of power to neighborhood-based democratic assemblies.

* * *

No one was thinking along those lines in what remained of the American antiwar movement. In April 1972, protesting the war in Vietnam, students at more than 150 colleges and universities shut down campuses and occupied administration buildings. Tens of thousands demonstrated in New York, Los Angeles, and San Francisco. Yet a week later, President Nixon went on to mine North Vietnam's ports. Students exploded with the old 1960s fury. Aghast, one Ivy League university president exclaimed, "I don't see how we can continue to run our universities if the war escalates. What will we face in September?" But as the semester came to a close, students turned their attention to final exams. On May 21, only a paltry fifteen thousand showed up for a march on Washington. Said activist David Dellinger, "Not enough people came. Not enough people were sure it made a difference."[73]

The ephemeral nature of antiwar marches persuaded Bookchin that demonstrations, as important as they were for protesting, could not in themselves create real popular empowerment. Instead of concentrating so much on marches, he thought, radicals should work to create permanent, stable alternative institutions, such as citizens' assemblies.

In his 1972 article "Spring Offensives, Summer Vacations," he advised them to set aside national Washington-based demonstrations: "Washington is their capital, not ours. Our 'capital' is the locality in which we live our everyday lives"—neighborhoods, workplaces, schools—"the world which their capital dominates and which we seek to liberate." Rather than march in Washington, radicals should stay in their home neighborhoods and try to create "popular assemblies and local action committees" there. And rather than repeatedly mouth ineffectual Marxist cant to one another, they should learn how to communicate with ordinary Americans.

Bookchin went on to advocate municipal electoralism. Radicals, he said, should run candidates in elections to town and city councils, on programs calling for "a radical restructuring of municipal institutions along directly democratic lines." They should try to persuade their fellow citizens to transform existing municipal councils and the like into democratic assemblies. In June 1972 *Anarchos* magazine, which had languished for a couple of years, was revived to publish the proposal.[74]

Bookchin felt sure that he could persuade anarchists to accept assembly democracy as an anarchist concept, as he had successfully done with ecology. Citizens' assemblies seemed to him, after all, their logical, decentralized political institution. But that hope proved to be overly sanguine. Even while the *Anarchos* issue was in production, trouble was brewing in the print shop. The anarchists in charge of printing the issue were shocked. Bookchin was saying that anarchists should run for elective office? They showed his piece to the theatrical anarcho-pacifist Judith Malina, who was appalled enough to pen a dissent. Anarchists must not "submit ... to the jurisdiction of the local constitutional government," she wrote, by running candidates. They must not even vote, because voting is "the tyranny of the many over the reluctant few." Tyranny is in fact "inherent in the democratic ethic." Anarchists should reject Bookchin's proposal and instead create alternative institutions, such as cooperatives and communes, by nonelectoral means.[75] The *Anarchos* printers typeset Malina's riposte and, without informing Bookchin, added it as an insert to that issue.

Once the issue was released, the "Spring Offensives" argument found few sympathizers among anarchists. Sam Dolgoff, for one, agreed with Malina that decision-making by majority vote was alien to anarchism. But up in Montreal, Dimitri Roussopoulos concurred with Bookchin from the outset that assembly democracy was compatible with anarchism, arguing that urban government should be "fundamentally re-organized into a federation of decentralized communities, each developing in cooperation and coordination with other community assemblies the process of social reconstruction."[76]

To persuade his anarchist friends, Bookchin saw that he would have to work harder to fill out his argument. Back in 1960, he had had to abridge his hefty, history-laden article "The Limits of the City" for publication in *CI*; now in 1973, he sent the full article to Harper Torchbooks, which agreed to publish it unabridged, as *The Limits of the City*. Delighted, he added some new chapters that suggested ways to use urban space communally and ecologically, drawing on experiments that he'd learned about in Berkeley. To reintegrate town and country, urban residents could dismantle the fences between their houses, and they could raise organic crops in communal backyard gardens. They could transform vacant lots into parks and community gardens. They could sell their produce in a "People's Market" (what we now call farmers' markets). They could limit their driving to necessary trips and even then carpool, otherwise get around by bicycling or walking; they could collectively recycle trash. They could establish cooperatives. They could tear down interior walls in their homes to create communal dining and meeting spaces. Shouldering community problems collectively in these and other ways could "reorient the individual" from "isolation and egotism and dependence." Commingling city, country, and wilderness could foster sharing and "the integration of individual and society, of town and country, of personal and social needs within a framework that retains the integrity of each." Roussopoulos, for one, welcomed the book as "indispensable."[77]

In the wake of the first Earth Day, the Nixon administration undertook a series of reforms. It established the Environmental Protection Agency, banned DDT, strengthened the Clean Air Act, and passed several clean water acts and the Endangered Species Act. Businesses, too, pledged to adopt pollution controls, while Continental Oil introduced "cleaner-air" gasolines. It was the kind of reform that, Bookchin often said, dealt with symptoms rather than systems and didn't get to the root of the problem.[78]

He labeled such reforms "environmentalist," in contrast to his own "ecological" proposal. Where ecology was organic, aiming to preserve the integrity of the biosphere as desirable for its own sake, environmentalism was instrumentalist, seeing the natural world in terms of useful resources. Environmentalism perpetuated the idea "that man must dominate nature" and sought merely "to facilitate that domination" by mitigating environmental insults; ecology, by contrast, was committed to eliminating all domination, both of nature and of people. Any solution to the ecological crisis must involve not Band-Aids but a fundamental social restructuring.[79]

Even within the ecology movement, certain tendencies not only deemphasized the social dimension but missed it altogether. In 1968 Paul Ehrlich's *The Population Bomb* had argued that the greatest danger to the biosphere was human overpopulation and called for "zero population growth." Ehrlich suggested that if voluntary methods of population control failed, compulsion would be necessary. (What kind of compulsion did he have in mind? Bookchin wondered. Forced birth control? Deliberate starvation?) Ehrlich was wrong, Bookchin said: the ecological crisis resulted not from sheer population numbers but from social arrangements—specifically, capitalism and hierarchy. The United States, for example, with just over 7 percent of the world's population, "wastefully devours more than fifty percent of the world's resources." But a society organized along eco-anarchist lines would not be rapacious toward the environment.[80]

People aren't the problem, he argued—they are the solution. Any form of social organization may be consciously altered. People are capable of recognizing that their current level of consumption is unsustainable, and they are capable of reversing it; humanity as a whole is imbued with the power to innovate social solutions. Bookchin didn't even care for astronaut Schweickart's as-seen-from-space photos of Earth—they didn't show the planet's inhabitants. So he would advocate not environmentalism but ecology; and his version of ecology would not include sociobiology or Malthusianism but rather would be a humanistic *social ecology*.

In 1972 the Club of Rome published *The Limits to Growth*, whose conclusions, like Ehrlich's, were dire: "If the present growth trends in world population, industrialization, pollution, food producing, and resource depletion continue unchanged, the limits to growth on this planet will be reached at some time within the next one hundred years."[81] A frightening specter of scarce resources and a war of all against all seemed, to its authors, in the offing. But Bookchin thought *Limits to Growth* was alarmist. Resources weren't running out—we were just overusing them, squandering them. If we were to choose to eliminate waste, we could make better use of resources. And if we were to choose to create an eco-decentralist, postcapitalist society, and redistribute the abundance created by technology, we could avert competition over scarce resources by providing for everyone. Raising alarms about scarcity did nothing but serve the interests of capital.[82]

Then in October 1973 the oil crisis erupted, seeming to verify the thesis of *Limits to Growth*: the country seemed already to be running out of at least one vital natural resource. Temperatures in public buildings were reduced to 68 degrees. At gas stations, drivers sat in long lines for hours, only to see, when they reached the pump, the price per gallon spike before their eyes.

But again Bookchin denied the general assumption that the United States was running out of oil—that impression was "simply false." Oil had become scarce for political reasons—OPEC had embargoed the supply to punish the United States for aiding Israel in the Yom Kippur War. But now the oil companies were concocting an artificial scarcity, to try to persuade everyone that oil was finite, so they could gin up their profits and gain permission to build a pipeline in Alaska and drill offshore.[83] They were enacting the capitalist imperative for competition over scarce resources.

Government and industry, Bookchin argued, were also invoking scarcity in this artificial "limits to growth" crisis to neutralize the ecology issue's revolutionary potential. In the wake of the so-called oil crisis, they were asking people to tighten their belts and consume less energy. And much to Bookchin's chagrin, environmentalists were singing the same refrain, rather than debunking the false claims about scarcity, rather than calling out the oil companies for their manipulations, and rather than denouncing large corporations for expending huge amounts of energy to produce useless goods and to power the megalopolis. To Bookchin, it sounded all too familiar—a few years earlier acolytes of Paul Ehrlich had been telling Americans they were having "too many babies."[84] In both cases, it was ordinary people—"we"—who were being blamed for ecological problems, first for having children, and now for consuming energy.

But who was this "we"? When it came to harming the planet, Bookchin argued, working people, people of color, and women were simply not in the same league as corporate executives. A "poor, exploited worker who must drive to a noisy assembly line" must not be judged by the same standard as "his fat-cat employer who is chauffeured to his comfortable office suite." Mainstream environmentalists lectured people to abandon their air conditioners and cars and televisions to conserve power, but for working people, "these pitiful amenities are all they've got after arduous toil." People of color, workers, and poor people had little reason to like or heed environmentalists' class-biased message and every good reason to "denounce the ecology movement as a cabal of privileged, elitist whites."[85]

Don't let the ecology movement be reduced to the lapdog of the corporate elite, Bookchin implored, to be stroked or kicked at their whim. The movement should be challenging the profit-oriented economy that is based on production for the sake of production, whose law of life is to grow or die, that turns people into buyers and sellers—and that views nature as a collection of resources to be devoured. A focus on individual lifestyle choices, or on the number of children people have, will only distract us from the necessity of creating the long-term solution we desperately need: an eco-anarchist society.

NOTES

1. Bookchin, "Introduction" (1970), in *PSA*, 18n; Bookchin, letter to the editor, *WIN Magazine* 7, no. 14 (Sept. 15, 1971), 32; Bookchin, *The Limits of the City* (New York: Harper & Row, 1974), 127.

2. Bookchin, "Listen, Marxist!" (1969), in *PSA*, 174.

3. The anthologies included Arthur Lothstein, ed., *"All We Are Saying...": The Philosophy of the New Left* (New York: Capricorn Books, 1970); Theodor Roszak, ed., *Sources: An Anthology of Contemporary Materials Useful for Preserving Personal Sanity While Braving the*

Great Technological Wilderness (New York: Harper Colophon, 1972); Henry J. Silverman, ed., *American Radical Thought: The Libertarian Tradition* (Lexington, Mass.: D. C. Heath, 1970); C. George Benello and Dimitrios Roussopoulos, eds., *The Case for Participatory Democracy: Some Prospects for a Radical Society* (New York: Grossman, 1971); among others.

4. Richard Merrill, "A Prosaic Ode to Murray Bookchin," Jan. 15, 2009, courtesy Richard Merrill.

5. Bookchin, "Introduction" (1970), 30. He inserted some new paragraphs about hierarchy into the essay "Ecology and Revolutionary Thought," which had not been present in the original version; he had not been writing about hierarchy in 1964.

6. Arthur Lothstein to author, July 7, 2010. Alternate U was located at 69 West Fourteenth Street. At this time Murray also taught social studies in an experimental program at Staten Island Community College.

7. I'm inferring what he told his students based on his writings from around this time.

8. "Murray Bookchin: Ecologist and Environmental Activist," *Mother Earth News*, July–Aug. 1971, 8, online at http://bit.ly/1fN2gog.

9. Bruce Schulman, *The Seventies: The Great Shift in American Culture, Society, and Politics* (New York: Da Capo Press, 2002), 88; Harry C. Boyte, *The Backyard Revolution: Understanding the New Citizen Movement* (Philadelphia: Temple University Press, 1980), 138.

10. Bookchin, "The Future of the Anti-nuke Movement," *Comment*, no. 3 (Nov. 1979), 7. For Bookchin's influence on the appropriate technology movement, see Andrew Kirk, "'Machines of Loving Grace': Alternative Technology Environment and the Counterculture," in Peter Braunstein and Michael William Doyle, eds., *Imagine Nation: The American Counterculture of the 1960s and '70s* (New York: Routledge, 2002), 361; Andrew Kirk, "Appropriating Technology: The *Whole Earth Catalog* and Counterculture Environmental Politics," *Environmental History* 6, no. 3 (July 2001), 374–94; David Dickson, *The Politics of Alternative Technology* (New York: Universe Books, 1975); Franklin A. Long and Alexandra Oleson, eds., *Appropriate Technology and Social Values: A Critical Appraisal* (Cambridge, Mass.: Ballinger, 1980); and Witold Rybczynski, *Paper Heroes: A Review of Appropriate Technology* (New York: Anchor Press / Doubleday 1980).

11. On electric cars, see Bookchin, "Ecology and Revolutionary Thought" (1964), in *PSA*, 75; Bookchin, "Social Anarchism," audiotape, Great Atlantic Radio Conspiracy, ca. 1972. On fuel cells, see Bookchin, introduction to Hans Thirring, *Energy for Man* (New York: Harper & Row, 1976), xiii.

12. "Murray Bookchin: Ecologist and Environmental Activist," *Mother Earth News*, July–Aug. 1971, online at http://bit.ly/0EN4Vn. On *Mother Earth News*, see Jeffrey Jacob, *New Pioneers: The Back-to-the-Land Movement and the Search for a Sustainable Future* (University Park: Pennsylvania State University Press, 1999).

13. Merrill, "Prosaic Ode to Bookchin."

14. Bookchin, "A Discussion on 'Listen, Marxist'" (1970), in *PSA*, 242.

15. Bookchin, "Response to Andrew Light's 'Rereading Bookchin and Marcuse as Environmental Materialists,'" *Capitalism Nature Socialism* 4, no. 2 (June 1993), 101–20.

16. Tom Athanasiou, *Divided Planet: The Ecology of Rich and Poor* (Athens: University of Georgia Press, 1998), 244; Allan Bérubé, "Ecology Green Paper," *WIN Magazine* 6, no. 9 (May 15, 1970), 16ff; Ecology Action, "The Unanimous Declaration of Interdependence," in Roszak, *Sources*, 388ff.

17. Ecology Action East was "a radical libertarian ecology group inspired by the ecology action movement that began in California." It was founded "at a class on ecology taught by

Murray Bookchin at Alternate U in the fall of 1969." Robert Palmer (as Grey Fox), "Squatters Park," *Roots*, no. 2 (ca. 1971), 22, MBPTL and copy in author's collection.

18. Bookchin, "The Power to Destroy, the Power to Create," (1970; rev. 1979), in *TES*, 45, 46. This manifesto was originally published in *Roots*, no. 1 (1970).

19. Ecology Action East, "Buy Now . . . Die Later," in Bookchin, *Ecology and Revolutionary Thought* (New York: Times Change Press, 1970), 61.

20. Bookchin, "Introduction" (1970), 16; "Toward an Ecological Solution," *Ramparts* (May 1970), 10, 12; "Toward an Ecological Society" (1973), in *TES*, 71. The ENACT Teach-In on the Environment took place on Mar. 10–14, 1970. See "Students Call It Dirty Business," *Business Week*, Mar. 21, 1970, 29; "Big Names Due at U-M Conference, Environmental Pollution is Topic," *News-Palladium* (Benton Harbor, Mich.), Mar. 5, 1970. On March 12, Bookchin joined Gaylord Nelson, Ralph Nader, Barry Commoner, and Walter Reuther for a panel called "The Bridge between Ideals and Action"; on Mar. 13 Bookchin participated on a panel with Reuther and René Dubos called "Root Causes of the Environmental Crisis."

21. Bookchin, "Post-scarcity Anarchism" (1970), in *PSA*, 40; Bookchin, "Introduction" (1970), 22; "Toward an Ecological Society," 70; "Toward an Ecological Solution," 15.

22. Alan Glenn, "The Turbulent Origins of Ann Arbor's First Earth Day," *Ann Arbor Chronicle*, Apr. 22, 2009, online at http://bit.ly/13XQgeu.

23. Quoted in Gladwin Hill, "Activity Ranges from Oratory to Legislation," *New York Times*, Apr. 23, 1970, 1; Elizabeth Kolbert, "In the Air," *New Yorker*, Apr. 27, 2009, 17.

24. Joseph Lelyveld, "Mood Is Joyful as City Gives Its Support," *New York Times*, Apr. 23, 1970.

25. Ecology Action East, "The Funeral of Garbage," reprinted in Bookchin, *Ecology and Revolutionary Thought* (New York: Times Change Press, 1970). Photos of the Earth Day action appear on the cover of *Roots*, no. 1 (1970).

26. Robert Palmer (as Grey Fox), "Squatters Park," *Roots*, no. 2 (ca. 1971), 20–33.

27. Timothy Lloyd Thomas, *A City with a Difference: The Rise and Fall of the Montreal Citizen's Movement* (Montreal: Véhicule Press, 1977), 24.

28. Editors, "Montreal Citizens Movement," *Our Generation* 10, no. 3 (Fall 1974), 3ff.

29. The struggle to save Milton-Park would continue for a decade and a half, under Roussopoulos's and Kowaluk's leadership. In 1984 this six-block urban village was not only rescued but transformed into the largest housing cooperative in North America. See Claire Helman, *The Milton-Park Affair: Canada's Largest Citizen-Developer Confrontation* (Montreal: Véhicule Press, 1987); and Lucia Kowaluk and Carolle Piché-Burton, eds., *Communauté Milton-Parc: How We Did It and How It Works* (Montreal: Communauté Milton-Parc, 2012), www.miltonparc.org.

30. Dimitri Roussopoulos, interview by author, July 2009.

31. "Report to Our Readers," *Our Generation* 7, no. 1 (Jan.–Feb. 1970), 120; "Radical Implications of the Ecology Question," *Our Generation* 7, no. 1 (Jan.–Feb. 1970), 3–4.

32. Conversation reconstructed from my interview with Roussopoulos and from Murray's accounts of it to me.

33. Todd Gitlin, "To the Far Side of the Abyss," *Nation*, Mar. 6, 1972, 309.

34. Brian Tokar, interview by author, Aug. 7, 2009.

35. Bill Koehnlein, interview by author, ca. 2008.

36. Howard J. Ehrlich, "Notes from an Anarchist Sociologist" (May 1989), in Martin Oppenheimer, Martin J. Murray, and Rhonda F. Levine, eds., *Radical Sociologists and*

the Movement: Experiences, Lessons, and Legacies (Philadelphia: Temple University Press, 1991), 235.

37. Peter Marshall, *Demanding the Impossible: A History of Anarchism* (New York: HarperPerennial, 1993), 602.

38. Peter Prontzos, interview by author, Aug. 16, 2009.

39. For anarchist criticism of *PSA*, see "Questioning the Premises," *Match* [Tucson, Ariz.], Oct. 1971, recapped in Marcus Graham, "Fifth Estate, Anarchism, Technology and Bookchin," *Fifth Estate* 15, no. 6 (Nov. 19, 1981), 2, online at http://bit.ly/nyocvj; Calvin Normore, review of "Post-scarcity Anarchism," *Our Generation* 8, no. 3, pt. 2 (1972), 33–36; and Robert S. Calese, letter to the editor, *WIN Magazine* 7, no. 13 (Sept. 1, 1971), 33. For Bookchin's replies, see "Reply to Normore," *Our Generation* 8, no. 3, pt. 3 (1972), 28; and reply to Calese, *WIN Magazine* 7, no. 14 (Sept. 15, 1971), 32.

40. See, for example, Theodore Roszak, *The Making of a Counter Culture* (New York: Doubleday, 1969).

41. Calese, letter to *WIN*, 33.

42. A transcript of the conference proceedings was published as Charles A. Thrall and Jerold M. Starr, eds., *Technology, Power, and Social Change* (Lexington, MA: Lexington Books, 1972), 35–48. Bookchin's informal conversation with Mumford wasn't part of the official proceedings and is not recorded in that book; he told me about it. On Kropotkin, see Bookchin, "Concept of Ecotechnologies and Ecocommunities" (1976), in *TES*, 104; and "The Postwar Period," interview by Doug Morris, *Anarchism, Marxism, and the Future of the Left* (San Francisco: A.K. Press, 1999), 57–58. See also Bookchin, "Deep Ecology, Anarchosyndicalism, and the Future of Anarchist Thought," in *Deep Ecology and Anarchism: A Polemic* (London: Freedom Press, 1993), 55–56.

43. James Herod, "Some Thoughts on the *Telos* Conference" (Nov. 1971), online at http://bit.ly/oPdOI5. The conference began on November 21.

44. Ibid. These words are not Gorz's but a paraphrase.

45. Bookchin, *Limits of the City*, 127. Bookchin's address to the *Telos* conference, "On Spontaneity and Organization" (Nov. 1971), was published in *TES*; the quotes are from 262, 263.

46. David Van Deusen, "Green Mountain Communes: The Making of a People's Vermont," Catamount Tavern News Service, Northeast Kingdom, VT, Jan. 14, 2008, online at http://bit.ly/n4rbbm; Robert Houriet, *Getting Back Together* (New York: Putnam, 1971); Robert Houriet, "Earthworks—the Franklin Commune," online at www.herbalenergetics.com/id12/html; and Roz Payne, interview by author, May 2008. Payne had been a member of the Third World Newsreel radical media collective and the Red Clover commune in Putney, part of Free Vermont.

47. Bookchin, "Socialism in One City? The Bernie Sanders Paradox: When Socialism Grows Old," *Socialist Review*, no. 90 (Nov.–Dec. 1986).

48. Bookchin moved to Vermont incrementally over the course of the 1970s. Holding on to the New York apartment turned out to be useful—when his peripatetic lecturing life brought him to New York, he could stay there. Sometime in 1972, while he was staying temporarily in the apartment, it was subjected to yet another break-in, one in which he feared for his life. That November he returned to Vermont but still kept the apartment, storing his possessions there, and in 1973 he sublet it to a student. He did not finally give up the apartment until 1977. *United States v. W. Mark Felt and Edward S. Miller*, US District Court, Washington, DC, Sept. 1980, transcript, MBPTL and author's collection.

49. The apartment was at 161 Main Street.

50. Michel Saint-Germain, "La liberté de chacun: Un café avec Murray Bookchin," *Mainmise* [Montreal], no. 76 (1978), 30–31.

51. Bookchin, "Introduction" (1979), in *TES*, 29.

52. Perhaps his old editor Angus Cameron helped Murray get the grant—he was on the advisory board of the Rabinowitz Foundation.

53. Dorothy D. Lee, *Freedom and Culture* (Englewood Cliffs, NJ: Prentice-Hall, 1959), 8, 40.

54. On the "irreducible minimum," see Paul Radin, *The World of Primitive Man* (n.p.: Henry Schuman, 1953), 106ff. Murray would invoke Lee and Radin on these points frequently in the coming decades.

55. Lee, *Freedom and Culture*, 21. See, e.g., Bookchin, "Toward an Ecological Society," 62–63; *The Ecology of Freedom: The Emergence and Dissolution of Hierarchy* (Palo Alto, CA: Cheshire Books, 1982), chap. 2.

56. Bookchin, *Ecology of Freedom*, chap. 3.

57. Ibid., 163.

58. Max Horkheimer and Theodor Adorno, *Dialectic of Enlightenment*, trans. John Cumming (New York: Herder & Herder, 1972).

59. Bookchin, "On Neo-Marxism, Bureaucracy, and the Body Politic" (1978), in *TES*, 234.

60. Bookchin, "Deep Ecology, Anarcho-syndicalism" (1993), in *Deep Ecology and Anarchism*, 53.

61. One early formulation is in "Toward an Ecological Society," in *TES*, 66–67.

62. Bookchin, "Introduction" (1979), 26; *Ecology of Freedom*, 124n, 304.

63. Bookchin, *Ecology of Freedom*, 177, 319, 322.

64. Ibid., 340.

65. Karl-Ludwig Schibel to author, Jan. 2010.

66. Paul Hockenos, *Joschka Fischer and the Making of the Berlin Republic* (New York: Oxford University Press, 2008), 112.

67. Schibel to author.

68. Ibid.; Bernd Leineweber and Karl-Ludwig Schibel, *Die Revolution ist vorbei—Wir haben gesiegt. Die Community-Bewegung. Zur Organisationsfrage der Neuen Linken in den USA und der BRD* (Berlin: Merve Verlag, 1975).

69. Daniel Blanchard's *Halte sur la rive orientale du lac Champlain, Vermont* (Paris: Julliard, 1990) is a novel based on his experiences in Burlington in 1970–71. A character who resembles Murray says, "Le capitalisme rend fou. Il est fou. Il me veut, tout comme l'esclavagiste; me traiter comme une marchandise, c'est son idéal, sa fin dernière" (184).

70. Bookchin, "Reflections on Spanish Anarchism," *Our Generation* 10, no. 1 (Spring 1974), 34.

71. "Robert J. Palmer from New York was so enthusiastic about the Amsterdam [Kabouter] movement that in the summer of 1971 he attempted to build a [Kabouter] movement in his own city ... during a full moon night in Central Park. For this he worked together with Murray Bookchin, editor of the ecological-anarchist paper *Anarchos* and author (under the pseudonym Lewis Herber) of *Stikkende steden* (*Crisis in Our Cities*), a book that was also referred to by Roel van Duijn in *De boodschap van een wijze Kabouter*." Coen Tasman, *Louter Kabouter: Kroniek van een beweging* (Amsterdam: Babylon/De Geus, 1996); I'm grateful to Rafa Grinfeld for this source and the translation. See Grey Fox, "The Dutch Kabouters," interview with Roel van Duijn, *Roots*, no. 2 (ca. 1971), MBPTL and author's collection.

72. Bookchin, interview by Jeff Riggenbach, *Reason*, Oct. 1979, 34–38.

73. Tom Wells, *The War Within: America's Battle over Vietnam* (Berkeley: University of California Press, 1994), 546–47.

74. Bookchin (as *Anarchos* Group), "Spring Offensives and Summer Vacations," *Anarchos*, no. 4 (Jan. 1972).

75. Judith Malina, "Anarchists and the Pro-hierarchical Left," insert to *Anarchos*, no. 4 (June 1972).

76. Sam Dolgoff, "The New Anarchism," *WIN Magazine*, Mar. 1, 1973; Dimitri Roussopoulos, "Debates: A Community Control Strategy," *Our Generation* 9, no. 1 (Jan. 1973).

77. Bookchin, *Limits of the City*, 129–34, recommending ideas from Berkeley Tribe, "Blueprint for a Communal Environment," in Roszak, *Sources*, 392–41; Dimitri Roussopoulos, review of *The Limits of the City*, in *Our Generation* 10, no. 3 (Fall 1974), 50–57.

78. Bookchin, "Toward an Ecological Solution."

79. Bookchin, "Toward an Ecological Society," 58–59, 77.

80. Bookchin, "Toward an Ecological Solution," 10; "The Power to Destroy, the Power to Create" (1970), in *TES*, 37.

81. Donella Meadows et al., *The Limits to Growth: A Report for the Club of Rome* (New York: Potomac Associates, 1972), 22.

82. Bookchin, "The 'Energy Crisis'—Myth and Reality, an Interview with Murray Bookchin," *WIN Magazine*, Dec. 20, 1973; "The Great Energy Crisis Hoax," *Crawdaddy*, Apr. 1974, 54–56; and "Economics as a Form of Social Control," *Liberation*, Jan. 1975, 6–8. Murray responded to Donella Meadows at a 1972 conference: see "Planned Parent Unit to Focus on 'Vermont 2000,'" *Bennington Banner*, Oct. 21, 1974, 3.

83. Bookchin, "'Energy Crisis'—Myth and Reality," 4.

84. Bookchin, "Great Energy Crisis Hoax," 55.

85. Bookchin, "There are disquieting signs . . ." letter to the editor, *University Review*, Mar. 1974.

8

Social Ecologist

THE OIL CRISIS had one positive effect—it raised popular awareness of renewable energy. Suddenly solar panels and wind turbines were hot news, and do-it-yourself manuals flew off bookstore shelves. In 1973 E. F. Schumacher published *Small Is Beautiful*, which argued that giant cities have undergone "pathological growth" and have led to unprecedented "anonymity, isolation, social atomization." What was needed was an "appropriate technology" that was small in scale and "compatible with the laws of ecology." Such technology, he said, would be "conducive to decentralization."[1]

If it sounds like Bookchin's ideas in *Our Synthetic Environment*, that's because they were, and Schumacher gave him credit. "I agree with Mr. Bookchin's assertion," he wrote, "that 'reconciliation of man with the natural world is no longer merely desirable, it has become a necessity.'"[2] Not that he shared Bookchin's radical social agenda—Schumacher wanted "a middle way" between socialism and capitalism, something he vaguely called "Buddhist economics." *Small Is Beautiful* became an international bestseller, quoted by everyone from Prince Charles to California governor Jerry Brown. The world was ready to hear these ideas; a year after Schumacher's book appeared, Harper published a new edition of *Our Synthetic Environment*, which got far more attention now than it had the first time around.

Bookchin's influence was felt elsewhere in the culture as well. In 1974 the science fiction author Ursula K. Le Guin published *The Dispossessed*, a utopian novel about anarchism and revolution, capitalism, and collectivism. "Bookchin's pre-1980 writings," Le Guin told me, "were one of the major sources for the thinking that went into my anarchist utopia *The Dispossessed*."[3] In 1975 Ernst Callenbach used Bookchin's word *ecotopia* as the title for another popular novel.[4]

Even Herbert Marcuse finally acknowledged the importance of the ecology issue, in his 1972 book, *Counterrevolution and Revolt*. "The discovery of the liberating forces of nature," Marcuse wrote, seeming to echo Bookchin, "and their vital role in the construction of a free society becomes a new force in social change." Moreover, social and ecological change should go hand in hand, he said, since "the liberation of man to his own human faculties is linked to the liberation of nature." Bookchin's efforts to interest Marcuse in ecology had finally borne fruit.[5]

In Burlington, some of the members of the Fresh Ground collective wanted to turn the coffeehouse into a conventional business. Others objected, but the first group prevailed, and the collective structure was discarded. The normalized coffeehouse would serve coffee, soup, and bread throughout the 1970s.[6]

Around this time, in the autumn of 1972, a graduate student from Goddard College approached Bookchin with a proposal to become a regular lecturer there. Dan Chodorkoff explained that a philosophy teacher had initiated a lecture series called "Technology and Society," then abruptly departed the college in mid-semester. A replacement was needed to coordinate the series.

Goddard College is located about fifty miles from Burlington, in the mountains of central Vermont, on a former dairy farm.[7] Chodorkoff, a onetime New Leftist, had attended as an undergraduate and read Bookchin's work; he planned to write his master's thesis about anarchism. When he learned that Bookchin was living in nearby Burlington, he wanted him for his adviser. He had lobbied Goddard to let him offer Bookchin the teaching post.

With his low-key but determined manner, Chodorkoff set out to persuade Bookchin to accept. The college, he surely explained, had an international reputation as a progressive school, innovating in independent study and off-campus learning. It used evaluations instead of grades to assess students' work. Its approach to education was based on principles of John Dewey: students decided what they wanted to study, devised their own curriculum, then drew conclusions from their experience. In early 1970s the school, with its alternative ambience, was attracting hundreds of students from the counterculture—open-minded, eager to initiate and manage their own studies.[8]

Bookchin hesitated as he listened to Chodorkoff's offer, since he had a heavy lecture schedule and two books to write.[9] But Goddard was appealing—it reminded him of the nonhierarchical, experimental Modern Schools, founded by a Catalonian anarchist in the 1910s.[10] So he accepted.

Arriving at Goddard in February 1973, he glimpsed for the first time the brown-shingled farm buildings nestled among landscaped gardens, against the stunning backdrop of the Green Mountains. As Chodorkoff told me, "Murray didn't want to coordinate the lecture series. He wanted to *be* the lecture series." Those first lectures were "a revelation," he recalled, "some of the best I ever heard him give." The students gave Murray excellent evaluations, and Goddard invited him to teach a course the next semester.[11]

When Bookchin lectured at colleges and universities, students would come up to him afterward and ask where they could study the kinds of "creative, visionary and imaginative approaches" to ecology that he had talked about. But he was perpetually stumped—he knew of no program to recommend. At the same time, as he traveled the lecture circuit, he had met quite a few cutting-edge thinkers in the eco-technics, organic farming, and neighborhood movements. In November 1973 he and Chodorkoff decided to bring these innovative thinkers together with idea-hungry, ecology-minded students for a three-day conference at Goddard.[12] Among them were Wilson Clark, a designer of solar and wind power systems; Day Chahroudi, a pioneer in solar architecture from MIT; Eugene Eccli, an alternative energy expert; and Sam Love, editor of *Environmental Action* magazine.

Also present at that first conference was John Todd, an oceanographer and aquatic biologist. In 1972, partly inspired by a close reading of Bookchin's "Towards a Liberatory Technology," Todd had begun applying his scientific training to the task of averting ecological disaster. He and several others had founded New Alchemy Institute on a farm in Falmouth, Massachusetts.[13] There they pioneered growing fruit and vegetables in greenhouses heated by the sun—a 1,700-gallon water-filled cylinder absorbed and stored solar heat throughout the day, then radiated it back to the building during the night. Todd and his associates stocked such cylinders with tilapia, a herbivorous, freshwater fish that thrives on algae and insect larvae; they had come across it in East Africa. Experimentation showed that they could grow edible tilapia in ten weeks. The cylinders' water, complete with fish wastes, was then used to irrigate the fruit and vegetable gardens. This closed-loop system, which they called a bioshelter, could produce fish, fruit, and vegetables year round, regardless of season, through even the bitterest New England winters, "with occasional backup from a wood-burning stove."[14] And the exchange of nutrients between plants and animals meant that little, if any, petroleum-based fertilizer was needed. Todd doubtless told the Goddard conference about this integration of solar energy, aquaculture, and organic gardening.[15]

Also present was a colorful, outspoken neighborhood organizer from Washington, DC. Bookchin had met Karl Hess late in 1968, at a supper hosted by the anarcho-capitalist Murray Rothbard. Hess had been a speechwriter for Barry Goldwater and in 1964 had penned his speech with the notorious line, "Extremism in defense of liberty is no vice." Subsequently Hess had become a New Leftist, associated with SDS and the Black Panthers, then identified himself as a left-libertarian—an anarchist. He refused to pay federal income tax, as an act of political resistance, whereupon the IRS put a 100 percent lien on his future earnings. So Hess learned to weld and eked out a livelihood by bartering welding work for food and other goods, outside the money economy. At Rothbard's dinner, Hess and Bookchin had recognized each other as kindred libertarian spirits: talented autodidacts, they both opposed big business as well as big government.[16]

In 1970 Hess was living in Adams-Morgan, an ethnically and economically mixed neighborhood in his native Washington, DC. Young counterculturists were moving in, opening food coops and communes of all sorts. Hess set up a shop in an old warehouse and taught neighborhood residents how to use saws and drills. As they achieved mastery, he found, they gained confidence—they "began talking about what 'we did' and what 'we can do.'" Experimentation became a community pastime in Adams-Morgan. A science teacher built a solar collector out of cat food cans, capable of heating household air to 120°F. A marine engineer set up an outside mirror to collect energy from the sun, then devised an indoor cooker to put that energy to use.[17]

Hess and his friends Gil Friend and David Morris experimented with ways to grow food in cities, in "self-sustaining and permanent loops . . . which we can directly control on a neighborhood level." They transformed vacant lots into community gardens. They experimented with French intensive gardening, a method particularly fitting for urban farming because it brings "two to eight times the yield of traditional gardens with half the water consumption." They collected vegetable wastes from local food coops, composted it in pits dug in abandoned driveways, and they used horse manure from a park police stable as fertilizer. They constructed hydroponic gardens. On the flat roofs of the neighborhood's three-story row houses, they set up boxes with organic soil, where they grew vegetables. Basements turned out to be "the ideal place to grow sprouts."[18]

Perhaps it was at the 1973 Goddard conference that Hess found out about New Alchemy's closed-loop "bioshelters." Whenever it was, he went to work building large fish tanks for Adams-Morgan basements. He would salvage discarded washing machines, remove the motors, and connect them with power pumps to create a current against which the fish could swim. They flourished—an urban basement, he found, could produce fish "at a cost of less than a dollar per pound."[19]

These projects proved, said Hess, that an urban neighborhood "could be self-sufficient in the production of its food and wealth." Their work showed Bookchin that eco-technics could contribute not only the technological infrastructure but the economic base for urban decentralism.[20]

At the 1973 Goddard conference, about fifty people exchanged ideas about fish farming and solar power and organic agriculture, about eco-decentralism and utopia. The give and take was lively, and the event was such a smashing success that afterward Murray and Dan decided to launch a regular summer program, taught by Goddard faculty, supplemented by these same guest lecturers.

They called it the Social Ecology Studies Program, using the name coined two decades earlier by E. A. Gutkind, to reflect the indissoluble connection between social problems and ecological problems.[21] The program, Bookchin wrote in the proposal, would integrate the practical with the intellectual. The hands-on curriculum would instruct students in organic gardening, solar panel and windmill construction, aquaculture, and community organizing. The intellectual curriculum would cover the politics of ecology, decentralism, and community; urban history; radical history; social theory; psychology; and much more.[22] Nothing like it existed in the United

States. Chodorkoff, savvy about bureaucracies, shepherded the proposal through the Goddard administration, which accepted it and hired him to coordinate it.

The first session was scheduled to run for three months in the summer of 1974, long enough that students could immerse themselves. The program had no advertising budget, but John Shuttleworth of *Mother Earth News* came to the rescue, donating a full-page advertisement in his magazine. Thanks to that ad, the session attracted a hundred students.[23]

Bookchin inspired them by talking about historical movements for freedom. Chodorkoff taught them about utopias. "Aquaculture," Bill McLarney told his class by that name, "can help us achieve stable, integrated, efficient food production systems and even regional or local economic self-sufficiency." Eugene Eccli told his solar class, "Solar energy is a capturing mechanism to spread out over a time period the availability of the sun."[24] The program was underway.

In 1970 the state of New Jersey had created Ramapo College, a four-year liberal arts school, on a former cattle farm in Bergen County. Too many of New Jersey's young people, in those days, were leaving the state for college elsewhere—to appeal to them, the school would be open to experimentation, its structure interdisciplinary, its faculty unconventional.[25]

Among the first faculty members was Wayne Hayes, a onetime community organizer who had helped defeat the notorious Lower Manhattan Expressway, which had been proposed to run through Greenwich Village.[26] Tall and feisty and as blunt-spoken as Bookchin, Hayes held degrees in economics and city planning. Another faculty member was Trent Schroyer, who taught Critical Theory and was writing a book on the Frankfurt School.[27] In 1974 Hayes and Schroyer created Ramapo's School of Metropolitan Studies as "a free academic space."

Hearing that Murray Bookchin might be available, they both wanted to hire him, but they had no tenure-track position to offer him. Hayes pieced together some courses and offered him a job anyway. Murray was interested. Hayes told him he needed to provide his résumé. Murray sat down at his typewriter, inserted a brown paper bag, and typed out his life experience. When he finished, he handed it to Hayes—who noticed a glaring absence. Where was the B.A.? Where, for that matter, was the high school graduation? Bookchin had only a high school equivalency degree from the RCA Institute.

Even if Bookchin's academic credentials had been in order, persuading Ramapo to hire him was going to be a problem, Hayes realized. "The founding administrators were open to things contemporary," he told me, "but not necessarily radical." Furthermore, when the college's vice president, Robert Cassidy, found out that Hayes had offered Bookchin a job, he was furious. "How could you represent that we had a job when we didn't?" Cassidy shouted at him. "It was basic fraud," Hayes told me. "I was in big trouble." What to do?

Hayes invited Bookchin, basically, to audition. For the mise-en-scène, he chose a large wood-paneled classroom; he brought in several classes' worth of students and made sure Cassidy was present as well. Hayes introduced Bookchin, who proceeded to discourse for several hours, with no index cards or notes, about "the medieval city, and *Stadtluft macht frei*, and the origins of capitalism." It was "a spellbinder," a fantastic talk, Hayes recalled. Afterward "everyone said wow, and Cassidy was blown away."

But Bookchin's lack of academic credentials still bothered Cassidy. So Hayes asked Murray to bring all his published writings out to the college. Bookchin showed up with a box of the just-published *Limits of the City*. Hayes brought it into Cassidy's office, turned it over, and emptied the contents onto the desk. "Take a look at this stuff and tell me we don't want this guy," he told Cassidy, "with or without the degree." Later that day Cassidy called Hayes: "Let's hire him." He offered him an assistant professorship at a good salary.[28]

Bookchin started teaching at Ramapo in the fall of 1974. For the rest of the 1970s, he would teach urban studies and environmental studies at Ramapo during the regular school year, then decamp to Vermont for the summer's social ecology program. Ramapo's environmental studies major soon attracted hundreds of students.[29] And all the while, Bookchin continued to barnstorm the alternative scene, lecturing indefatigably on hierarchy and domination, on decentralization and technology, and on urban gardening with solar collectors and aquaculture.[30] He was riding the historical moment.

In 1962 Bookchin had been certain that American cities had reached their limits: "Megalopolitan life is breaking down, psychologically, economically, and biologically," he had written in *Our Synthetic Environment*.[31] A decade later the urban crisis had only intensified, due to ever greater pollution, racial strife, unresponsive city halls, police brutality, dismal housing, crumbling schools, crime, and poor social services. White middle-class families, and the business enterprises for which they worked, fled to the suburbs or to the Sunbelt.

Those who remained—the less affluent, often people of color—were left to fend for themselves. Local banks would not give them loans. City governments reduced or cut off essential services to minority-dominated neighborhoods, like police patrols, garbage removal, and firefighting. Urban elites (following a policy called planned shrinkage) were actively trying to make poor neighborhoods unlivable, to drive the impoverished residents away, so that developers could come in and the middle class would return.

On the Lower East Side, some of those urban residents whom society had abandoned took matters into their own hands. A rough Puerto Rican street gang decided to go straight. Somehow the engineer-futurist Buckminster Fuller taught them how to build geodesic domes. The domes were sturdy yet cost little to build and were easy to transport—and they had many possible applications in poor communities: as shelter; as greenhouses for urban gardens; as meeting places; and as playground structures. The gang, led by Chino Garcia, named itself CHARAS (after the six members' initials) and began building geodesic domes, east of Tompkins Square Park, in the area they took to calling Loisaida. They recruited street kids to help and within a few years had two dozen volunteers. Dan Chodorkoff, who was studying CHARAS for his Ph.D. in anthropology, said that creating the domes gave the members a "means of empowerment."[32]

Meanwhile, in other cities, young radicals were moving into low-income neighborhoods, where, as in Adams-Morgan, they organized communes and cooperatives. When banks wouldn't lend them money for their food coops or community health clinics, they formed community-based credit unions. And implementing the new techniques of community organizing developed by Saul Alinsky, they mobilized residents to resist police brutality and to challenge oppressive zoning policies and urban renewal projects.

Community activists began to think further, too, in terms of empowering neighborhoods—devolving political power from unresponsive city governments to popular neighborhood councils. As they escalated their pressure, city governments agreed to establish community boards, mainly to handle citizens' complaints but sometimes to allow them to participate in planning and zoning. Between 1970 and 1974, New York, Boston, Seattle, Atlanta, St. Paul, Pittsburgh, and other cities created these "little city halls" or changed their charters to create neighborhood councils. Sociologists called it the "community revolution."[33] Observed the democratic theorist Benjamin Barber, "The experiment evolved into substantial decentralization."[34]

Decentralization—once again a proposal of Bookchin's seemed prescient. But he and other proponents of neighborhood radicalism whom he befriended around this time thought "little city halls" didn't go far enough—they were just vehicles for municipal governments to pretend to seek citizens' views while continuing to accommodate moneyed interests. Milton Kotler, for one, called for effecting a wholesale "political transfer from existing units of government" to neighborhood assemblies. And David Morris and Karl Hess wanted neighborhoods to secede from cities altogether and devolve power to "an assembly of citizens within the community who directly participate in decisions." In their 1975 handbook, *Neighborhood Power*, Morris and Hess showed step by step how it could be done. A crucial concept was that neighborhood activists would have to "gain power in order to decentralize power"—that is, they would have to participate in municipal elections, gain elective local office, and then use their power to devolve municipal power to the neighborhoods—as Bookchin had been advocating.[35]

Hess and Morris and their friends in Washington showed the way by creating a practical example. When the city stopped collecting trash in Adams-Morgan and garbage piled up in the streets, a neighborhood committee organized a volunteer cleanup. Other committees sprang into being to address problems of affordable housing, pest control, and recreation. To tie themselves to the community, these committees created a community government, "a neighborhood assembly, based on the town meeting model," as Hess wrote. Here "people in the neighborhood could get together, discuss their problems, discuss solutions, and then actually decide what they themselves could do." At its peak, participation in the assembly reached three thousand.[36]

In addition to constructing political self-government, Adams-Morgan residents built economic self-sufficiency as well. They farmed trout in their basements and reduced their energy bills by using solar and wind power. The stereotype, as Chodorkoff observed, was that poor people were too concerned with survival to be interested in the environment. But concern with survival had the opposite effect—it was what "led many of the groups to begin working on alternative technology."[37] Renewable energy, urban gardening, and aquaculture could potentially provide the basis for a self-reliant, cooperatively owned and managed neighborhood economy.

In Anacostia, another Washington, DC, neighborhood, residents also "instituted a town meeting of the entire neighborhood," and when it got too big, covering too large an area, "they sensibly divided into six parts and instituted six separate town meetings that met monthly and one federated meeting that met twice a year to discuss problems relative to all the neighborhoods." Anacostia residents built a bicycle factory (in an abandoned garage) and a cooperative furniture factory. Solar collectors supplied enough energy to meet half the neighborhood's heating requirements during winters as well as for cooling during summers. Enough trout and tilapia were cooperatively farmed in geodesic domes and basements to supply the neighborhood's needs for protein. Residents grew tomatoes hydroponically on rooftops, harvesting several tons from the roof of the Roosevelt Hotel alone. These Washington neighborhoods became relatively self-reliant with vibrant local economies, while their town meetings, observed Hess, were "forms of direct participation unequaled in any city."[38] Bookchin praised the Washington efforts for achieving "urban community based on popular control of urban resources and institutions" through the use of these "ecotechnologies."[39]

In the spring of 1974, citizens of Montreal, still chafing under the oligarchical rule of Mayor Jean Drapeau, yearned to participate in city government, to bring reality to the slogan "power to the people." Community and labor organizers, peace activists and feminists, independent radicals and socialists and environmentalists, immigrants and homeless advocates came together to mobilize the city to finally unseat the mayor in the coming November election. In May they elected delegates to an assembly that founded the Montreal Citizens' Movement, or MCM. Its program, *Une ville pour nous*, proposed that every two years, in each neighborhood of Montreal, a citizen assembly should elect delegates to a neighborhood council. The delegates would be mandated and recallable, to ensure their accountability to the assemblies.[40]

Bookchin's writings, published in *Our Generation*, influenced this proposal: as one historian observed, "much of the political current that fuelled [the MCM] was partially rooted in anarchism, in social ecology and in forms of libertarianism." Black Rose Books publisher Dimitri Roussopoulos, working with Hess, Morris, and Kotler, as well as Bookchin, formulated a call for neighborhood power that was widely read.[41]

Six months after the MCM's founding, on November 10, Montrealers handed the MCM over 40 percent of the vote and an astonishing eighteen out of fifty-five city council seats. The citizen groups were jubilant. Drapeau's stranglehold on city hall was not entirely broken, but it was significantly loosened. Thereafter, said Roussopoulos, "the movement of urban insurgency was like a dammed-up force and could not be held back."[42]

The MCM, both Bookchin and Roussopoulos acknowledged, was not the perfect embodiment of assembly democracy. Its general assemblies met only once a year, which was "not frequent enough," Bookchin thought, and the two-year span between municipal elections was too long. Its program, although radical, was "far from revolutionary." Roussopoulos, for his part, worried about the fact that among the MCM's founders were some pragmatically oriented politicians who were less interested in devolving municipal power or fighting capitalism than in creating a conventional, albeit municipally based, political party. "Sooner or later more opportunists and careerists" might descend on the MCM.[43]

Nonetheless, Bookchin found the MCM overall to be an "impressive achievement." It had a robust internal democracy and "a libertarian ambience." It defined itself programmatically as "consciously committed to a decentralized political structure for the city."[44] And it had good prospects—as good as one could hope for in 1974—for replacing the municipal government with a network of popular assemblies.

Bookchin made frequent trips to Montreal, to inspire the citizenry with "ideas of where things can go from there," as Roussopoulos recalled. In 1975 a half-dozen anarchist affinity groups joined the MCM, which could only strengthen its libertarian ambience. "Social ecologists and anarchists . . . were far more numerous and influential" in Montreal than in any other Canadian city, observes historian Timothy Lloyd Thomas. They even "toyed with the idea of freeing Montreal from the constraining jurisdictional control of both the federal and provincial governments."[45]

In November 1975 an MCM party congress elected a new executive committee. Six of its eight members were libertarian socialists, committed to establishing effective neighborhood councils. One member, Stephen Schecter, was an associate of Roussopoulos and advocate of "a revolutionary strategy." His election, said Roussopoulos, "was a symbolic victory of the MCM's radical libertarian element."[46] With its decentralist program, its bottom-up structure, and its libertarian executive, the MCM was turning Montreal into a laboratory for radical municipal politics.

In Vermont, the Social Ecology Studies Program needed land in order to experiment with alternative technology and organic farming. Fortuitously, in 1975 a forty-acre parcel owned by Goddard became available. Cate Farm, located in a bend of the Winooski River, adjacent to the main campus, had a nine-room brick farmhouse, a big uninsulated barn, several outbuildings, and glorious open fields. Chodorkoff's negotiating skills did the trick: Goddard agreed to let Social Ecology use it.

The Cate Farm grounds needed a lot of work, but the hundred-plus students who arrived for the 1975 summer session were more than eager to pitch in. They'd listen to Bookchin discourse on hierarchy and radical social theory, then go outside and do construction work.[47]

Their first task was to retrofit the farmhouse and heat its interior air, to create a year-round office for the program.[48] When innovators in the alternative energy field arrived to lecture, they shared their expertise on how to combine solar arrays with traditional building materials.[49] Author Wilson Clark contributed his encyclopedic knowledge of solar technology. New Mexico inventor Steve Baer explained his solar-heated drumwall houses. Day Chahroudi detailed how a greenhouse could harness solar radiation and store it as heat in the dirt below. For the Cate Farmhouse, the students and faculty finally decided to install a four-foot solar panel in its south-facing wall. But would the interior air stay warm enough during the long, overcast New England winters? To find out, they would measure the indoor temperatures that winter.

Meanwhile, students and faculty built an earthen shelter into a hillside, to see if a house could be warmed solely by the sun and insulated naturally. It was the first solar building in Vermont. They went on to design and build a greenhouse, a variation on the New Alchemy bioshelter, to house year-round closed-loop fish farming. A flat-plate solar collector installed atop this so-called Sunhouse warmed the interior and the thousand-gallon fish tank inside, stocked with tilapia, catfish, clams, and more. Students fed the fish alfalfa from the garden, as well as larvae and worms, while the fish waste (rather than chemicals) was used to fertilize vegetables.[50]

Another faculty-student collaboration was a horizontal-axis Jacobs wind machine. Mounted atop a sixty-foot steel tower, it could capture wind blowing from any direction. The energy was used to pump the Sunhouse's tank water through a biological filtration unit, which aerated it and also filtered the fish wastes.[51] A brigade from CHARAS came up from Loisaida and constructed a geodesic dome, to function as another greenhouse, with an aquaculture tank inside (figure 8.1).

The students built an open-sided shed, for compost. There they layered hay, kitchen wastes, dirt, and animal manure in four-foot piles, which they let stand; when the compost was ready, they applied it to the soil in the open fields, to enrich it without the use of chemical fertilizers.[52] In the farmhouse basement, they constructed a composting toilet, to recycle human wastes.

These small-scale eco-technics projects, using minimal resources, were simple enough that students with little technical training could build and manage them. In an urban context, similar eco-technics could allow neighborhood residents, even as amateurs, to exercise control over the energy that powered their communities. Eco-technics at the human scale could thereby transform (in a riff on Marxian terminology) technology from "instruments of domination and social antagonism" into "instruments of liberation and social harmonization."[53]

While its sister projects, New Alchemy in Massachusetts and Farallones in California, emphasized research, the Social Ecology Studies Program offered three months of total immersion for beginners in these hands-on techniques. In addition, it gave them a grounding in social theory. That first session, Dan Chodorkoff taught a class on utopia, "those images of the good life which have shimmered on the horizon throughout the whole of history."[54] The feminist activist Ynestra King taught a class

FIGURE 8.1 Institute for Social Ecology at Cate Farm, ca. 1976. During the 1970s, the ISE educated students in solar and wind energy, organic farming, composting, aquaculture, and permaculture, as well as radical social theory and revolutionary history.
Goddard College Archives.

called Women and Ecology, which discussed the domination of women in relation to the domination of nature. King hoped to develop a feminist ecology movement that would solve the ecological crisis by generating "truly human relationships," as opposed to relations of domination. Guest speakers included the feminist anthropologist Rayna Rapp, the author and poet Grace Paley, the antinuclear activist Anna Gyorgy, and others. A women's coffeehouse was held on Sunday nights.[55]

Karl Hess, in his classes, described the Adams-Morgan project to the students. At the town meetings, he said, people's personalities changed: "People who had been shy spoke out. People who had seemed without hope sparked to new life." Longtime cynics and naysayers aired their gripes, but when they saw others take up their ideas and turn them into actions, their lives gained "new meaning, new excitement, and a new sense of dignified purpose." The Adams-Morgan assemblies "were the most exciting political experiences I have ever had," Hess said. "After tasting a participatory democracy, I would never want to trade it for a merely representative one."[56]

And Bookchin, of course, taught social ecology, which by now straddled history, philosophy, anthropology, and sociology, as well as social theory and politics. The students revered him for his moral imagination, his ebullience, and his generous open-heartedness. He was irrepressible as he lectured, pacing before them and gesticulating, insisting on the urgent need for a large-scale social and ecological transformation, reiterating that all the organic farming and eco-technics they were doing could not be separated from a liberated social context. Afterward, he shared his words and his time unstintingly, spending hours talking to students informally and frankly. Sharing ideas did not deplete him—on the contrary, it energized him. As a venue for lively political discussion in a natural setting, the Social Ecology Program recapitulated aspects of his childhood utopia, Crotona Park.

Visiting lecturers would arrive and stay for a week at a time, living among the students, advising them on their experiments, letting them pick their brains. Recalled Richard Merrill, "John Todd, and I, and Murray Bookchin would stay up for hours just talking." Bookchin "pushed all of us to develop a more radical critique of the culture," Sam Love told me. "How he could keep it up all those years amazed me."[57]

But no matter how much everyone respected him and were inspired by him, one thing remained baffling. Even as he extolled organic farming, he was unabashedly fond of junk food. And even as he championed solar and wind energy, he drove around the Goddard campus in his car, even for short distances.[58] "Everybody saw it, and everybody commented on it," former student Barry Costa-Pierce recalled. The students even formed a Walk the Talk Working Group, so they could "call each other's shots on personal behavior." Much the same thing had bothered Joyce Gardner back in 1964–65, and it bothered the Ramapo students. "The thing that got people was the Twinkies," Wayne Hayes told me. "He ate a lot of them."[59]

Bookchin sometimes got defensive about it—in 1978 he exclaimed to an interviewer, "Always the judgments 'What are you eating there?'" Such intolerance had broken up many a commune, he pointed out, pleading for forbearance.[60] In my view, his unusual food choices were a way of asserting his proletarian background—even amid organic-minded hippies, he remained culturally working class and proudly so. His personal choices in general were also a way of pointing out that the ecological crisis was not caused by individual lifestyles—rather, its causes were systemic, and so its solutions must be collective, not individual. In any case, said Costa-Pierce, "nobody ever held it against him, because he gave so much. It made whatever he needed to do as a person secondary." Despite his eccentricities, agreed Richard Merrill, "for those of us who saw into his heart, he was a revelation, a loyal friend, a mentor."[61]

In their downtime the social ecology students "did their fair share of partying" in the Green Mountains, swimming in the streams and pools and camping in the lush forests.[62] Conscious that the fourteen-month program was a rare experience, one to be savored, they bonded as a community. When it ended on August 22, the leave-takings were often traumatic.

So successful was the 1975 session that Murray and Dan proposed to turn the program into a year-round Institute for Social Ecology (ISE). It would offer an undergraduate program for Goddard students and a summer session for anyone interested.[63] Goddard approved the proposal, and Bookchin became the ISE's first director.

As for Ramapo College, it too turned out to be a good match for Bookchin—it was "an extremely left-wing place," where Marxists and anarchists squabbled, but at least "they took ideas seriously," former student Jim Morley recalled. "It was a very idealistic time."[64] Bookchin found the students in New Jersey especially congenial because of their working-class backgrounds—many of their parents worked at the huge Ford assembly plant in Mahwah.

Yet when it came to eco-technics, Ramapo students and faculty alike felt they were "blazing new trails away from an oil-based overconsuming." The college, said Mike Edelstein, "probably offered more innovative courses on renewable energy, social ecology, and the like than any other undergraduate institution of the area."[65]

Bookchin taught social ecology as well as the history of social thought. In classrooms whose windows looked out onto the Ramapo Mountains, he traced the ebb and flow of utopian ideas from Plato to Rousseau to Marx and the socialists, situating them all in the Western intellectual tradition. "What Murray could do with that!" marveled Wayne Hayes.

Self-conscious about his lack of formal education, he immersed himself in urban history, to understand the rise, decline, and limits of the city. Historically, he observed, the city was the place where people born in a village society could migrate in order to escape the oppressions of the blood tie—of tribe and ethnicity. Village society tended to be organized according to clan, where one's status was a matter of birth. Some were privileged over others. But those who held low status in a clan-based village and chafed at it might leave those local inequities behind by departing for a city, where they could join other former provincials and gain a new status, equal with others. In small cities and urban neighborhoods—in the public arena—the stranger "becomes transformed into the citizen." In the cosmopolitan city, the blood tie—not only ethnicity but potentially also gender and sexual identity—can give way to a concept of universal humanity. For Bookchin, this stranger-welcoming-and-integrating, citizenship-bestowing city "represented the triumph of society over biology, of reason over impulse, humanity over folkdom."[66]

He left his students, many of whom were children of immigrants like himself, "hungering for more," Morley told me. In exploring history's grand themes, he worked in the tradition of nineteenth-century generalists, systematizers who straddled disciplines. A set of ideas, he believed, had to be coherent. He was "enormously erudite," recalled Hayes, "and deployed that knowledge brilliantly, and with heart."[67]

Once Ramapo realized that it had hired somebody special, said Hayes, "there was never any question that he'd get tenure." Robert Cassidy, the cautious vice president who'd approved Bookchin's hiring, actively invited him to apply to become an associate professor. "That's very rare," Hayes told me. "He was a superb teacher."[68] In 1976 Bookchin was promoted to associate professor in the School of Environmental Studies.

In Vermont, the winter of 1975–76 was severe, but the ISE staff diligently logged the temperatures inside the Cate Farmhouse. The solar installations, they found, did indeed heat the inside air sufficiently, and the well-insulated structure retained the heat even at night. If solar energy was practicable in sun-deprived central Vermont, then it could work pretty much anywhere.

The 1976 summer session attracted a whopping 180 students, eager to become ecological pioneers. While Bookchin lectured on the domination of nature and its relation to the domination of human by human, the "biological agriculture" class set up experimental garden plots to test how well compost functioned as fertilizer compared to chemicals. The soil in the composted plots, they found, had higher and steadier rates of nitrogen uptake, while the chemical plots were more vulnerable to cucumber beetle infestations. So they blended compost with soil and sand and used it to grow cucumbers, tomatoes, and bell peppers in the Sunhouse, where the aquaculture tanks, meanwhile, yielded approximately seventy pounds of fish twice a year.[69]

Other students constructed several dozen six- by twelve-foot French intensive beds in the open fields. They dug soil two feet deep and mixed humus into it, then added layers of compost and topsoil, to form a low mound. These raised beds, which yielded the most produce in the least area, turned out to be eminently suitable for urban gardening—say, on the Lower East Side or in Adams-Morgan.[70]

"It was about food for the poor and communities," aquaculturist Barry Costa-Pierce told me. "A lot of people call it *integrated* aquaculture—integrated with communities and with agriculture." Combining aquaculture and solar and wind power, it was now clear, could allow low-income urban neighborhoods to achieve self-reliance. Dan Chodorkoff envisioned the ISE as, in part, an alternative agricultural extension service, sharing practical knowledge with people in urban ghettos as well as rural poor and Native Americans. The ISE set up a training program for a union of small farms in Puerto Rico, and it developed a strong connection with the Akwesasne Mohawk nation, in upstate New York: John Mohawk and Ron LaFrance came to the ISE to teach, while ISE students traveled to Akwesasne to join political protests.[71]

But the poster child for achieving self-reliance through alternative technology was CHARAS, on the Lower East Side. In 1976, as a result of the city's deliberate neglect, trash was yet again building up in the neighborhood, especially in abandoned buildings and vacant lots. CHARAS's cadre of high school students were by now adept at spotting promising vacant spaces, and they found one on Ninth Street (between Avenues B and C). They cleared away the rubble and weeds and rats and established La Plaza Cultural there, a cultural center, "an oasis of color in an otherwise bleak cityscape." Poets, musicians, and dancers now had a place to perform, and visual artists a place to display their work.[72]

A few blocks away, an abandoned five-story tenement at 519 East Eleventh Street (between Avenues A and B) had been gutted by fire; the city acquired it, barely more than a shell, and boarded it up. In 1976 a group of young architects decided to buy and renovate it. CHARAS helped them obtain financing on the basis of their own labor—"sweat equity" loans, they were coming to be called—then helped them clear the building of garbage and rubble. On the roof, they installed a bank of flat-plate solar collectors—the first rooftop solar panels in Manhattan. They found an old farm turbine, and about forty people from CHARAS carried it up to the roof and set it up as a windmill—the first roof-mounted urban windmill in the United States.[73]

By 1977 the solar collectors and windmill were providing more than enough power to heat and illuminate 519 East Eleventh—the excess energy was going into the city's power grid. The residents demanded that Con Ed reimburse them, but it refused and took them to court. The 519 group won: for the first time, a small-scale power generator received credit from a power company.[74]

Inspired by 519's success, urban homesteaders took over forty more neglected buildings in the neighborhood and renovated them, creating cooperatives, in what was called the Eleventh Street Movement. CHARAS helped many of them create rooftop gardens, hand-fashion solar panels and windmills from scrap, and install them. One co-op set up a three-hundred-gallon fish farm in the basement—plywood tanks stocked with trout, carp, catfish, tilapia, and crayfish. CHARAS cleared a vacant lot on the south side of East Twelfth Street (between A and B), where Linda Cohen, an ISE alumna, used growing and composting techniques she'd learned in Vermont to found El Sol Brillante, the first organic community garden on the Lower East Side. Wastewater from the neighborhood's fish tanks fertilized the garden, while vegetable wastes and worms from the garden nourished the fish.[75]

Through the use of such eco-technics, Bookchin observed, CHARAS and its fellow Loisaida residents gained "control over the material conditions of their lives" and thereby acquired a sense of neighborhood solidarity, self-empowerment, and proficiency.[76]

In April 1976 the neighborhood power movement coalesced into the Alliance for Neighborhood Government and the National Association of Neighborhoods. Meeting in Philadelphia, the two groups adopted a "Bill of Rights for Neighborhoods" that declared that people in neighborhoods "can and should govern themselves democratically and justly." They should be able to "determine their own goals, consistent with the broad civic ideas of justice and human equality," and to "decisively influence ... all actions of government and private institutions affecting the neighborhood."[77]

During this bicentennial year of the American Revolution, Bookchin took a close look at American history and realized that that revolution's institutional engine had been, in fact, a kind of neighborhood government: the New England town meeting.[78] The town meeting had originated in the seventeenth century, when English Puritans established small villages in order to be able to practice their religion based on scripture alone. Religious autonomy became political autonomy, as these congregations doubled as citizens' assemblies, vested with authority to handle civil affairs. Over the course of a century, New England townspeople gained the heady experience of face-to-face democratic self-governance. By 1776 the town meeting, along with the plethora of committees that sprang into existence throughout the colonies, had become a "form of freedom" comparable to the revolutionary sectional assemblies of 1793 Paris. After the Revolution, the new, independent nation's elites turned against what they considered tumultuous civic overparticipation and instituted a centralized government. Still, during the two hundred years since then, town meeting self-government had persisted in northern New England, especially in Vermont.

In the bicentennial year, Bookchin sought out Frank Bryan, an expert on the town meeting who taught at the University of Vermont. Sitting in his Burlington office one day, Bryan "heard this clumping from the two staircases, *clomp clomp*, and in walks this guy. He had on a black leather jacket. It was apparent from the minute he walked in the door that he was not—and I don't say this disparagingly—an academic." Although Bookchin was then an associate professor, he "wasn't one of that crowd." Verily, Bryan was impressed by Bookchin's interest in the town meeting—not only did he take it seriously, he "had a paradigm for it."[79]

Bryan and other Vermont libertarians at the time were concerned that the state government was eroding the powers of the town meeting. Coming from both the political Left and Right, they shared a belief that "centralized power is the enemy of individual liberty, self-reliance, and voluntary cooperation."[80] In March 1977 they joined forces to form the Decentralist League, to protect and if possible strengthen the towns' powers. "The Right fears big government, and the Left fears big business," Bryan told me, but they both understood that "it's really scale that matters."[81]

Bookchin joined the group, comfortable associating with right-wing libertarians, believing that they shared common ground; after all, his friendship and collaboration with Karl Hess, the onetime Goldwater speechwriter, remained close. And starting in the mid-1970s another onetime Goldwater supporter, John Clark, entered Bookchin's orbit and quickly became his protégé, soaking up all Murray had to teach about social ecology, anarchism, and Left social theory generally.[82]

Once Bookchin was awakened to the crucial role of the town meeting in the American Revolution, he came to realize that all the great Western revolutions had been based in cities, in places where ordinary people were concentrated, where they could share experiences, learn from one another, hear speeches, read local newspapers, discuss issues in clubs and cafés, organize political action—and become empowered. From Boston to Paris, St. Petersburg, and Barcelona, the great revolutionary movements had been urban in nature. Marx had been blind to this fact, preferring the factory as the revolutionary arena, and under his influence generations of leftists afterward had been similarly unseeing. But cities could recapture that role, by recreating town meeting assemblies in their neighborhoods.

In Montreal, the MCM, with its decentralist program, was committed to such a devolution of power. But Dimitri Roussopoulos was worried. The new party had so far failed to clarify how the neighborhood democracies were to be created and structured. It had not even instituted structures of democratic accountability between the movement and the eighteen MCM city councilors. The pragmatists in the party were preparing a rebellion, insisting on transforming the MCM into a conventional parliamentary party.

The next battle would be fought, both sides knew, at the following party congress, to be held in December 1976. The *Our Generation* editorial board mobilized itself as a kind of think tank for the MCM's radical-socialist left wing, and Roussopoulos's living room became its hub.[83] The phone lines between Ramapo and Montreal buzzed, and

Bookchin came up from New Jersey often. In November 1976 he told a McGill audience that the citizen must replace the worker as the "elusive 'historical agent' . . . that will effect revolutionary social change."[84] While Bookchin was the charismatic and inspiring speaker, Roussopoulos was the experienced organizer. Between the two of them, they developed and tested the program for creating neighborhood democracy. Montreal was their laboratory.

The left wing, inspired by Bookchin, drafted a radical program to submit to the coming congress. It committed the MCM to organizing "committees on a street-by-street or block level. Each street or block committee would send delegates to the neighborhood council." The councils would become "the organizational instruments" through which the citizens could create a "truly democratic power." Once the councils gained power, they must become, not part of the state apparatus, but "an alternative power to the present state at all its levels."[85] Murray praised this document as "the most radical program of all."[86]

Once the congress convened, the pragmatists tried to wrest control of the party executive, but the Left fought back and stayed at the helm. In the end, the congress adopted Roussopoulos's radical-socialist program. In the next two years, neighborhood organizing in Montreal's nineteen districts would reach a crescendo.

NOTES

1. E. F. Schumacher, *Small Is Beautiful: Economics as if People Mattered* (New York: Harper & Row, 1973), 106, 145.

2. Ibid., 107.

3. Ursula Le Guin to author, Apr. 19, 2009.

4. Murray had used the word *ecotopia* in "Toward an Ecological Society" (1973), in *TES*, 70–71.

5. Herbert Marcuse, *Counter-revolution and Revolt* (Boston: Beacon Press, 1972), 59, 69. Marcuse credits Bookchin's "Ecology and Revolutionary Thought" and "Towards a Liberatory Technology" on page 61.

6. Dan Chodorkoff, interview by author, May 12, 2009; Michel Saint-Germain, "La liberté de chacun: Un café avec Murray Bookchin," *Mainmise* [Montreal], no. 76 (1978), 30–31.

7. "Goddard College's Roots Deep in Vermont History," *Burlington Free Press* (hereafter *BFP*), Sept. 4, 1980.

8. Ibid.; Marilyn Adams, "Goddard College Struggles to Survive," *BFP*, Jan. 4, 1981, 1B.

9. Bookchin, "Reflections: An Overview of the Roots of Social Ecology," *Harbinger* 3, no. 1 (Spring 2003), 6.

10. Bookchin, *The Spanish Anarchists: The Heroic Years, 1868–1936* (New York: Free Life Editions, 1977), 4–5. The founder of the Escuela Moderna was the Catalan anarchist educator Francesc Ferrer i Guàrdia.

11. Chodorkoff interview; Bookchin, "Overview of Roots," 6f.

12. "Goddard Sessions to Focus on Ties of Society, Ecology," *Bennington Banner*, Oct. 13, 1973.

13. Todd and his wife Nancy Jack Todd credited Murray as an inspiration for New Alchemy in *Bioshelters, Ocean Arks, City Farming: Ecology as the Basis of Design* (San Francisco: Sierra Club Books, 1984), 11. On New Alchemy, see Nancy Jack Todd, "New Alchemy: Creation Myth and Ongoing Saga," *Journal of the New Alchemists*, no. 6 (1980); Ted Morgan, "Looking for Epoch B," *New York Times Magazine*, Feb. 29, 1976; "New Alchemy Institute Starts Its Second Decade," *Mother Earth News*, Jan.–Feb. 1980; and The Green Center, www.thegreencenter.net.

14. Jeffrey Jacob, *New Pioneers: The Back-to-the-Land Movement and the Search for a Sustainable Future* (University Park: Pennsylvania State University Press, 1999), 197.

15. Bookchin praised New Alchemy's work in "The Concept of Ecotechnologies and Ecocommunities" (1976), in *TES*, 108.

16. Hess affirmed libertarianism in "The Death of Politics," *Playboy*, Mar. 1969, online at http://bit.ly/nToFam.

17. Karl Hess, *Community Technology* (New York: Harper & Row, 1979), 45, 236; "Gil Friend and David Morris of the Institute for Local Self-Reliance," *Mother Earth News*, Nov.–Dec. 1975; David Morris and Karl Hess, *Neighborhood Power: The New Localism* (Boston: Beacon Press, 1975), 137.

18. "Gil Friend and David Morris of the Institute for Local Self-Reliance," *Mother Earth News*, Nov.–Dec. 1975.

19. Karl Hess, *Mostly on the Edge* (Amherst, NY: Prometheus Books, 1999), 239; Hess, *Community Technology*.

20. Hess, *Mostly on the Edge*, 237.

21. Bookchin had used the term *social ecology* in passing in "Ecology and Revolutionary Thought." In August 1970 he had written that as anarchism and ecology converge, "natural ecology becomes social ecology." Bookchin, "Introduction" (1970), in *PSA*, 21. In "Concept of Ecotechnologies and Ecocommunities," he noted that "the late E. A. Gutkind . . . coined [the term] a quarter of a century ago" (108). In 1974 he referred to René Dubos as "that grand old man of social ecology." *Our Synthetic Environment*, rev. ed. (New York: Harper Colophon, 1974), xiv.

22. Bookchin to Resident Undergraduate Faculty, "Proposal for Social Ecology Studies Program" (ca. 1973), ISE Archive, Marshfield, VT.

23. The flyer for the Social Ecology Studies Program described it as "an intensive twelve-week program in social ecology and environmental sciences exploring alternative technologies, a no-growth economics, organic agriculture, urban decentralization, the politics of ecology, and the design and construction of experimental models of wind, solar and methane-powered energy production." For that first session, among Goddard faculty, Charles Woodard and James Nolfi (a marine biologist) taught agriculture; Scott Nielson, economics; Jules Rabin, ecological anthropology; and John Mallory, architecture. The visiting faculty were Todd, McLarney, and others from New Alchemy, as well as Hess, Eccli, Love, and Clark. ISE Archive, Marshfield, VT. See also Bookchin, "Overview of Roots," 6f; Chodorkoff interview.

24. "Bill McLarney, Aquaculture, 7/2/24," and "Eugene Eccli, Solar Energy, 7/26/74," audiotapes, MBPTL.

25. Michael R. Edelstein, "Sustaining Sustainability: Lessons from Ramapo College," in Peggy F. Barlett and Geoffrey W. Chase, eds., *Sustainability on Campus: Stories and Strategies for Change* (Cambridge, MA: MIT Press, 2004), 271–92.

26. Wayne Hayes, interview by author, Aug. 25, 2009.

27. Chodorkoff interview.

28. Hayes interview.

29. Edelstein, "Sustaining Sustainability."

30. Bookchin, introduction to the second edition of *The Limits of City* (Montreal: Black Rose Books, 1986).

31. Bookchin (as Lewis Herber), *Our Synthetic Environment* (New York: Alfred A. Knopf, 1962), 238.

32. Daniel Elliot Chodorkoff, "Un Milagro de Loisaida: Alternative Technology and Grassroots Efforts for Neighborhood Reconstruction on New York's Lower East Side," Ph.D. diss., New School for Social Research, Mar. 1980.

33. On the community revolution, see Benjamin Barber, *Strong Democracy: Participatory Politics for a New Age* (Berkeley: University of California Press, 1984); Daniel Bell and Virginia Held, "The Community Revolution," *Public Interest* 16 (Summer 1969); Harry C. Boyte, *The Backyard Revolution: Understanding the New Citizen Movement* (Philadelphia: Temple University Press, 1980); Douglas Yates, *Neighborhood Democracy: The Politics and Impacts of Decentralization* (Lexington, MA: Lexington Books, 1973); and Joseph F. Zimmerman, *The Federated City: Community Control in Large Cities* (New York: St. Martin's Press, 1972).

34. Barber, *Strong Democracy*, 205.

35. Milton Kotler, *Neighborhood Government: The Local Foundations of Political Life* (Indianapolis: Bobbs-Merrill, 1969), xii, 33, 41; and Morris and Hess, *Neighborhood Power*, 99, 101, 148. According to Douglas Yates, writing in 1973, "decentralization holds the promise of creating more direct democracy than has ever existed before in American cities." Yates, *Neighborhood Democracy*, 161.

36. Karl Hess, *Dear America* (New York: William Morrow, 1975), 237; Hess, *Community Technology*, 40.

37. Chodorkoff, "Un Milagro de Loisaida," 120, 121.

38. Hess, *Community Technology*, 54, 55, 60.

39. Bookchin, "Concept of Ecotechnologies," 108.

40. Dimitri Roussopoulos, interview by author, July 6, 2009; "The MCM Manifesto," *Our Generation* 10, no. 3 (1974), 10ff. See also "The Montreal Citizens' Movement," *Our Generation* 10, no. 3 (Fall 1974), 3ff.

41. Timothy Lloyd Thomas, *A City with a Difference: The Rise and Fall of the Montreal Citizens Movement* (Montreal: Véhicule Press, 1977), 3; Dimitri Roussopoulos, "Debates: A Community Control Strategy," *Our Generation* 9, no. 1 (Jan. 1973).

42. Dimitri Roussopoulos, "Neighborhood Councils," *City Magazine* 3, no. 8 (1978), reprinted in Roussopoulos, ed., *The City and Radical Social Change* (Montreal: Black Rose Books, 1982), 214.

43. Bookchin, "The Urban Vacuum: Prospects and Possibilities," *Liberation* 19, nos. 8–9 (Spring 1976), 118; Roussopoulos quoted in Thomas, *City with a Difference*, 50, 54.

44. Bookchin, "Urban Vacuum," 118.

45. Roussopoulos interview; Thomas, *City with a Difference*, 14.

46. Roussopoulos interview; Stephen Schecter, "Urban Politics in Capitalist Society: A Revolutionary Strategy," *Our Generation* 11, no. 1 (Fall 1975). Schecter emphasized to me that he no longer holds these views. Thomas, *City with a Difference*, 58.

47. The 1975 session ran fourteen weeks from June 2 to August 22. The core faculty were Dan and Murray, plus Goddardites Jim Nolfi, Scott Nielson, and Charles Woodard. The New

Alchemists came to lecture, as did Wilson Clark, Steve Baer, and Robert Reines, on solar and wind energy. Other visiting lecturers spoke on organic agriculture, urban decentralization, the politics of ecology, and the "no-growth economy," including Sam Love, Karl Hess, John Shuttleworth, and Stewart Brand. See poster for the 1975 Social Ecology Studies Program, ISE Archive, Marshfield, VT. For a detailed description of the mid-1970s program, see Judith Edwards, "Pioneering for the Future: The Social Ecology Institute," *Vermont Summer* [magazine], *Bennington Banner*, Aug. 27, 1976.

48. Judith Edwards, "Space Patrol Station Set Up, around an Old Farmhouse," *Bennington Banner*, Mar. 26, 1976.

49. These projects are discussed in Bookchin, "Concept of Ecotechnologies"; and in his introduction to Hans Thirring, *Energy for Man* (New York: Harper & Row, 1976). See also Wilson Clark, *Energy for Survival: An Alternative to Extinction* (New York: Anchor Press, 1974). For photos of some of the projects, see Bruce Anderson and Michael Riordan, *The Solar Home Book: Heating, Cooking, and Designing with the Sun* (Palo Alto, CA: Cheshire Books, 1976), 38.

50. Bookchin, introduction to Thirring, *Energy for Man*.

51. The Jacobs produced 1,800 watts of 32-volt DC electricity. Two other wind machines were set up: a Windcharger (a 12-volt DC unit with an output of 2,000 watts at 700 rpm) and a Savonius air pump. ISE Archive, Marshfield, VT.

52. Ted Mackey et al., "Composting," *Wanekia: The Cate Farm Journal* (Plainfield, VT: Institute for Social Ecology, 1977), 44, ISE Archive, Marshfield, VT.

53. Bookchin, "Energy, 'Ecotechocracy,' and Ecology," *Liberation*, Feb. 1975, 32.

54. Dan Chodorkoff, "Towards the Future," *Wanekia*, 19.

55. Nesta King, "Women's Perspective," *Wanekia*, 23.

56. Hess, *Community Technology*, 41, 51, 54.

57. "Richard Merrill, writer and editor," n.d., online at http://bit.ly/qeHmxL; Sam Love to author, June 13, 2009.

58. Steve Baer to author, June 5, 2009.

59. Barry Costa-Pierce, interview by author, June 6, 2009; Joy Gardner interview, Apr. 5, 2008; Hayes interview.

60. Saint-Germain, "La liberté de chacun."

61. Costa-Pierce interview; Richard Merrill, "A Prosaic Ode to Murray Bookchin," June 15, 2009, courtesy Richard Merrill.

62. Bookchin, "Overview of Roots," 6–11.

63. Chodorkoff interview. For the proposal, see Bookchin to Resident Undergraduate Faculty, "Proposal for Social Ecology Studies Program" (ca. 1973), ISE Archive, Marshfield, VT.

64. James Morley, interview by author, July 12, 2009.

65. Edelstein, "Sustaining Sustainability," 274.

66. Bookchin, "Toward a Vision of the Urban Future" (1977), in *TES*, 174.

67. Stephanie Mills to author, June 14, 2009.

68. Hayes interview.

69. "Biological Agriculture at Cate Farm," *Wanekia*, 33; Rick Beck, "Creating Fertile Greenhouse Soil," *Wanekia*, 34; Alvin Atlas and Rick Beck, "Greenhouse Veggies," *Wanekia*, 34; Alvin Atlas, "Insects in the Greenhouse," *Wanekia*, 35. See also "Student Charts Own Course," *BFP*, Sept. 1, 1976. The windmill was a 1.8 KW Jacobs, which generated electricity for the pumps and lights in the aquaculture complex. See Edwards, "Pioneering for the Future"; Bookchin, "Concept of Ecotechnologies," 108; Chodorkoff, "Un Milagro de Loisaida," 117.

70. Leslie Sproule, "French Intensive Beds," *Wanekia*, 36.

71. Costa-Pierce interview. Pierce described Bookchin and Nolfi as "early pioneers of the social ecology of food systems, alternative energy, and social strategies" in Barry A. Costa-Pierce, *Ecological Aquaculture: The Evolution of the Blue Revolution* (New York: John Wiley & Sons, 2002), xv. See also Boyte, *The Backyard Revolution*, 145; "At Goddard: Getting from Ego to Eco Is the Goal of Social Ecology Institute," *Bennington Banner*, Feb. 18, 1976, 6; Chodorkoff interview; Constance Holden, "NCAT: Appropriate Technology with a Mission," *Science*, Mar. 4, 1977, 857; "1974," memo, n.d., ISE Archive, Marshfield, VT.

72. Chodorkoff, "Un Milagro de Loisaida," 118, 152.

73. Ibid., 156ff; see also Josh Weil, "The Wind Farmers of East 11th Street," *New York Times*, Aug. 3, 2008.

74. The process has since become known as net metering. See Abigail Rao, "Adding the Final Touch: A Windmill and Solar Panels, Feb. 1975: An Experiment in Green Living Leaves a Mixed Legacy," *City Limits Magazine* Nov. 2001.

75. Chodorkoff interview; Chodorkoff, "Un Milagro de Loisaida." In May 1976 on the Lower East Side, the ISE held a conference, called "Urban Alternatives: Toward an Ecological City," that brought neighborhood organizations together with technical innovators. Bookchin gave the opening address, followed by Karl Hess, Travis Price of the Eleventh Street Movement, Gil Friend of the ILSR, Liz Christy of Green Guerrillas, Jim Nolfi on agriculture; Milton Kotler, on neighborhood government; Eugene Eccli on local energy production; Ramon Rueda, on sweat equity in the South Bronx; and more. Robert Palmer papers, courtesy Bob Erler.

76. Bookchin, "Self-Management and New Technology" (1979), in *TES*, 130. For a video about CHARAS's work on Loisaida, see Marlis Momber's ten-minute film *Viva Loisaida 1978*, at http://bit.ly/13QYku0.

77. Quoted online at http://1.usa.gov/qIdYDC (accessed 2008) and in Boyte, *Backyard Revolution*. The document was renamed Neighborhood Bill of Responsibilities and Rights.

78. Bookchin, "On Neo-Marxism, Bureaucracy, and the Body Politic" (1978), in *TES*, 246.

79. Frank Bryan, interview by author, July 2009. Bryan is a leading authority on the town meeting. See his *Real Democracy: The New England Town Meeting and How It Works* (Chicago: University of Chicago Press, 2004). Murray also recommended to me Jane Mansbridge, *Beyond Adversary Democracy* (Chicago: University of Chicago Press, 1983).

80. Greg Guma, "Breaking Out of the Empire Box," *Toward Freedom*, Sept. 1, 2008; online at http://bit.ly/qSdG47.

81. Bryan interview. The Decentralist League disbanded when its left wing opted for electoral politics and the right signed on for the Reagan "revolution." Guma, "Breaking Out." Guma says Murray wasn't a member, but Bryan says he was.

82. Bookchin, "Turning Up the Stones: A Reply to Clark's October 13 Message," Oct. 1998, online at http://bit.ly/qc1Yw6.

83. Dimitri Roussopoulos, "Beyond Reformism: The Ambiguity of the Urban Question," *Our Generation* 11, no. 2 (Winter 1976).

84. Bookchin, "Urban Vacuum," 114. On Nov. 26, 1976, he was in Montreal speaking to a general audience on "A Radical Analysis of the City," and to an anarchist study group on "The Development of Anarchism"; in 1977 he returned to lecture on the development of anarchism and Marxism. All are part of the *Our Generation* Tape Series, Labadie Collection, Ann Arbor, MI.

85. The MCM's 1976 program is quoted in Dimitri Roussopoulos, "Neighborhood Councils" (1978), reprinted in Roussopoulos, ed., *City and Radical Social Change*, 216. See also Marc Raboy, "The Future of Montreal and the MCM," *Our Generation* 12, no. 4 (Fall 1978).

86. Bookchin, "Toward a Vision," 190. He told me, years later, that he would have preferred that it called for assemblies rather than councils.

9

Antinuclear Activist

⁂ ───

IN THE SUMMER of 1977, the renowned anthropologist Margaret Mead visited the ISE as a guest speaker. As she toured Cate Farm, taking in the solar buildings, the windmills, the dome, and the raised-bed gardens, she nodded approvingly, but then as she was leaving, she turned and waved her stick at it all. "You know," she said, "it won't do a bit of good if we don't stop nuclear power."[1]

Mead had reason to be concerned: about seventy-five nuclear reactors had been built since 1966, arousing ever more unease about the likelihood of accidents and the absence of a safe way to dispose of radioactive wastes. A year before her visit to the ISE, the Public Service Company (PSCO) of New Hampshire had gained permission from the Nuclear Regulatory Commission to construct two reactors next to a broad expanse of salt marshes, estuaries, and tidal mudflats in the sleepy coastal town of Seabrook, New Hampshire. PSCO planned to run a pair of giant two-mile-long pipes from the reactors through the marshes to the Atlantic. The pipes would suck in a billion gallons of seawater every day to cool the reactors and turn the turbines, then return the water to the ocean, forty degrees warmer. The huge soft-shell clam bed in the shoreline's delicate ecosystem could not possibly survive those elevated temperatures.[2]

In July 1976, as bulldozers began clearing the site, thirty-two activists gathered in nearby Rye to form the Clamshell Alliance to fight the project.[3] They were inspired by a remarkable action at Wyhl, in southwestern Germany, the year before. In February 1975 a nuclear power plant had been slated to be constructed at Wyhl. By the time the German bulldozers arrived, groups of protesters were already in place, peacefully occupying the construction site. Police dragged them away, but tens of thousands occupied it again—not only activists and students but local citizens. They set up an encampment of huts and tents—an alternative village. This time the police left it in place, and in March a court withdrew the reactor's construction license. The reactor was never built. The Clamshell Alliance, many of whose founders were Quakers, intended to follow Wyhl as its model. They would occupy the Seabrook construction site peaceably, using nonviolent direct action, and rally the seacoast community against the reactors.[4]

At their early organizational meetings, the "Clams" agreed that they would make decisions by consensus—that is, in making decisions, they would try to achieve unanimity. Consensus process, which they learned from the Quakers, seemed superior to traditional majority-rule voting: consensus was inclusive, accommodating everyone's ideas, whereas majority voting was win/lose—it permitted only one view to prevail, and the minority had to put up with it. The original Clams were a small group of friends brimming with goodwill, their meetings relatively informal, so they easily achieved consensus among themselves and forged ahead.[5]

In preparation for occupying the Seabrook site, a group of eighteen Clams practiced doing civil disobedience and role-played possible encounters with police. They schooled one another to avoid doing anything that might provoke a confrontation. On August 1, 1976, some six hundred Clams gathered at Seabrook, and the eighteen who had undergone nonviolence training made their way to the construction site, where they planted some pine and maple saplings. When PSCO requested that they leave, they sat down and refused. A squadron of state police then dragged them, limp, to the station and charged them with criminal trespass.[6]

The next site occupation was to take place three weeks later. About 180 people volunteered to do civil disobedience and underwent training in six- to eight-hour sessions. Quaker Clam Suki Rice decided to train them in small groups—she called them affinity groups. It's unclear where she got that phrase; one historian says it was her "informal innovation," but by 1976 Bookchin's much-read *Post-scarcity Anarchism* had long since popularized the term. In any case, Rice's affinity groups were organized for the practical purpose of nonviolence training.[7]

When August 22 arrived, the 180 Clams occupied the site; they were arrested, went limp, and got dragged into buses. They declined bail and were taken to the National Guard armory in Portsmouth. Most were released the next morning.[8]

This second action generated headlines, and new members flooded into the Clamshell Alliance, from the feminist, Native American rights, labor, and environmental movements. They showed up at meetings eager to put their bodies on the line. The next action was scheduled for April 1977, and the Clams organized them into affinity groups for nonviolence training.

During the winter of 1976–77, Clamshell's congresses adopted an organizational structure to accommodate the growing alliance. It set up a Coordinating Committee (CC), consisting of representatives from the various regions of New England. But the CC was not empowered to make decisions. Rather, all Clams were to participate in decision-making and work to achieve consensus. When an issue arose requiring a decision, the CC would discuss it and develop a position or positions on it; the representatives would take the CC's position(s) back to their respective regions, which would discuss and arrive at their own positions; the reps would then bring the regions' positions to the next CC meeting. To harmonize the regions' positions, the reps would hammer out a compromise, which they would then take back to their regions for discussion and, they hoped, approval. Usually an issue would go back and forth between the regions and the CC for weeks, or even months, before consensus could be reached. The CC adhered scrupulously to this process, even as it became lengthy and cumbersome.[9]

The creation of Clamshell gratified Bookchin, who had opposed nuclear power since the 1950s and had helped prevent construction of the Ravenswood reactor in the 1960s. Now here was a whole movement dedicated to challenging it through an ethically charged revolt. Furthermore, the movement wasn't Marxist—it had a libertarian structure, and the members were even forming affinity groups. Bookchin joined eagerly, with high expectations.

He understood that the Clamshell's use of the term *affinity group* was not like his own or the Spanish anarchists'. Clam affinity groups were ad hoc, made up of strangers who underwent training together to carry out a predetermined action, then disbanded. Bookchin's affinity groups were more autonomous, ongoing groups of "brothers and sisters" who knew each other well and coalesced to be politically creative together, to conceive and carry out actions collectively. He and some friends at the ISE formed the Spruce Mountain Affinity Group (named after a mountain near Goddard) in order to be part of Clamshell. Bookchin's friends in Boston (who had adopted the name Black Rose) formed an affinity group in this vein as well; it was separate from the main, Quaker-dominated Boston group. Other groups were formed around Vermont, and on July 30, 1977, they met at Goddard and formed the Green Mountain Alliance.[10]

As the Clamshell continued to grow, Bookchin knew, its democratic ethos would likely be challenged. Often in loosely structured, egalitarian organizations, some members who covet control take on de facto leadership roles, and the very absence of structure allows them to accumulate informal power without accountability to the rest.[11] Clamshell would have to prevent that from happening. It needed a well-defined structure, yet one that was compatible with a libertarian ethos.

Bookchin thought of the manuscript that had been lying in his drawer for the past few years. He hadn't yet finished writing *The Spanish Anarchists*, partly because the story's denouement was so heartbreaking: the CNT leadership had betrayed anarchism by joining the Popular Front government, which had contributed to the movement's destruction. How could he present the Spanish anarchists as some kind of model, when their prominent members had in the end trampled on their own principles?

But the manuscript—originally written to rebut New Left criticisms of anarchism as disorganized—described a large-scale, democratic, libertarian organizational structure that had persisted for decades. As this very question was looming over

Clamshell, the times cried out for the Spanish story to be told. So he decided to finish the manuscript, call it volume one, and defer the story of the 1936–37 betrayals to a second volume. He finished volume one and dedicated it to Russell Blackwell, "mi amigo y mi compañero." Free Life Editions published it in 1977.[12]

In preparation for Clamshell's next planned occupation of the Seabrook site, nonviolence training proceeded fast and furious. Then on the morning of Saturday, April 30, 1977, two thousand protesters converged at Seabrook, among them a contingent from the ISE. Organized into affinity groups, they pitched their tents in the construction area in orderly rows, dug latrines, and laid out "roads," creating a village they called Freebrook. By consensus, they made a decision to ban all nuclear material within its limits.

The occupiers slept there overnight, then awoke on Sunday to find National Guards arrayed around them. The state governor warned them to evacuate within twenty minutes or face arrest. Most stayed put. Police arrested 1,414 and herded them onto buses that took them to the Portsmouth Armory. It wasn't big enough to hold such a large group, so the National Guard dispersed them to several more armories around the state.[13]

During the two weeks of their incarceration, as Brian Tokar, one of the detainees, told me, the affinity groups took on a new function: they "structured how people incarcerated together could make decisions collectively while enduring harassment by police." In the process, the incarcerated Clams reconceived the affinity group as a unit much more like what Bookchin had intended: one that "could form the basis for a much more widely participatory, directly democratic form of movement organization."[14]

As it happened, in advance of the occupation, Clamshell cofounder Howie Hawkins had invited Murray to speak at Dartmouth the first week in May. (Murray had asked, "Have you read my writings? I'm a revolutionary anarchist! Are you sure you want me?") When the date came, the fourteen hundred Clams were still in the armories. "He agreed to speak on the spot," Hawkins recalls, at a demonstration in solidarity with them. "He held the crowd with a rousing five-minute speech off the top of his head on the dangers of nuclear power and the power of popular direct action. He could have kept the crowd listening intently all that evening if we had let him go on."[15]

After two weeks, most of the charges were dismissed, and the Clams were released. The expanding antinuclear movement was jubilant, considering the action a triumph for nonviolent mobilization. Bookchin lauded the "return to the élan of the civil rights and antiwar movements" of the mid-1960s. Indeed the site occupations represented an advance over those movements, he said, because they constituted anarchist direct action, in which "the individual recovers her or his social being, a sense of power over events, and meaning as an active human being." All in all, the occupations "produced a thrilling sense of hope" among radicals nationwide.[16]

That summer the ISE attracted 150 students for its twelve-week session. New practical workshops taught the students skills in electrical wiring and plumbing, soldering and welding, and carpentry. One class constructed an entire garage using only native materials and hand tools. Structures were designed and redesigned, built and rebuilt, several times, so students could gain hands-on experience. Karl Hess marveled, "The 'hippies' have dropped their flowers and picked up wrenches and hammers."[17]

They plunged into eco-technics work. One group built flat-plate solar collectors from scratch and installed them in the Sunhouse for year-round aquaculture. Another designed a parabolic concentrating solar collector, which moved water on the principle that hot water rises. An instructor bought up some old wind generators from midwestern farms and brought them to the ISE, where the students refurbished them into usable turbines. The class designed a sailwing wind-power system, with twenty-two-foot Dacron sails, and set it up on a thirty-foot steel tower near the Sunhouse. Used to pump water for solar heating into a storage tank, it "spins in even the gentlest breeze."[18]

The biological agriculture students dug thirty-three French intensive beds—as well as ordinary beds, for controls, so they could observe the difference. The produce grown in the intensive beds, they found, outshone the controls "in vitality, health and taste." Nearby they planted marigolds, nasturtiums, sage, and peppermint around cabbage plots, to see whether these companion plants could fend off insects. Student Calley O'Neill designed what she called a Solar Shield Garden, using algae-laden water from five tilapia ponds as fertilizer. "The garden just went ballistic, crazy, wild, . . . like a garden on steroids," she said. Students planted an herb garden, constructed sprouting frames for alfalfa sprouts, and built a root cellar in the farmhouse basement; they built cold frames, along a wall of the compost shed, where plants could survive into the winter.[19]

In their downtime, they'd swim in the Winooski River. Murray, who stayed on dry land, was a "big Russian teddy bear," O'Neill told me. Sometimes he'd come up to the students' dorm with some vodka, "And we'd share a shot with him." O'Neill marveled at how polite and gracious he was. "If he had to interrupt, he was so patient, he was 'Excuse me please,' 'Thank you so much.' He was the most courteous conversationalist I'd ever met."

During the three-month session, the students bonded as a community. "A hundred fifty people came together and fell in love," O'Neill told me.

But no matter what project people were working on, when it was time for Bookchin to lecture, everybody—faculty and students—went to listen. "We'd all get to the barn early—we didn't want to miss a word," O'Neill said. "And he'd mesmerize everyone. He'd lecture for three hours, nonstop, no notes, full throttle, people on the edge of their seats." He'd build up his ideas on democracy and the history of civilization "very methodically," presenting them "in a spellbinding way. . . . He was a force of nature."[20]

Even the technically oriented people appreciated his historical perspective. "He'd always blow everyone away," recalled Joseph Kiefer, who specialized in organic farming.[21] "We realized it's not just about systems," said Costa-Pierce. "It's about people, it's about gender, families, community." One night "We all went to hear Murray talk at the University of Massachusetts at Amherst. Murray was the man! There were thousands of people. By the time he was done, people were yelling 'Amen, brother!' People were screaming and laughing and crying. And there was a giant roar, and everybody got up and we roared again. That speech was one of the greatest I've ever seen anyone give."[22]

At summer's end, many of the students agreed that the ISE session was "the best, most productive, metamorphosizing experience we had ever had."[23] It was "transformational for many of us," recalled Kiefer. "Without question he inspired what I do today," he told me, which was to connect local food growers with local human service agencies.

"But you know, Murray would drive everywhere," Kiefer told me. "He'd eat at McDonald's."[24] It was that proletarian thing.

Jim Morley entered Ramapo College as an angry young freshman in 1976. "I was really pissed off as a kid," he recalled. But when he sat down in Professor Bookchin's classroom, ready to act out against authority, the man who stood at the front of the room drove all such ideas from his mind. He was "dressed like a janitor," with baggy green pants and a tire air-pressure gauge in his front pocket. When he opened his mouth, he "spoke as a New Yorker, in our working-class idiom," to "children of parents who did not go to college." As he listened, Morley thought, "This is in my blood, this is me."

Bookchin took anger like Morley's and "gave it intellectual voice." He rarely bothered to consult a syllabus or look at any notes. He would orate extemporaneously, eloquently, "and he made perfect sense." He spoke in earnest and passionate monologues. "I was in absolute awe of Murray," said Morley. "He was my ideal of a human being." For his part, Bookchin enjoyed mentoring a generation of ecological revolutionaries.[25]

To these children of Ford assembly plant workers, Bookchin narrated the history of capitalism, as interpreted less by Marx than by the economic historian Karl Polanyi. His book *The Great Transformation* (1944) taught, as Bookchin put it, that contrary to popular belief, the rise of capitalism was not inevitable: "People were dragged into capitalism screaming, shouting and fighting all along the way, trying to resist this industrial and commercial world."[26] They resisted being reduced from communal beings to monadic egos. They resisted the reduction of social relations to exchange relations. Now today, he told his students, we must resist the dissolution of neighborhoods into suburbs, of retail shops into shopping malls. The market economy is becoming the market society, transforming everything into a commodity. And in the process capitalism is disassembling the biosphere, reversing the evolutionary process.

"A lot of teachers just want forty-five minutes," recalled Wayne Hayes, but at the forty-five-minute mark, "Murray was just getting warmed up." The academic mind, Hayes explained, "is trained to be narrow and to focus on particular subfields, and you're not supposed to roam beyond the boundaries." But Bookchin had no patience for that: his mind was "synoptic," able to pull things together.

He didn't take grading seriously—anyone who showed up regularly in class got an A. If a student didn't do the work, he or she would just get an incomplete. While the students loved his grading policy, the dean did not. But otherwise the Ramapo administration was delighted with Bookchin. On September 1, 1977, the onetime high school dropout achieved the rank of full professor.[27]

That fall the antinuclear movement was ballooning. Activists around the country organized groups, modeled on Clamshell, to occupy nuclear sites nonviolently: the Abalone Alliance in California, the Catfish Alliance in Florida, the Shad Alliance in New York, and so on.[28] The movement had verve and nerve, and its peaceful yet feisty protests attracted members in droves.

In New England, the Clamshell, too, expanded rapidly. In November 1977 a congress of two thousand Clams met at Putney, Vermont, to decide on the next step. Some wanted to do another site occupation, this time installing windmills and solar panels. But others thought Clamshell should step away from direct action for now and concentrate on educating other groups. The plenary, searching for consensus on the issue, got bogged down; the meeting dragged on for twelve hours. Many got tired and left. At 6:00 a.m. the rump agreed to go the route of a new site occupation that would not only protest but actually try to prevent construction. They scheduled it for June 24, 1978.[29]

In preparation, the Clamshell produced a handbook that affirmed that affinity groups are "the basic decision-making structure for the occupation," allowing for "decentralized tactical decision-making through the use of consensus," in which all participants have a voice. Bookchin's 1969 article "A Note on Affinity Groups," from *Post-scarcity Anarchism*, was copied and inserted. Clamshell was made up of affinity groups, he noted with satisfaction, networked egalitarian style (like the spokes of a wheel), rather than hierarchically (like the steps of a ladder). In other words, "its practice is anarchist."[30]

In the fall of 1977 *The Spanish Anarchists* was published. This "first history of Spanish anarchism to be written in English," historian George Woodcock observed in a review, provided a sense of the "humane" in anarchism. The book would "make many readers revise considerably" their views of anarchism, he predicted, which was what Bookchin intended.[31] Finally and most important, the Left now had a documented study of anarchist organization, and the antinuclear movement could grasp the Spanish roots of its affinity groups.

Clamshell's occupations of the Seabrook site had hitherto been logistically uncomplicated: the demonstrators had simply stepped into the wide-open construction area. But this time, in anticipation of the June 24 occupation, PSCO surrounded the site with a chain-link fence, topped with barbed wire.

Some Clam affinity groups, like Bookchin's Spruce Mountain and the radicals in Boston, saw the fence as no obstacle—they'd simply cut it (as Ecology Action East had done at Squatters' Park), proceed with the occupation, and get arrested as planned. But other Clams, especially those influenced by Quakerism, objected to fence cutting as an act of destruction and hence of violence. They would not go along with it. Several members of the CC held this view. The argument raged across the Alliance. "Much logistical planning of the occupation can't proceed until this question is settled," noted the CC. But the question was far from settled—on this issue, the Clams could not achieve consensus, not even close.[32]

A few weeks before the June 24 date, the New Hampshire attorney general Tom Rath took to the airwaves and made a proposal. The Clamshell Alliance should set aside its plan for an illegal site occupation, leave the fence intact, and instead hold a legal rally in an area just outside the fence. It could protest nuclear power, then leave the site before the construction workers arrived on Monday morning.

Clamshell now had to decide whether to accept Rath's proposal. Most of its fifty-five affinity groups across New England opposed it, on the grounds that the Putney Congress had consensually agreed on the site occupation. They wanted to proceed as planned and cut the fence. But the Quaker-oriented members, including those on the CC, preferred accepting Rath's proposal to cutting the fence.

Counterproposals and negotiations flew back and forth between center and periphery, but time was running short, and consensus remained remote—both sides had dug in. Finally, the CC members who supported the legal rally made an end run around the process—they went directly to the media and simply announced that the Clamshell Alliance had canceled the illegal site occupation and accepted Rath's offer. They had no power whatsoever to make such a decision—they created a fait accompli simply by announcing it. "In Vermont we learned about the 'decision' over the radio," said Bookchin.[33]

The legal rally, held on June 24–25, 1978, attracted eighteen thousand people. When Bookchin and his affinity group arrived, they rubbed their eyes in disbelief. The Seabrook marshland had been transformed into an energy fair, with tents, geodesic domes, small windmills, solar cookers, and bicycle generators. There were displays and food stands and concession booths, jugglers and Sufi dancers—it was like a carnival midway. On an impromptu stage, Jackson Browne and Arlo Guthrie sang, and Pete Seeger and Dick Gregory spoke.[34] "As I wandered around Seabrook during the 'Woodstock' of 1978," said Bookchin, people were "rocking along with the stars." It wasn't even a rally—it was an event to be consumed, "a star-studded 'legal' festival ... [a] spectacle."[35]

Stunned, two hundred or so New England radicals huddled to share their feelings of betrayal and to take stock of what had happened. The CC had unilaterally overturned a legitimate decision. Did the committee claim to be running the Alliance now, rather than merely coordinating it? Were the affinity groups from now on merely to follow its orders? The dissenters constituted themselves as a caucus within the Clamshell, called Clams for Democracy, to challenge the CC and fight for the power of the affinity groups.

On Sunday evening the rally organizers tore down the energy fair and duly carted all the exhibits away. Governor Meldrim Thomson, who had once referred to the Clamshell Alliance as "a lawless mob," pronounced it "regenerated and rehabilitated."[36]

The CC, fearing a hemorrhage of Clamshell membership, sent reps around New England to try to soothe the outraged feelings. At a meeting in Montpelier, the reps urged the Vermonters to remain in the Alliance. And they asked the Vermonters, by the way, to stop talking so incessantly about hierarchy. The Clamshell should focus solely on shutting down nuclear reactors, they said, in order to win mass support. Leave everything else alone. Nonsense, said Bookchin. Nuclear power wasn't a problem of technology; it was a problem of society. Any viable antinuclear movement must challenge "the social institutions and the domineering sensibilities that led to nuclear power plants in the first place."[37]

That July, Bookchin drafted a constitution for the Clamshell, one that would empower local affinity groups as the Alliance's base. Rather than be temporary, ad hoc training units for civil disobedience, they would be the Alliance's permanent fundamental unit, embedded in a democratic structure. Every month the affinity groups would send spokespersons to a regional coordinating meeting; when that meeting had a decision to make, the spokes would vote only as mandated by the affinity group that sent them, and should the spokes fail to do so, they would be recalled. An Alliance-wide committee would be elected at annual congresses. This structure all but guaranteed that power would rise from the bottom up. The Clamshell would be, in effect, a confederation of affinity groups—inspired by the Spanish anarchists' history. "I brought in stuff from the CNT and FAI," Bookchin recalled, and the Spruce Mountain Affinity Group signed it.[38]

In July 1978 Clams for Democracy gathered for a conference at Hampshire College and considered, among other things, this proposed constitution. While they agreed that affinity groups should be the Clamshell's basic unit, they objected to the levels of regional and Alliance-wide coordination. Affinity groups alone should constitute the Alliance, they insisted, with no coordinating tiers. Second, they rejected nonviolent civil disobedience as the means of opposing Seabrook—it was too passive. Confrontational direct action should replace it. (In other words, Tokar told me, "We said, 'We're going to cut the fence.'") Third, Clamshell should oppose nuclear power alone—it should oppose all hierarchy and domination and seek a large-scale transformation of the society that was trying to impose nuclear power regardless of the popular will.[39]

In advance of this Hampshire conference, Boston's anarchist affinity group Hard Rain had recruited as a potential ally the Red Balloon, a Marxist collective based in Brooklyn.[40] Bookchin was alarmed: if his life experience from the YCL to SDS had taught him anything, it was *Don't work with Marxists*. But now even as late as 1978, naive, well-intentioned young people were once again succumbing to the fatal attraction and even welcoming Marxists into a promising movement. He objected vociferously to Red Balloon's presence.

His young friends were bewildered. "Instead of talking about what we had in common, Murray spent a lot of time arguing with them," Tokar recalled. He "didn't even want to discuss more theoretical issues," John Lepper told me. But for Bookchin, theoretical issues became meaningless when Marxists were around. He told Tokar that if Clams for Democracy admitted Red Balloon, it would go nowhere. "I didn't agree," said Tokar, to which Murray responded, "You're too nice—you get along with too many different people."[41]

Bookchin dropped out of Clams for Democracy. Over the next year, these radical Clams would go on to scale the Seabrook fence using ladders and ropes; when they climbed down on the other side, police arrested them. When they went at the fence with wire cutters, police drove them back with clubs, mace, tear gas, and fire hoses. Their direct actions, while confrontational, failed to energize the movement or even generate media coverage.[42]

In the summer of 1978, the ISE's study of eco-technics surged. A wind power instructor conceived of an elaborate sensor to monitor the comparative temperatures in the two greenhouses and outdoors; the students designed and built it. Students set up a Windcharger atop the chicken coop, and they installed solar panels for a hot-water system on the farmhouse. They grew watercress, tomatoes, and green peppers hydroponically; a biofilter was used to remove fish wastes from the water. Master's student Barry Pierce studied food production by aquaculture in passive solar greenhouses. Between October 1977 and February 1979, he cultured rainbow trout, carp, tilapia, yellow perch—and achieved a growth rate of 106 pounds per year. Even New Alchemy, where the technique had originated, was impressed. The short film *Karl Hess: Toward Liberty* shows the ISE's wind machines and solar greenhouse in their mountain setting. On the Lower East Side, CHARAS and its youth cadres continued to manufacture geodesic domes for use as greenhouses and closed-loop aquaculture systems on rooftops, and they transformed an abandoned building on Avenue B for their headquarters. On Loisaida, eco-technics could indeed provide a base for self-reliance.[43]

Much to Bookchin's amazement, the United States had engendered an anarchist political party: the Libertarian Party, founded in 1971. It was procapitalist, to be sure, but its very existence suggested to Murray that an indigenous American streak of antistatism—or antigovernmentalism—was growing in popularity. In 1978 he keynoted a Libertarian Party meeting in Boston. When he told the 150 right-wing libertarians that America needed to become a society free of bureaucracy and centralization and hierarchy, they gave him a standing ovation. Afterward, an interviewer asked him if he thought it inconsistent for libertarians to try to achieve their ends through a political party; he agreed that it was, but he defended a libertarian use of the political process at the local level, based on neighborhood groups. "I find it perfectly consistent," he replied, "for libertarians to operate on the municipal or county level, where they are close to the people."[44]

As they had done, for example, in Amsterdam and Montreal. In 1977, however, the Montreal Citizens' Movement had undergone a split. The conventional pragmatists within it had had enough of the party's libertarian socialist program, the one calling for neighborhood empowerment. They left and formed their own new party, the Municipal Action Group (MAG), with a conventional structure and the conventional aim of attracting business to Montreal. In the next municipal election, in 1978, the

MCM, the MAG, and Drapeau's old Civic Party would square off. The socialists, anarchists, and community activists went all out for the MCM.[45] And on election day, the Montreal citizens went to the polls—and voted for a restoration of King Drapeau, giving his Civic Party fifty-three out of fifty-five council seats. The MCM and the MAG got one seat apiece.

The MCM radicals were shocked: the citizens had massively—and democratically—repudiated all their work to build neighborhood democracy. "We did a postmortem," Roussopoulos told me. "We had a packed meeting at a Spanish restaurant in downtown Montreal. 'Where do we go from here?' was the question." Most of the anarchists concluded that electoral politics was hopeless and left the MCM. Roussopoulos chose to stay and try to salvage something from the disaster, but in the next months the MCM dropped the radical elements in its 1976 program, then abandoned it altogether.[46]

The defeat in Montreal was a symptom of changing times. Over the course of the 1970s, once-alternative institutions had gradually given up their revolutionary aims and gone straight, when they did not perish altogether. Food and other cooperatives, in order to survive, adopted the ways of conventional business enterprises. Community organizations shed their demands for neighborhood power and applied for the Carter administration's community development block grants. "The government wanted enterprise rather than political action in the neighborhood," Milton Kotler observed caustically. "It would move the people out of the meeting hall and put them behind cash registers."[47]

Bookchin had long insisted that eco-technics was inseparable from eco-communities, but it turned out that solar panels and wind turbines could indeed be divorced from a radical social program. The highest-profile advocate of alternative technology and renewable energy, Amory Lovins, shared his advice with a gaggle of corporations (Bank of America, Dow, Lockheed Martin, Monsanto, among many others) and even with the U.S. government. "If Amory Lovins and his ilk can place their 'AT' know-how in the service of the Pentagon," Murray lamented, "then the environmentalist 'movement' . . . might just as well incorporate and sell itself to Exxon."[48]

When Jerry Brown was elected governor of California in 1975, he hired Bookchin's friend and renewable energy expert Wilson Clark as his principal energy adviser. The onetime ultrarevolutionary Tom Hayden (whom Bookchin could not interest in ecology in 1968) got on board with Brown to promote solar energy and social justice causes. Eco-technics, Murray was dismayed to admit, could be decoupled from decentralism and stripped of its revolutionary potential—and even be put to use by corporations and the state, to shore up domination, with "solar power utilities, space satellites, and an 'organic' agribusiness."[49]

Meanwhile the incipient "neighborhood power" movement for urban decentralization was being reversed. In New York, the new mayor, Edward Koch, called on young suburbanites to come back to the city and help rebuild it. Investment returned to neighborhoods in New York and other cities, and gentrification began, under the euphemism "urban renaissance." Outside the large cities, the postwar suburban boom was producing a distorted kind of decentralization that was not what Bookchin had meant at all. Urban sprawl was a mere diffusion of the metropolis, dependent on automobiles and highways and typified by visual blight, shopping malls, residential subdivisions, tract housing, and industrial and office parks—places that made community and civic participation all but impossible. In the Sunbelt, new cities were constructed that lacked any neighborhoods at all. The new urban agglomerations ("urbanization without cities," Bookchin would call them) negated the humanly scaled town and city and replaced neighborhood solidarity with worship of the cult of business. Residents of suburb and megalopolis alike, far from being active citizens, were ever more a passive "client population," paying taxes in exchange for services.[50]

As for the environmental movement, it continued along the road to reformism that Bookchin had criticized, concentrating its efforts on lobbying in Washington. In his widely reprinted 1980 "Open Letter to the Ecology Movement," Bookchin pointed out once again that attempting to rectify specific ecological problems, as environmentalists were doing, was mere tinkering; to address the ecology crisis at its root causes, a majoritarian movement had to emerge to challenge the market economy as well as the social system of hierarchy and domination.[51]

During the 1970s, various individual elements of his program had been detached from the whole, only to be absorbed into the existing system in distorted form: ecology had been narrowed to environmentalism, the cooperatives that survived had become enterprises, decentralization had been warped into sprawl, and eco-technics had become interesting to government and corporations. "It was the unity of my views," he lamented, "that gave them a radical thrust."[52]

In 1978 a new "voluntary simplicity" movement arose, advocating an antimaterialist lifestyle: it considered living simply, with minimal goods and services, to be the proper ecological choice, one that would bring inner richness. But as always, proposals for material renunciation struck the proletarian-minded Bookchin as elitist, demanding sacrifices of those least able to make them. You could impose bridge tolls or raise the price of gasoline, he objected, but that would only hurt low-income and poor people, while the rich would sail over those toll bridges and fill up their gas tanks unscathed. "They have chauffeurs! They have the means to pay!" As for organic food, it was too expensive for most working-class people to buy with any regularity—"I can't afford [it]. I buy cheap food at the supermarket."[53]

In these years, too, the New Age—a Western adaptation of Asian spirituality—had arisen as an alternative culture and was winning the hearts and minds of onetime ecological radicals. Valuing a sense of cosmic oneness or unity with nature, it fed into a new ecological spirituality. Bookchin thought any spirituality was a step backward, since it represented a renunciation of reason and action (as in the Taoist concept of *wu wei*) in favor of mysticism, political passivity, and withdrawal into private life. He preferred the Western disposition for rationally and actively attempting to understand the unknown rather than passively worshipping it, and for actively attempting to change society for the better rather than passively accepting an unjust or unfree status quo. The dynamism of the ancient Greek tradition held far more appeal for him than the Indian and Chinese ideas that fed the New Age: it offered "rational, technical and ethical inspiration for developing human-oriented technologies and communities."[54] He undertook a study of Greek political and nature philosophy.

As for the young radical intellectuals who had entered the universities, they were finding ever more satisfaction in exploring the copious and sometimes contradictory writings of Karl Marx. In 1973 the *Grundrisse* (written in 1858) got its first English translation. Works of the so-called Western Marxists were translated, too, like Georg Lukacs's *History and Class Consciousness* (1923, English translation 1972) and Karl Korsch's *Marxism and Philosophy* (1923, English 1970). But in Bookchin's view, these "new" Marxist writings not only exhibited the familiar limitations of Marxism; they were burdened with theoretics so abstract and arcane that they were an obstacle to actual political organizing among ordinary people. The Situationist Guy Debord talked of the "theory of the spectacle," Murray often observed, but it was becoming more accurate to speak of "the spectacle of theory."[55]

Bookchin respected Marxism for what it had been in its time, but that time had come and gone, and these neo-Marxist academics were working with a corpse. Worse, in their attempts to revive it, they were adorning the corpse with ideas that were foreign to it—like feminism and communitarianism—and thereby reducing gender and community to matters of economics.

Worst of all, in Bookchin's eyes, they were trying to present Marx as some kind of early ecologist. This was more than Murray could bear. Yes, Marx's passages about "town and country" had influenced him long ago, but they were marginal to Marx's thought; Marx had fundamentally regarded the conquest of nature as a prerequisite for the socialization of humanity. He had commended bourgeois society for rendering nature "purely an object for humankind, purely a matter of utility," and he praised science's discovery of nature's laws, the better "to subjugate [nature] under human needs, whether as an object of consumption or as a means of production."[56] In other words, Marx had welcomed the domination of nature as a prerequisite for human progress.

Moreover, Marxism was indelibly authoritarian. Marx and Engels had regarded authority as essential for disciplining the proletariat and enforcing obedience—Engels had explicitly praised the factory as "a school for hierarchy, for obedience and command." Marx had even thought capitalism, by destroying earlier economic forms and developing technology, had played a historically progressive role. He thought class society had been historically necessary to achieve humanity's ultimate liberation. Such notions, Bookchin wrote, made Marxism, all appearances to the contrary, "the most sophisticated ideology of advanced capitalism."[57]

Neo-Marxists could not have it both ways; Marxism was quite simply incompatible with ecology. By trying to fit the expansive new social movements—ecology, feminism, and community—onto the archaic procrustean frame of Marxian economics, they were obstructing the development of a successor ideology, one that would be ethical and antihierarchical with "a truly revolutionary conception of freedom."[58]

He carried the battle to academic conferences, challenging academic Marxists to their faces. In May 1980, at UCLA, he discoursed to some of the other professors on anarchist history. "Murray's command of political history had us all spellbound," Carl Boggs recalled, until the wee hours. After everyone else went to bed, he stayed up writing a twenty-five-page talk. That afternoon fifty-nine-year-old Bookchin strode to the podium and "delivered a breathtaking tour de force on the long and tortured relationship between Marxism and anarchism—showing no signs of travel, the late-night soliloquy, or the vodka." As the day was winding down, "Murray was quite prepared to extend the theoretical debate for many more hours just as the rest of us were completely overwhelmed by exhaustion."[59]

In March 1979 the partial meltdown of a reactor core at Three Mile Island in Pennsylvania spurred tens of thousands to join the antinuclear movement. But in Bookchin's view, as the movement grew in popularity, its leaders were neutralizing its radical message. The same people who had high-handedly quashed the 1978 Seabrook action were now going on to become star-tripping celebrities and were turning the potentially revolutionary antinuclear movement into a harmless frolic. "No Nukes" concerts were held at Madison Square Garden, then at Battery Park, where the performers called for ending nuclear power but said nothing about hierarchy or capitalism or the nation-state. The huge audiences passively consumed these spectacles rather than taking action. Antinuclear politics was being transmuted into concerts, albums, and movies.[60]

When Bookchin's 1980 "Open Letter" was published in *WIN* magazine, Pete Seeger wrote a rebuttal. Fund raising through concerts was part of movement building, he scolded Bookchin; entertainers-as-activists were just trying to keep "the poor who may still inherit the earth" from inheriting "more than a poisonous garbage dump." He accused Murray of being a purist and a sectarian.[61] Bookchin didn't mind being dismissed as a purist, he said in reply, if that was the label attached to those who upheld a principled commitment to radical change. And coming from those who were "virtually dissolving [their] principles in order to work with almost everyone," it was practically a compliment.[62]

In contrast to the present's amorphous ecumenism, he yearned for the kind of "embattled commitment" and social idealism that had marked revolutionaries of the past. In 1793 Danton had memorably roared, "L'audace! L'audace! Encore l'audace!" By contrast, today's "self-styled radicals ... demurely carry attaché cases of memoranda and grant requests into their conference rooms." Many were joining Barry Commoner's Citizens Party, which called for economic democracy and environmentalism, but it was merely an effort to channel radical impulses into the mainstream; like all conventional parties, it offered "a maraschino cherry to everyone" while defusing serious challenges to the system.[63]

Where radical politics once stood for full citizen empowerment, it now stood for the empowerment of professional politicians in state and national government; where it once endorsed democratic assemblies, it now recommended "the numbing quietude of the polling booth, the deadening platitudes of petition campaigns"; instead of complex social theory, its new métier was bumper-sticker slogans; and instead of stirring demands for revolution, it meekly begged for paltry reforms. People no longer wanted to dedicate themselves to a revolutionary project that might "require the labors and dedication of a lifetime." Instead, they craved instant gratification and were willing to surrender their long-term ideals to get it. Indeed, the "world of 'fast politics' . . . closely parallels the world of 'fast food.'"[64]

But the preference for instant success and immediate gratification leads to trade-offs and compromises, he warned. It leads to choosing the lesser evil over the greater good. It decouples politics from ethics. The way to rejoin them was to do what the ancient Greeks (for whom politics was infused with ethics) had done: build an assembly democracy, in which all adult citizens could participate in decision-making processes, and to build a humane society in which the material means of life were communally owned, produced, and shared according to need.[65]

Amid the disappointments, Bookchin found ways to advance his project regardless. The teaching at Ramapo was satisfying—Bookchin had developed new courses on urban utopias, alternatives to modern medicine, and nature philosophy. Black Rose Books published his 1970s essays to acclaim from eco-anarchist reviewers. *Toward an Ecological Society* was "a work of crucial importance," wrote his protégé John Clark. "Practically every paragraph is filled with mind-expanding ideas," observed Bob Long. Bookchin was recasting "the emancipatory traditions of human history" in contemporary form, wrote John Fekete of Trent University.[66]

And although Karl Hess's community technology projects in Washington had succumbed to rising crime, CHARAS was thriving in New York. In 1979 the group transformed an abandoned turn-of-the-century school building on East Ninth Street into El Bohio Cultural and Community Center. For the next few decades, it would be an uplifting force in Loisaida (figure 9.1).[67]

One day in September 1979, Bookchin opened his mailbox to find a subpoena from the US government, summoning him to testify at a trial in Washington. It took him a while to understand what it was all about.

Since the 1950s, FBI agents had been spying on the American Left by entering people's homes surreptitiously, without warrants, looking for evidence of wrongdoing. In 1972 the Supreme Court had declared these illegal searches unconstitutional. But at that moment the FBI had been intent on continuing the searches in order to get at the Weather Underground, whose members were then planting bombs in government and corporate buildings. But the Weather people had gone underground, their whereabouts unknown—the bureau couldn't get at them. In 1973 the second- and third-ranking FBI officials, W. Mark Felt and Edward S. Miller, instructed agents to resume "black bag jobs," in defiance of the Supreme Court's ruling. Agents were to break into the homes of friends and families of the Weather people, to look for evidence of the fugitives' activities.

Antinuclear Activist

FIGURE 9.1 Chino Garcia and CHARAS helped empower the Hispanic community on the Lower East Side, at a time when city government abandoned it. Ca. 1985. Mark Ivins.

In February 1973, FBI agents had surreptitiously entered Bookchin's apartment at 235 Second Avenue and searched it.[68] Evidently, someone had fed the FBI some false information about him.

In the late 1970s, the US Senate committee investigated the illegal resumption of black bag jobs. On April 10, 1978, a federal grand jury charged Felt and Miller with conspiracy to violate the constitutional rights of American citizens by searching their homes surreptitiously and without warrants. The two FBI directors were to be tried in US district court in Washington. The special prosecutor, John Nields, subpoenaed Bookchin to testify on behalf of the prosecution.[69]

On September 22, 1980, Bookchin entered the federal courthouse. He had to walk past hundreds of FBI rank and file who were lined up in military formation, loyal to their two chieftains, scowling at him. When he took the stand, Felt's and Miller's attorneys produced exhibits that made his hair stand on end, photos of his personal and political papers of all kinds: the *Anarchos* mailing list; the address book that he had compiled in Europe in 1967, with contact information for expatriate Spanish anarchists; his list of anarchist contacts from the 1969 Chicago SDS conference, where the Weather Underground was born; and more. This last one they offered as compelling evidence of his connection to the Weather people.

But Bookchin replied calmly, "All of these factions were Marxist, Leninist, and I believe Maoist."[70] His own faction, the Radical Decentralist Caucus, he said, "was adamantly opposed to them." The defense lawyers, unaware of any distinction between anarchists and Marxists, were taken aback. Are you saying, one asked, "that the people that are on this list and attended this caucus would not be people that were affiliated or supporting the RYM 1 or the Weatherman Faction of the SDS?" Bookchin replied, "I would hope that they weren't, otherwise they were in the wrong faction."

But look here, said one bewildered lawyer. Your so-called anarchist list contains the name Jeff Jones. And Jeff Jones, as we all know, went on to become a leader of the Weather Underground.

Bookchin replied that there were two people named Jeff Jones. The one on his list lived in Austin and was a student leader at the University of Texas. "He is not . . . the Jeff Jones who was normally associated with the Weathermen."

A defense attorney asked him questions about Weather people. Had he ever been at a meeting with Cathy Wilkerson or Kathy Boudin? ("I was never at any meeting of the Weathermen or Weather Bureau, whatever they choose to call themselves"). Were you ever at a meeting attended by Mark Rudd? ("Oh hell no—excuse me, I'm sorry.") Do you know a Leibel Bergman? ("A what?")

Next, a government lawyer questioned him. "You are neither a friend or a relative of any Weatherman fugitive, isn't that correct?" *Yes.* And "if the indictment in this case alleges that you are a relative or an acquaintance of a Weatherman fugitive, it is just flat wrong, isn't it?" "I would say so," Murray replied. Across the courtroom, John Nields teared up.

Had he ever written about the Weathermen? Murray said no, but he had written often against the Marxism that was their ideology. His "Listen, Marxist!" was "vehemently anti-Marxist, anti-Leninist, anti-Maoist."

"Mr. Bookchin, were your views on that subject well known in the community in 1972 and early 1973?"

"Very well known."

"I have no further questions, your honor."

By this time, Nields was sobbing tears of joy—the trial had barely begun, and the defense's case was in debris. During that recess, a defense lawyer actually went over to the government lawyers and congratulated them.[71]

Bookchin returned to Burlington, and the trial continued for another few weeks. Five former attorneys general, as well as former president Nixon, testified on behalf of Felt and Miller.[72] On November 6, the jury returned its verdict: Felt and Miller were guilty of violating citizens' constitutional rights. It was a victory for the Fourth Amendment and the Bill of Rights. Nields told Bookchin that his testimony had been primary in gaining that verdict.

At the 1980 summer session of the ISE, student Leo Chaput constructed a solar heating system that could warm a house as well as an aquaculture system. Introducing algae into the system's water made it dark green, which boosted the water's heat absorption, which made it warm enough to support fish. He raised fish and used their waste as fertilizer for the gardens. He raised worms for fish food, in a medium containing rabbit droppings, and he raised the rabbits and grew the rabbit food in the solar greenhouse. "The whole thing is interconnected," he told an interviewer.[73]

Meanwhile, Bookchin kept the goals of social ecology clear: it wasn't about the technology alone; it was about getting rid of the domination of nature by eliminating the domination of human by human.

For several years now, Ynestra King had been teaching her Women and Ecology class (figure 9.2) at the ISE, featuring a version of cultural feminism known as "ecofeminism" (a term Bookchin invented). According to ecofeminism, the domination of women is connected to the ecological domination of the planet. The patriarchal society that demeans women is the same society that has brought about acid rain and global warming. Strikingly, patriarchy demeans women in terms of nature: it regards them as "closer to nature" than men, as more "natural," less rational, and so on. In the age of ecological resistance, ecofeminists decided to recast the demeaning patriarchal analogies between women and nature in a positive light and present women as ecologically enlightened. Emotionality, compassion, peacefulness—qualities associated with women—are a necessary part of the solution to the ecological crisis, ecofeminists argued. If women are more "natural" than men, they said, that is not a defect but a boon—it means they can mediate our society's return to nature. The perpetuation of gender stereotypes might be troubling, but King and other ecofeminists argued for embracing this imagery for positive political purposes.[74]

FIGURE 9.2 Women and Ecology was a popular program at the ISE in the late 1970s and early 1980s. Poster ca. 1980.
Courtesy Goddard College Archives.

In March 1980, King and her colleagues joined together to form Women for Life on Earth to oppose nuclear power and nuclear weapons. Identifying warmaking as male and patriarchy as militaristic, they argued that women had a strong interest in and link to peacemaking. On November 17, 1980, two thousand female activists assembled in Washington for a four-stage Women's Pentagon Action. The Pentagon, their statement read, is "the workplace of the imperial power which threatens us all." Using large puppets and cardboard tombstones and brightly colored yarn, banging on cans and chanting "Take the toys from the boys," they encircled the Pentagon, hand in hand, singing peace songs.[75]

With its defiant radical energy and imaginative symbolism, ecofeminism impressed Bookchin. Its proponents, he said, were now at "the forefront of anti-militarist as well as ecological activities," demanding an end to hierarchy and domination in all their varieties. Ecofeminism, if fused with an anarchist movement, he thought, could "open one of the most exciting and liberating decades of our century."[76]

But what anarchist movement? Since no dedicated eco-anarchist organization existed, Bookchin set out to create one, with a broad goal of eradicating hierarchy and domination. In the summer of 1980, he and several northern Vermont affinity groups invited other groups throughout the region to join them in creating a New England Anarchist Conference (NEAC). It would be structured, once again, as a confederation of affinity groups, in which decisions were made from the bottom up; it would focus on creating "direct personal and collective empowerment" at the local level so that as the movement grew in numbers and strength, citizens would have the consciousness and experience necessary to manage their own communities through popular assemblies, leading to "a truly democratic and non-hierarchical society."[77]

By now, Bookchin was wary of movements whose members were uneducated in radical social theory and history. He had seen SDS lose its moorings when the hitherto nonpolitical hippies had flooded in, and he had seen the antinuclear movement, once it became popular, deteriorate into alternative energy fairs and rock concerts. To his mind, the movement NEAC spawned could remain small, at least initially: the number of its members was less important to him than their quality as activists and teachers. Indeed, he actively preferred a handful of members to a large movement, for only members who were committed and educated would be capable of moving history forward.[78]

More than 175 anarchists from New England, New York, New Jersey, and Quebec attended NEAC's founding conference, held on October 17–19, 1980, at Goddard. They discussed anarchism in relation to ecology, feminism, militarism, and urban alternatives and adopted the proposed structure and resolves. It was a promising start. In January 1981 the second NEAC conference convened, in Somerville, Massachusetts, but the conference showed itself to be not up to the task of creating a movement. The 125 participants "got lost in identity politics," as Brian Tokar told me. Bookchin would have to look elsewhere for the kind of educated activists necessary to create the movement he had in mind.[79]

* * *

After the ISE's summer 1980 session ended, Bookchin resigned as director, saying he'd had enough of dealing with the Goddard administration and wanted more time to write. Dan Chodorkoff, who had by now earned his Ph.D. in cultural anthropology at the New School, took the institute's helm (figure 9.3).

At that moment Goddard was having serious financial troubles. Declining enrollment was one reason: the kind of self-designed, independent study that Goddard had pioneered could by 1980 be found at other schools, at lower cost. But the college had also mismanaged its finances—it had a $1 million debt, and the bank was threatening to foreclose unless it produced a balanced budget. Rumors flew that the school was going to lose its accreditation.[80]

The ISE itself, however, was going strong. "There's nothing like it in the world," Bookchin told a reporter around this time. The ISE "has been imitated," agreed Chodorkoff, "but never really equaled." If Goddard were to go under, the ISE could continue—its finances were in the black. Perhaps it could even buy Cate Farm and establish a living-learning program there.[81] In December 1980, Chodorkoff let Goddard know that the ISE was interested in purchasing Cate Farm, and negotiations began.

In early 1981, Goddard teetered on the brink. In May its accreditation went on hold. So its board members bit the bullet: they sold its off-campus programs to a

FIGURE 9.3 Dan Chodorkoff, who cofounded the Institute for Social Ecology with Bookchin, in 1980 took over as its director. Ca. 1985.
Mark Ivins.

nearby university, laid off most faculty and staff, and reduced the student body by 80 percent. And they decided to sell off some of its 450 acres.[82]

The ISE negotiated with the bank for a mortgage, but in the midst of the negotiations, Goddard suddenly decided the interest rate was too high and terminated the discussion. Instead of selling it to the ISE, it put Cate Farm, along with other land parcels, on the market. It canceled the ISE's 1981 summer program, two weeks before it was to begin. It seized the ISE's equipment and put the buildings and other structures under lock and key.

Shocked, the ISE asked a local wealthy friend to buy Cate Farm and lease it to them. The buyer agreed, and in September 1981 his group made a 10 percent down payment. But once the sale went through, he ignored the agreement with the ISE and established a private farm on the land.[83]

On September 30, 1981, its collective radical heart breaking, the ISE bade farewell to the solar panels, the Sunhouse, the geodesic dome, the windmills, the compost shed, the solar food dehydrator, and the thermosiphoning air panel.[84] It bade farewell to Cate Farm, the mountains, and the natural sauna. The ISE was now homeless.

From the YCL's ideology of proletarian insurgency of the 1930s to Weber's theory of retrogression in the 1950s to the revolutionary youth of 1968–69, Marxists had been predicting that the demise of capitalism in the industrialized West was imminent. But looking back in retrospect from 1980, Bookchin saw that capitalism in those years had only just been beginning to consolidate itself. Far from collapsing from some internal contradiction, and far from being toppled by a mass movement of the oppressed, it had flourished, boomed, explosively, even globally. It had mocked every prediction that Marxists had flung at it since the end of World War II. Now it was already so consolidated that it had even become more than an economy—it had become both socially and psychologically a "given."[85]

That being the case, Bookchin had to admit that the revolutionary era, too, had almost certainly come to an end. The era of proletarian revolution had begun on the barricades of June 1848, continued through the Bolshevik Revolution in 1917, and reached its zenith in the Spanish Revolution of 1936—the greatest of them all. But it had ended on the barricades of Barcelona in May 1937, the "last great historic confrontation."[86] Castro in Cuba, Mao in China, Ho in Vietnam, so admired by the Left, had been peasant revolutions aiming for modernization, not proletarian ones aiming for socialism. The 1960s upheaval had been furious and turbulent, but it had not been genuinely revolutionary. During the 1970s, revelations about Marxist bloodbaths in China and Cambodia and elsewhere had redefined revolution from potentially liberatory to actually catastrophic. But the greatest desecrator of the revolutionary remained Stalin, who had made the very idea of revolution fearsome, synonymous with dictatorship, murder, and gulags. The red flag had been "stained more by the blood of its own revolutionary bearers than its reactionary opponents."[87]

Yet despite its triumph, capitalism continued to gnaw away at the biosphere, continued to undermine the foundations of all life on the planet. It continued to exploit workers, and it continued to inject itself into every aspect of social life, commodifying even interpersonal relations, personal character structure, demeaning people from citizens active in their communities to isolated consumers. The election of Ronald Reagan in November signaled that individualism had triumphed as the national ideology. Nothing requiring collective action or consciousness—let alone environmentalism, and let alone social ecology—would be on the agenda anytime soon.

Bookchin consoled himself that his own failure to create a social ecology movement to this point did not negate the problem at hand or invalidate social ecology as an idea. "To be in the minority is not necessarily testimony to the futility of an ideal," he observed. On the contrary, a humanly scaled, communal, ecological society remained as necessary as ever to avert ecological catastrophe, regardless of circumstances.

Contrary to generations of radicals, capitalism would not perish from internal contradictions, he concluded. If it were to die, it would die only "like a cancer"—that is, only when, having grown out of control, like a tumor, it "destroys its host," the host being society and the biosphere. And when it perished, it would take the rest of humanity, and probably most other species, down with it.

The only possible alternative was for an eco-anarchist movement to arise that would "corrode, weaken, and hollow [it] out" by warning humanity of the social and ecological peril it faced and by inducing the majority of people to take action and create an ecological, decentralized, and rational society.[88]

But the end of the revolutionary era deprived social ecology of the activist tradition upon which he had always thought a social ecology would depend. Given that reality, he knew he must dwell "on the margins of experience and practice" and try to make the best of it, "even as the center seems triumphant." He told people, "I'm one of those characters that lives on the dark side of the moon."[89]

From the margins, he could still demand the impossible, demand utopia. And untethered from current developments, he need make no apologies for it. He comforted himself, too, with a quote from Marx's *Eighteenth Brumaire*: "Men make history, but not under conditions of their own choosing."

Still, quietly, he made a subtle shift in his writing. In the past he had often distinguished between the *what is* and the *what could be*. The phrase *what could be* implies a possibility of practical attainment, but now he no longer used it. Instead, he replaced it with *what should be*—a moral standpoint, making no reference to the practical attainment. Living on the margins would mean writing in the subjunctive.

NOTES

1. Quoted in Nancy Jack Todd, "No Nukes," *Journal of the New Alchemists*, no. 6 (1980), 18. I'm guessing that Mead's visit to the ISE was in 1977, but it could have been a year earlier. (Mead died in 1978.) The story is also told in John Todd and Nancy Jack Todd, *Tomorrow Is Our Permanent Address: The Search for an Ecological Science of Design as Embodied in the Bioshelter* (New York: Harper & Row, 1980), 40–41.

2. Federation of American Scientists and Washington and Lee University, *The Future of Nuclear Power in the United States* (2012), http://bit.ly/1cMjVoe; Marc Mowrey and Tim Redmond, *Not in Our Backyard: The People and Events That Shaped America's Modern Environmental Movement* (New York: William Morrow, 1993), 169.

3. "Clamshell Alliance," founding statement, July 1976, Box 1, Clamshell Alliance Papers, Milne Special Collections, University of New Hampshire, Durham (hereafter CPUNH).

4. Ibid.

5. Barbara Epstein, *Political Protest and Cultural Revolution: Nonviolent Direct Action in the 1970s and 1980s* (Berkeley: University of California Press, 1991), chap. 2.

6. Mowrey and Redmond, *Not in Our Backyard*, 171.

7. Etahn Cohen, *Ideology, Interest Group Formation, and the New Left: The Case of the Clamshell Alliance* (New York: Garland, 1988), 119, 141 n. 20.

8. Mowrey and Redmond, *Not in Our Backyard*, 171.

9. Cohen, *Ideology, Interest Group Formation*, 109.

10. Bookchin, "Anarchism," *WIN Magazine*, May 1, 1971, 20; CPUNH.

11. As Jo Freeman warned in her much-circulated "The Tyranny of Structurelessness," http://bit.ly/1F3UbXX.

12. Bookchin, *The Spanish Anarchists: The Heroic Years, 1868–1936* (New York: Free Life Editions, 1977). When A.K. Press republished the book in 1998, someone at the press, without consulting Murray, removed the dedication to Blackwell and replaced it with another dedicatee. In fact, Blackwell was the only person to whom Murray Bookchin ever dedicated this book.

13. Mowrey and Redmond, *Not in Our Backyard*, 171; Epstein, *Political Protest*, 66.

14. Brian Tokar, interview by author, July 18, 2009.

15. Howie Hawkins, "Remembering Murray Bookchin," March 18, 2013, unpublished, author's collection.

16. Bookchin, "The Seabrook Occupation . . ." *Liberation*, July–Aug. 1977, 7, 8; "The Plot Thickens: MUSE and the Demise of the Clamshell Alliance," *It's About Times*, Aug. 1981, 7.

17. Jeanne Major and Wes Miles, "The Institute for Social Ecology: A Master Plan for Land Use: The Cate Farm," master's thesis, Goddard College, 1981; Gloria Goldberg, "Three Techniques for Organizational Development: A Case Study of Intervention at the Institute for Social Ecology," master's thesis, Goddard College, 1982, III-9; Hess quoted in Leslie Sproule, "The Universe Is Coming Together," *Wanekia: The Cate Farm Journal* (Plainfield, VT: Institute for Social Ecology, 1977), 29.

18. Barry, Doug, Joe, Sally, Bob, "Steel Sandwich Plate Collector," *Wanekia*, 61; "Aquaculture Collector," *Wanekia*, 61; Alvin, "Greenhouse Aquaculture," *Wanekia*, 49; "Thermosiphoning Hot Water Collector," *Wanekia*, 62–63; "Solar Test Facility," *Wanekia*, 86. The Savonius rotor was hooked up to a 12-volt generator for lights and tools in the workshop. See Judith Edwards, "Pioneering for the Future: The Social Ecology Institute," *Vermont Summer* [magazine], *Bennington Banner*, Aug. 27, 1976; Bookchin, "The Concept of Ecotechnologies and Ecocommunities" (1976), in *TES*, 108; Dan Chodorkoff, interview by author, May 12, 2009; Maia, Rob, et al., "Sailwing Wind-Power System," *Wanekia*, 73.

19. Leslie Sproule, "French Intensive Beds," *Wanekia*, 36; Calley O'Neill, "Maia and Calley's French Intensive Bio-dynamic Bed," *Wanekia*, 37; Major and Miles, "Institute for Social Ecology"; Leslie Sproule, "Cabbages and Companions," *Wanekia*, 39; Ron Zweig, "Solar Aquaculture: Historical Overview," *Journal of the New Alchemists*, no. 6 (1980), 93–95; "The Solar Shield Backyard Garden," CalleyONeill.com, http://bit.ly/1JjSoxo (with map and

photos); Calley O'Neill, interview by author, Nov. 21, 2009; Goldberg, "Three Techniques for Organizational Development," III-8, 9; Phil, Leslie, and Jay, "The Perfect Cold Frame," *Wanekia*, 87.

20. O'Neill interview.

21. Joseph Kiefer, interview by author, Feb. 15, 2010.

22. Barry Costa-Pierce, interview by author, June 6, 2009.

23. Geri Berger, "The Best I Ever Had," *Wanekia*, 104.

24. Kiefer interview.

25. Jim Morley, interview by author, July 12, 2009; Thomas Pawlick, "A Return to First Principles," *Harrowsmith* 43 (1982), 35. Around this time, a student, Jane Coleman, became enamored of Murray and moved in with him. For a professor to live with a student was a breach of ethics, so he married her in 1978. Their personalities were mismatched, and they divorced in 1980. Coleman declined to be interviewed for this book.

26. "Interview with Murray Bookchin," *Open Road* [Vancouver], Spring 1982.

27. Wayne Hayes, interviews by author, June 9 and Aug. 25, 2009; George T. Potter to Bookchin, Mar. 10, 1977, MBPTL.

28. Epstein, *Political Protest*, 195.

29. Ibid., 73.

30. "Seabrook 78: A Handbook for the Occupation/Restoration Beginning Jun. 24," Box 8, CPUNH; Tokar interview; Bookchin, "The Seabrook Occupation," 8.

31. George Woodcock, "The Libertarian Virtues," *TLS*, Apr. 28, 1978.

32. Clamshell Alliance Coordinating Committee Minutes, Feb. 19, 1978, CPUNH.

33. Mowrey and Redmond, *Not in Our Backyard*, 197–99; Bookchin, "Plot Thickens," 7.

34. Beer and Parsons, "Clamshell Alliance Fights," *Ubyssey*, Sept. 28, 1978, online at http://bit.ly/9ZFI1v (accessed March 2014).

35. Bookchin, "Plot Thickens"; Bookchin, "Future of the Anti-nuke Movement," *Comment* 1, no. 3 (1979), 4.

36. Mowrey and Redmond, *Not in Our Backyard*, 200; Thomson quoted in Beer and Parsons, "Clamshell Alliance Fights."

37. Bookchin, "Future of the Anti-nuke Movement," 1.

38. Spruce Mountain Affinity Group, "On Process and Policy: A Position Paper by the Spruce Mountain Affinity Group of the Green Mountain Alliance, Plainfield, Vermont," July 1, 1978, MBPTL and author's collection. The proposal was signed by Bookchin and nine others. The quotation is from MBVB, part 38.

39. Agreed July 1978, Amherst: "We affirm that the affinity group is the fundamental unit of the Clams for Democracy network. These affinity groups will be organized into a network of interlocking relations. . . . There will be no superstructure built over the affinity group base. Clams for Democracy decisions can only be made at conferences of the entire caucus. There can be no decision of any kind made in the name of this caucus without the approval of the affinity groups making up the caucus." "Clams for Democracy: Points of Agreement," *Clam Journal* 1, no. 1 (Jan. 1979) (Burlington), Box 2, CPUNH; Tokar interview.

40. Cohen, *Ideology, Interest Group Formation*, 296; Red Balloon, "Got a Dime?" ca. Jan. 1979, Box 2, CPUNH. In this document Red Balloon defines itself as "Marxist-Leninist in orientation." See also Beer and Parsons, "Clamshell Alliance Fights."

41. John Lepper to author, June 1, 2010; Tokar interview.

42. Cohen, *Ideology, Interest Group Formation*, 175–76. See William E. McKibben, "Seabrook Protest—a Victory of Sorts," *Harvard Crimson*, Oct. 13, 1979, online at http://bit.ly/pCoBF7.

43. Goldberg, "Three Techniques," III-8; Barry A. Pierce, "Water Reuse Aquaculture Systems in Two Solar Greenhouses in Northern Vermont," *Proceedings of the World Mariculture Society* 11 (1980), 118–27; John Wolfe, "The Energetics of Solar Algae Pond Aquaculture," *Journal of the New Alchemists* 6 (1980), 106–7; Daniel Elliot Chodorkoff, "Un Milagro de Loisaida: Alternative Technology and Grassroots Efforts for Neighborhood Reconstruction on New York's Lower East Side," Ph.D. diss., New School for Social Research, March 1980, 155ff. *Karl Hess: Toward Liberty* won the Academy Award for best documentary on a short subject in 1980.

44. "Hess, Bookchin Team up at NJLP convention," *New Jersey Libertarian*, Jan. 1980, 3; Bookchin, interview by Jeff Riggenbach, *Reason*, Oct. 1979, 34–38.

45. Dimitri Roussopoulos, interview by author, July 6, 2009; Timothy Lloyd Thomas, *A City with a Difference: The Rise and Fall of the Montreal Citizen's Movement* (Montreal: Véhicule Press, 1977), 67, 38; "Rassemblement de citoyens de Montreal, Rapport du comité de stratégie électorale, mars 1978," quoted in Marc Raboy, "The Future of Montreal and the MCM," *Our Generation* 12, no. 4 (Fall 1978).

46. Karen Herland, *People, Potholes, and City Politics* (Montreal: Black Rose Books, 1992), 12–13.

47. Harry C. Boyte, *The Backyard Revolution: Understanding the New Citizen Movement* (Philadelphia: Temple University Press, 1980); Milton Kotler, "The Politics of Community Economic Development," *Law and Contemporary Problems* 36, no. 1 (Winter 1971), 3–12.

48. Amory Lovins, *Soft Energy Paths: Towards a Durable Peace* (San Francisco: Friends of the Earth, 1977); Bookchin, "Notes and Reflections of an Old New Yorker," in *Raise the Stakes* (1981), 7. *Raise the Stakes* was a publication of the Planet Drum Foundation in San Francisco.

49. Bookchin, "Self-Management and New Technology" (1979), in *TES*, 131.

50. Bookchin, "Toward a Vision of the Urban Future" (1977), in *TES*, 179. On the failure of twentieth-century urban decentralism, see Mark Luccarelli, *Lewis Mumford and the Ecological Region: The Politics of Planning* (New York: Guilford, 1995).

51. Bookchin, "An Open Letter to the Ecology Movement" (1980), in *TES*.

52. Bookchin, *The Ecology of Freedom: The Emergence and Dissolution of Hierarchy* (Palo Alto, CA: Cheshire Books, 1982), 3.

53. Michel Saint-Germain, "La liberté de chacun: Un café avec Murray Bookchin," *Mainmise* [Montreal], no. 76 (1978), 30–31.

54. Bookchin, "Concept of Ecotechnologies," 101.

55. Bookchin, "Social Anarchism or Lifestyle Anarchism," in *Social Anarchism or Lifestyle Anarchism: An Unbridgeable Chasm* (San Francisco: A.K. Press, 1995), 51.

56. Karl Marx, *Grundrisse*, Notebook IV, mid-Dec. 1857 to Jan. 22, 1858, online at http://bit.ly/qSanfP.

57. Friedrich Engels, "On Authority," quoted in Bookchin, "Self-Management and New Technology," 127. See also Bookchin, "Marxism as Bourgeois Sociology" (1979) in *TES*, 196, 209.

58. See Bookchin, "Introduction" (1979), "Spontaneity and Organization" (1971), "On Neo-Marxism, Bureaucracy, and the Body Politic" (1978), and "Marxism as Bourgeois Sociology" (1979), all in *TES*; quote at 209.

59. Carl Boggs, "The Life and Times of an American Anarchist," MBPTL and author's collection. The article referred to is "Anarchism: Past and Present," lecture delivered to the Critical Theory Seminar of the Conference on Marxism and Anarchism, University of California at Los Angeles, May 29, 1980; published in *Comment* [n.s.] 1, no. 6 (1980).

60. Bookchin, "Future of the Anti-Nuke Movement," 1–8.

61. Bookchin, "Open Letter," published as "The Selling of the Ecology Movement," *WIN Magazine*, Sept. 15, 1980; responses by Pete Seeger, *WIN Magazine*, Dec. 1, 1980, and Dave McReynolds, *WIN Magazine*, Dec. 15, 1980.

62. Bookchin, reply to Seeger, *WIN Magazine*, Feb. 1, 1981, 33.

63. Bookchin, "Introduction," 12; "Open Letter"; "Statement of Purpose," *Comment*, Jan. 14, 1979, 6.

64. Bookchin, "Introduction," 11, 20; "Statement of Purpose," 5.

65. Bookchin, "Introduction."

66. John Clark, review of *Ecology of Freedom*, *Telos* 57 (Fall 1983), 226ff.; Bob Long, "An Important Source for Social Ecologists," *Synthesis: A Newsletter for Social Ecology* [San Pedro, CA], no. 8 (May 1981); John Fekete, "Ecology as Master Concept," *CAUT Bulletin*, Dec. 1981.

67. Chodorkoff interview. The old school was P.S. 64, at 605 East Ninth Street, between Avenues B and C.

68. "Declassified U.S. Government Intelligence Information, Regarding the Communist and Foreign Connections of the Weather Underground, presented as Evidence . . . in the Trial of W. Mark Felt and Edward S. Miller," MBPTL and author's collection. Adding to the confusion, Larry Weiss, a Weather Underground member whose name is mentioned in the indictment, was also the name of an academic colleague of Murray's at Ramapo.

69. The case was *United States v. W. Mark Felt and Edward S. Miller*, US District Court, Washington, DC, Sept. 1980, transcript, author's collection and MBPTL.

70. The quotes in these paragraphs are from the transcript at 1776–1806.

71. On Bookchin's testimony, see Gregory Gordon, "Prosecution Witnesses Testify at FBI Trial," UPI, Sept. 27, 1980; "Professor Testifies at FBI Trial," UPI, Sept. 27, 1980.

72. Laura A. Kiernan, "Weatherman Link to Foreign Power Held a 'Judgment Call,'" *Washington Post*, Sept. 20, 1980; and Laura A. Kiernan, "Weatherman Prober Grilled by Prosecution," *Washington Post*, Sept. 25, 1980; Robert Pear, "Testimony by Nixon Heard in F.B.I. Trial," *New York Times*, Oct. 30, 1980. On April 15, 1981, the newly elected president, Ronald Reagan, issued a pardon for the two convicted FBI directors.

73. The 1980 program ran from June 2 to August 22. Murray was still director and taught social ecology; the visiting lecturers included Peter Barry Chowka, Eugene Eccli, Karl Hess, Richard Merrill (of New Alchemy Institute West), feminist anthropologist Rayna Reiter, Lee Swenson (of Farallones), John Todd and Nancy Jack Todd (of New Alchemy), among others. Advertisement for ISE in *Mother Jones*, Apr. 1980; Robert Palmer papers, courtesy Bob Erler. See also "Goddard College's Roots Deep in Vermont History," *BFP*, Sept. 4, 1980. On the specific solar and wind installations, see Goldberg, "Three Techniques," III-8; Major and Miles, "Institute for Social Ecology." On Leo Chaput, see Neil Davis, "Social Institute Could Survive Goddard," *BFP*, Jan. 25, 1981, 1a.

74. Ynestra King, "Toward an Ecological Feminism," in Judith Plant, ed., *Healing the Wounds: The Promise of Ecofeminism* (Gabriola Island, BC: New Society, 1989), 23.

75. "Women and Life on Earth: A Conference on Eco-Feminism in the 1980s" was held at the University of Massachusetts at Amherst in March 1980. Six hundred female antinuclear and environmental activists discussed the links between militarism and patriarchy. See Rosemary Skinner Keller, Rosemary Radford Reuther, and Marie Cantion, *Encyclopedia of Women and Religion in North America* (Bloomington: Indiana University Press, 2006), 3:1057; Harriet Hyman Alonso, *Peace as a Women's Issue: A History of the U.S. Movement for World*

Peace and Women's Rights (Syracuse, NY: Syracuse University Press, 1993); Epstein, *Political Protest*.

76. Bookchin, "New Social Movements: The Anarchic Dimension," in David Goodway, ed., *For Anarchism: History, Theory and Practice* (New York: Routledge, 1989), 266; "Open Letter," 82.

77. "NEAC Manifesto," in "Anarchists Unfurl the Black and Green," *Open Road*, Summer 1981, 13.

78. "Policy Statement Resolves" (unsigned), c. 1981, unpublished ms., 4, MBPTL and author's collection.

79. Brian Tokar to author, June 30, 2009.

80. Marilyn Adams, "Goddard College Struggles to Survive," *BFP*, Jan. 4, 1981, 1B. Some accounts say Goddard was $3 million in debt, e.g., Susan Green, "Goddard's Rebound," *BFP*, Oct. 14, 1982, D1.

81. Neil Davis, "Social Institute Could Survive Goddard," *BFP*, Jan. 25, 1981, 1A.

82. Goldberg, "Three Techniques," III-6; Norma Jane Skjold, "Goddard Barely Survives the Eighties," *Vermont Vanguard Press*, Sept. 4–11, 1981.

83. Goldberg, "Three Techniques," III-18.

84. David Goska, "Graduates Experiment in Farming," *BFP*, Feb. 7, 1982.

85. Bookchin, "Were We Wrong?" *Telos*, no. 65 (Fall 1985), 63.

86. Bookchin, "Reflections on Spanish Anarchism," *Our Generation* 10, no. 1 (1974), 24.

87. Bookchin, "1984 and the Specter of Dememorization," in Marsha Hewitt and Dimitrios Roussopoulos, eds., *1984 and After* (Montreal: Black Rose Books, 1984), 36. See also "Between the Thirties and the Sixties," in Sonia Sayres et al., eds., *Sixties without Apology* (Minneapolis: University of Minnesota Press, 1984).

88. Bookchin, "Were We Wrong?" 70–71.

89. Bookchin, "Utopianism and Futurism" (1979), in *TES*, 277; "Cities: Salvaging the Parts," *Planet Drum* 1, no. 3 (1979), 13.

10

Municipalist

⁂―――――――――――――――――――――――――――――

IN 1980 BOOKCHIN'S friend and translator Karl-Ludwig Schibel organized a speaking tour of West Germany for him in January 1980, and when he set out, his hopes for the political scene there may not have been high. But as he moved through Frankfurt, Kassel, Hanover, and West Berlin, he found it to be buzzing with action and hope. In Europe the postwar baby boom had begun somewhat later than in the United States, and now as American youth had done ten years earlier, many European young people were rejecting consumerism and power politics, militarism and the patriarchal family, racism, and environmental destruction, and were seeking to form a communal, cooperative, nonhierarchical culture as an alternative.[1]

Throughout the 1970s, citizens' action groups in West Germany had been organizing around issues of the environment, gender equity, minority rights, housing, social services, and public transportation. After the successful site occupation at Wyhl in 1975, opposition to nuclear power had spread. "We weren't just trying to do symbolic actions," recalled the socialist journalist and activist Jutta Ditfurth. "We were trying to actually block the construction of the plant."[2]

Now in Frankfurt a culture of political protest was emerging. The international airport was planning to build a new runway extension through an old-growth forest—and activists were opposing it. The chemical conglomerates Hoechst and Merck had buried toxic waste near city water supplies and dumped pollutants straight into the Main and Rhine rivers, so demonstrators took to the streets. Faced with a housing shortage, students and youth squatted abandoned houses communally and fought off police attempts to evict them. Frankfurt bookstores, like the Karl Marx Bookshop, spilled over with volumes of radical theory and analysis, including Bookchin's, which had been published in German translation since 1974–75. In Germany, it seemed, Bookchin need not live on the dark side of the moon after all.

In December 1979, NATO had announced that it intended to station more than two hundred Pershing II and Cruise missiles on West German soil, nuclear-capable missiles able to reach Moscow in minutes. If a nuclear war broke out between the United States and the Soviet Union, West Germany would be the battlefield—yet these terrifying weapons would be outside any effective democratic control. Instantly, a concentrated peace movement sprang to life.

But no one was trying to tell this movement to concentrate on nuclear weapons alone, Bookchin realized, the way the Clamshell reps had tried to tell him to concentrate on nuclear energy alone. On the contrary, these young Germans understood that nuclear weapons were merely a symptom of a dysfunctional society. They were creating the kind of broad, multi-issue, antihierarchical movement that he had hoped that Clamshell and NEAC could generate, linking issues of ecology, feminism, energy, and peace. Their movement, moreover, was decentralized, emphasizing grassroots democracy in structure and direct action in practice. And finally, German youth seemed to have a greater appreciation for social theory than did pragmatic Americans, which made their movement all the more impressive. To them it was axiomatic that ecological problems stemmed from social problems—hence their movement was tantalizingly social-ecological. It was close to what Bookchin had been looking for in the United States.[3] Perhaps he could mediate a cross-cultural interchange, he thought, combining German theoretical rigor with the American libertarian impulse.

Due to his teaching commitments at Ramapo, Bookchin could not linger in Europe. But in the spring of 1981, when he had a sabbatical, he returned and spent several weeks traveling through West Germany, Belgium, the Netherlands, Switzerland, and Italy, once again soaking up the radical scene.

Squats were important nodes around which this European youth movement was forming, and the squatters' movement, a loose network of communes, had evolved into a struggle for free, alternative spaces generally. In Freiburg and Zurich, squatters had formed autonomous youth centers that were "basically anarchistic in character." At weekly meetings at the Zurich youth center, the microphone was open and people could "just walk up . . . and say whatever they want to say," without a moderator, yet the meetings were functional. The circled A, a symbol of anarchism, was ubiquitous. Impressed, he told the communal squatters about the America libertarian tradition that was grounded communally, in town meetings, dating back to the Revolution.[4]

Strikingly, this admirable youth movement was based not in large urban centers but in small cities like Freiburg and Nuremberg. Such cities, he realized, still had "some sense of community"; one could be politically creative there, in a way that one could not in big cities like New York or Frankfurt. This robust European scene helped him realize that living in small-scale Vermont didn't have to mean giving up political life or being sidelined to margins: the margins could have an unexpected importance as the birthplace of "the rich variety of forms, sensibilities, and institutions that are likely to supplant and transcend the given 'centers' of today."[5] Living far from an urban core, forty miles from the Canadian border, could thus actually be an advantage, opening up new possibilities for action. His grandmother Zeitel, after all, had chosen to live on the margins, in the borderlands of the Russian empire, and had found a way to be politically creative there.

He returned to the United States to his teaching job at Ramapo, but he moved his primary residence to Burlington, where the political possibilities beckoned.[6] The commute was grueling—during the school year he would drive the 266 miles to Ramapo, check into a motel, teach his classes, then drive back to Burlington.[7]

But Vermont, with its political culture that emphasized civic participation and community, was worth it. Each of its 242 towns had a town meeting, which assembled on the first Tuesday in March every year. The Vermont town meeting, he thought, bore a striking resemblance to the *ekklesia*, the citizens' assembly of the ancient Athens. Vermont towns were small in scale, like the Athenian poleis, small enough "to be taken in at a single view," as Aristotle had prescribed. The Vermont citizens who governed themselves through the town meetings were, like ancient Athenians, amateurs at politics who met face to face and made decisions about the shared, communal elements of their lives. They too were a *demos* (people) who enjoyed *philia* (solidarity) and *koinonia* (community life). They shared, in Bookchin's eyes, a reverence for *aretē* (virtue) and *dikē* (justice). Their political experience taught them *politikē technē* (political judgment). And on certain matters, like local infrastructure and schools and budget, they enjoyed *autonomia* (self-rule) or independence from the state, possessing "direct, unmediated control of society."[8]

Ancient Greek political thought, imbued with such ethical concepts, was closely tied, he saw, to a nature philosophy that emphasized that the *kosmos* (natural world) had an orderly structure that was comprehensible by *nous* (mind) or *logos* (reason). A nature philosophy based on these ideas could "guide us toward a deeper sense of ecological insight into our warped relationship with the natural world," he thought—and would be far more relevant to Western ecology movements than Asian philosophy.[9]

Particularly in the work of the pre-Socratic philosophers, he found a sense of reality that was not only orderly and comprehensible but "pregnant, fecund, and immanently self-elaborating." The Marxists had trained Murray to think dialectically, in terms of unfolding and emergence. As he studied Aristotle's concepts of *dunamis* (potentiality) and *entelechia* (actuality, the fulfillment of potentiality), he recognized in them the roots of his familiar philosophy. His very cast of mind was dialectical: the source of his optimism was his ability to recognize potentialities for freedom and progress in both present and past. Human beings had the *dunamis* to create an eco-decentralist society, governed by assembly democracy, as its *entelechia*. Hayes recalled that, while carpooling with him to Ramapo, Bookchin "took you out of today's newspaper, and you entered a world of human potential that could be actualized and recovered. His attitude was, *Hey, this is real, it happened, it can happen again.*"[10]

To begin to fulfill Burlington's potentialities, Bookchin brought together a small group of anarchists to work with, individuals who were educated in social ecology and committed to putting it into practice. He found them in Clamshell veteran Brian Tokar, antinuclear activist Alan Kurtz, social ecology student David Block, and others who followed him to Burlington. Their study group read Bookchin's by-now-standard texts—Buber's *Paths in Utopia* and Horkheimer's *Eclipse of Reason* and *Dialectic of Enlightenment*—as well as current work in evolutionary theory and nature philosophy. At the same time they would get involved in local politics.

Burlington's mayor, Gordon Paquette, was intent on developing the city's waterfront. Through various quirks of history, the land on the shore of Lake Champlain had been reduced to an industrial dumping ground, home to oil storage tanks, an old grain mill, and warehouses, as well as debris and weeds and broken bottles.[11] (What CHARAS might have done with it!) In its nineteenth-century heyday, it had been a busy port, transshipping trees felled in Canada onto railcars. But that industry had vanished almost a century earlier, whereupon the waterfront—owned by the Central Vermont Railway—had fallen into disuse.

In 1971 Burlington's city council (then called the board of aldermen) formed a committee to decide what to do with the waterfront, which led to studies, development proposals, financing schemes, and zoning changes. Most recently, local developer Antonio Pomerleau had proposed to construct a high-rise luxury enclave at the water's edge, with three condo towers, as well as a marina, a 150-room hotel, restaurants and offices, a parking lot, and some shops.[12] Mayor Paquette, catering to the local business establishment, was solidly behind the $35 million project.

But many city residents demurred: Pomerleau's luxurious plan would create an enclave for the wealthy, at a time when the city lacked adequate housing for working people. A neighborhood movement sprang to life, in part to oppose it, and the battle would be joined in the next mayoral election.[13] Mayor Paquette would face the voters in March 1981. A thirty-nine-year-old writer, filmmaker, and socialist activist, Bernard Sanders, stepped forward to challenge him, running as the voice of the neighborhoods. Sanders opposed the Pomerleau plan and pledged, if elected, to "establish an alternative waterfront development policy—one which will bring jobs and prosperity for all of Burlington rather than a few wealthy individuals."[14]

Campaigning on the slogan "Burlington is not for sale," Sanders slogged door to door through five-foot-high snowdrifts. On March 4, 1981, Burlington elected him mayor—by a margin of ten votes out of more than 9,600 cast.[15] ("Ten *anarchist* votes!" Murray would say. "And I know who they were!")

While Burlington's political establishment reeled at the prospect of a socialist mayor in the Georgian-style city hall, the neighborhood groups rejoiced. Bookchin commended the socialist awakening in the city's political culture: "The Vermont citizenry seems to have risen up with a political vitality that we have not seen for years."[16]

The city's neighborhood movement was planning a conference to transform itself into a cohesive political force. In advance of the conference, neighborhood groups were invited to submit resolutions on a whole spectrum of social and environmental issues. Bookchin's anarchist friends, under his tutelage, submitted three resolutions. "Waterfront and Downtown Development" argued that the waterfront should be held in a public trust and developed in accordance with Burlingtonians' desires. "Energy and the Environment" urged the city to make more use of renewable energy.

The third resolution, "On Neighborhood Democracy," was the most remarkable. It proposed replacing the existing city council with "a system of open democratic neighborhood assemblies." Each of the city's six wards would form an assembly, to hold their respective city councilors accountable through mandate, rotation, and recall. If a councilor refused to submit to such accountability, the assembly would run its own candidate in the next election. "When a sufficient number of the Alderpeople [city councilors] represent the neighborhood assemblies, the city charter should be revised so that the board [council] is officially composed of mandated, recallable representatives of the neighborhood assemblies." The council's function would then be merely to coordinate the assemblies, which would constitute a neighborhood government.[17]

On May 16, 1981, the Conference of Neighborhoods endorsed these and thirty-odd more resolutions.[18] Burlington had gained not only a socialist mayor but a dynamic popular movement.

Since Bookchin had become a teacher in 1973, his book on hierarchy had languished in a state of incompletion. In the spring of 1981, after he came home from Europe, he used the rest of his sabbatical time to finish it. When it was complete, he gave the manuscript to his friend Michael Riordan to publish at the small press he had founded called Cheshire Books.[19]

The Ecology of Freedom was Bookchin's fullest exposition of radical social ecology, using material drawn from history, anthropology, dialectical philosophy, and science.[20] The first chapters, as we have seen, described an original "organic society," with a mutualistic, egalitarian social organization, then went on to trace the rise of hierarchies: gerontocracy, warriors, patriarchy. Domination was the next phase, with states, tyrannies, inquisitions, empires. The eighteenth-century Enlightenment had made the domination of nature part of the civilizatory enterprise: humanity, if it were to progress, must subdue and conquer nature.

Fortunately, the West also has a long-standing antihierarchical tradition, dating from uprisings in the ancient world to medieval heretics and peasant revolts to the democratic revolutions of England, the United States, and France, to the Paris Commune to the Spanish Revolution. The common, if unspoken, dream of all these movements was to revive the principles of organic society—usufruct, the irreducible minimum, and the principle of complementarity—but within the framework of modernity.

The liberation of humans from exploitation and domination, Bookchin argued, is a precondition for creating a society in harmony with nature. Only a free, emancipated society could create an ecologically sound planet. A social ecology movement would advance a moral economy as an alternative to capitalism, replacing competition with values of reciprocity and interdependence, of responsibility and integrity. It must seek to construct a cooperative economic life, producing not for profit but for excellence and for the sake of the community.

In *Ecology of Freedom*, supplemented by a series of articles in the early 1980s on nature philosophy and ethics, Bookchin rejected the dualistic idea, derived from Descartes, of a radical dichotomy between humans and the rest of nature and embraced instead the idea of a graded continuum, stretching from the simplest life-forms to human society, as described by evolutionary theory.[21] Human beings, like all other organisms, are an integral part of that continuum. Natural evolution (what Bookchin called "first nature") has also given rise to human society ("second nature"), and people necessarily inhabit that social environment as well as that biological one. With our symbolic faculties and our capacity for cooperation, *Homo sapiens* is uniquely social as well as natural; our behavior is conditioned not only by biology but also by society, language, psychology, and culture. Part of our dual nature is creative agency. Through our labor and our imagination, we are structured to interact with nonhuman nature, even to modify and transform it.

This distinctiveness, however, does not give us the right to subordinate the rest of nature or to wield dominion over the earth. Least of all does it give us the right to treat it as an assemblage of resources available for our use. Capitalism, with its market economy geared to producing goods for profit rather than need, views it this way, but that system is social not biological, a part of our second nature (culture), not first nature (our physiological makeup), and hence alterable. We are capable of changing that system and replacing it with a different system, one that relates to the rest of nature with respect and intelligence. And we can orient our use of science and technology toward humane and life-affirming purposes, rather than toward the enhancement of profit and environmental destruction.

Removing ourselves from the ideologies of domination and submission, we can cultivate an "ecological sensibility": respect for natural phenomena, sensitivity to the interdependence of life-forms, and a feeling of responsibility for the natural world. In so doing, we can potentially even improve on it, promoting the flourishing of people as well as the biosphere.

That sensibility must be underpinned by an ethics. In order to have substance, Bookchin believed, ethics must be grounded in tangible reality. What was to be the ground for an anticapitalist ethics? Moral relativism, with its fleeting, changeable values, offered none. Nor could religion, with its tablets of commandments, provide an ethical foundation: such authoritarian systems, demanding of their adherents unquestioning obedience to "Thou shalt not" injunctions, had no place in a modern, rational movement.

Not even the Frankfurt School had a good answer, in Bookchin's view. Its members thought instrumental rationality had reduced ethics to the utilitarian calculation of risks versus benefits, of greater evils versus lesser ones. But the rejection of rationality opened the door to the demonic. Horkheimer and Adorno were, Bookchin found, unable to anchor an emancipatory ethics in nature philosophy and ended with a "dark pessimism about the human condition."[22] As always, while he admired their analysis, he couldn't accept their pessimism.

The objective basis for ethics, he concluded, lay in nature itself. By that he did not mean simplistic parallels between natural and social phenomena, say, by reductively explaining human behavior in terms of biology—especially genes—rather than by considering the social factors at work. Historically, the privileged and powerful had tried to justify hierarchy—oligarchy, slavery, sexism, imperialism, and the state—and had dominated women, people of color, and colonized ethnic groups and nationalities by asserting that they were "naturally" inferior based on their biology. Social Darwinism had tried to justify capitalism by maintaining that since competition exists in nature ("survival of the fittest"), competition, rather than cooperation, must exist in society.

No, if nature were to be a ground for ethics, it would have to be understood not in terms of parallels but for its ontology: fecund and creative, ever-changing, evolving toward greater complexity, dialectical. Life-forms are not only competitive but cooperative—and their mutual cooperation and symbiosis have helped advance natural evolution at least as much as competition, if not more. According to the evolutionary biologist William Trager, "Mutual cooperation between different kinds of organisms—symbiosis—is just as important [as competition], and . . . the 'fittest' may be the one that most helps another to survive."[23] Cellular life itself may have begun with a symbiotic cooperation between viruses and bacteria, as the biologist Lynn Margulis showed.

But to assert that we must have a cooperative society because cooperation exists in nature would itself be a kind of reductionism: we must avoid "projections of our own social relationships into the natural world," Bookchin observed, even cooperative ones. Nature is neither cruel nor kind; it has no morality: "Nature is a ground for ethics, [but is] not ethical as such."[24]

Rather, Bookchin argued instead that mutuality, cooperation, and complementarity are potentialities in the natural world, ones that have historically been expressed and, one hopes, will continue to be. As a result of their expression, natural evolution has had a cumulative history of increasing complexity and diversity, toward ever more elaborate and conscious life-forms. Over eons, along that graded evolutionary continuum, creatures' neural and sensory systems became more differentiated, resulting in consciousness and culminating in the layered human brain.

Tantalizingly, the paleontologist Elisabeth Vrba's "effect hypothesis" suggested that evolution "includes an immanent striving," or directionality. That implied, Bookchin thought, that evolution could be self-directive, even "participatory."[25] He denied that this notion amounted to teleology, asserting that the directional process is neither inexorable nor preordained. Potentialities do not inexorably achieve actualization. "Potentiality is not necessity; . . . No specific stage of a process necessarily yields a later one."[26]

But as natural history shows, more neurologically complex organisms did evolve, and so did their capacity to make choices—and so did the possibility of freedom.

As the twentieth-century German philosopher Hans Jonas pointed out, even the most rudimentary organism, the cell, makes an effort to preserve itself—to maintain its identity—through metabolism, and cellular metabolism is evidence of "germinal freedom."[27] Human beings have the potentiality for both freedom and self-consciousness, not by analogy to natural evolution, not as a projection onto it, but by virtue of being a continuation of natural evolution.

Selfhood and reason and consciousness are products of natural evolution: in this respect, evolutionary biology accorded with dialectical philosophy, with Hegelian ideas of potentiality and actualization: "The history of the world," Hegel had said, "is none other than the progress of the consciousness of freedom."[28] Reinterpreted in terms of natural evolution, natural evolution seemed to Bookchin to constitute a dynamic nature philosophy.

And it also constituted the basis of an ecological ethics. The evolution of nervous systems and consciousness, of choice and freedom, are what Bookchin meant, in my opinion, when he said nature is a ground for ethics. His ethical *ought* is grounded in the objective reality of potentialities for freedom and consciousness. The purpose of an ecological ethics, he said, was "to help us distinguish which of our actions serve the thrust of natural evolution and which of them impede it." Human beings have the objective potential to bring natural evolution to a new level of freedom and consciousness; they are objectively capable of actualizing the potentiality of natural evolution to produce a rational, ethical, cooperative ecological society (which Bookchin named in advance "third" or "free nature"), although they may or may not actually do so. The German idealist philosopher Johann Fichte had remarked that humans are "nature rendered self-conscious";[29] Bookchin agreed only by adding the qualifier "potentially."

The Ecology of Freedom (as well as the philosophical essays that followed its publication) was dense and erudite, rich and sweeping, both polemical and speculative. It navigated multiple disciplines, connected them in original ways, and tied them all to concrete social praxis.

As the pages emerged from his Selectric typewriter, he shared them with the Burlington anarchists. "We would sit in groups and read it out loud," Tokar recalled. "It was fantastic."[30] His friends understood that it was a milestone in the history of anarchism. John Ely commended it as anarchism's "first comprehensive and cohesive theory," while John Clark called it "the first elaborated and theoretically sophisticated anarchist position."[31]

Bookchin finished the manuscript in October 1981, and upon its publication a few months later, some reviewers pointed to shortcomings: anthropologist Karen Field, for one, thought it selectively painted an "overly homogenized—even sanitized—picture of preliterate peacefulness and egalitarianism." Nonetheless she lauded the book as "the kind of wide-ranging and impassioned synthesis that is all too rare in this age of scholarly specialization" and "a graceful model of what scientific synthesis can be."[32]

Most reviewers accorded it high praise. John Fekete noted that it urged "a cosmic evolutionary ethic" that was "favourable to the survival and life interest of the human race, and authentically grounded in the potentialities and actualizations of nature." He commended Bookchin for integrating ethical philosophy with natural evolution: "Even to have raised an agenda [this] complicated and significant . . . testifies to the courage, dedication and intelligence of the author." Theodore Roszak called it "perhaps the most important contribution to environmental thought we will see in our generation. With it, Bookchin takes his place with Thoreau, Lewis Mumford and Paul Goodman as a major American political philosopher." Stanley Aronowitz placed Bookchin "at the pinnacle of the genre of utopian social criticism, the successor to the many generations of Diggers, Levellers, and Ranters." Science writer Robin Clarke remarked, "Bookchin's relationship to Marx can be paralleled with that of Albert Einstein to Isaac Newton."[33]

In 1980–81, as Bookchin was finishing the manuscript, the movement against nuclear weapons was growing. NATO, as we have seen, had announced its intention to deploy Euromissiles in central Europe, and now President Reagan was signaling a willingness to fight a "tactical" and "limited" nuclear war—to actually use the heinous weapons. To many Americans and Europeans, nuclear war seemed not only possible but imminent.

Karl Hess was one. Living off the grid in West Virginia, he started a newsletter called *Survival Tomorrow* and invited Bookchin to participate. Paging through an issue, Murray found articles advising libertarian readers on how to survive after a thermonuclear bomb blast: they were to stockpile weapons and seek "a 'survivalist' refuge in the hinterlands of America, be it a remote commune in Oregon that is fleeing the Apocalypse or a sand-bagged family fortress where the kids pack 45-calibre automatics." When the neighbors broke into your garden to take your produce, you could ward them off with lethal violence. Bookchin turned Hess down: he wouldn't even want to survive a thermonuclear war, he said, and even if he somehow did, far from driving his neighbors away, he would share whatever he had with them. He was an anarchist communist, not an individualistic libertarian. Least of all would he try to survive at the expense of others—what would be the point? He turned Hess down.[34] So did the two old friends part ways.

Far more to Bookchin's liking, the peace movement in Europe, opposing NATO's Euromissiles, had rapidly achieved mass proportions. Nonviolent, symbolic actions were being carried out everywhere, it seemed: peace initiatives, die-ins, nuclear-free zones. Hundreds of thousands of people demonstrated in Bonn. In Britain, Women for Life on Earth—the group that had originated with Ynestra King and her colleagues—established a peace camp outside the US airbase at Greenham Common, with protesters linking arms in human chains.[35]

In March 1982 Vermonters assembled in their town meetings to consider their local school budgets and road repairs, but this year many of them considered a national policy question as well: Should the United States and the USSR freeze the testing, production, and deployment of nuclear weapons and delivery systems? Activists throughout the state had put the question on as many town meeting agendas as they could. By day's end, a total of 177 towns endorsed the so-called nuclear freeze. Those endorsements, from a majority of Vermont towns, catapulted the freeze to a national issue. Organizers seized the momentum and went to work over the next months to plan a demonstration in New York. On June 12, almost a million people marched through the streets of Manhattan to demand a nuclear freeze—it was the largest political demonstration to date in American history (figure 10.1).[36] To Bookchin, the episode proved that ideas and actions originating on the margins—in Vermont—could have an impact on the country's political and financial centers. And beyond that, it proved that citizens in assemblies could have enormous ethical power—even to the point of challenging US foreign policy.

Bookchin continued his weekly commute between Burlington and Ramapo, but with West Germany bursting at its seams to protest the missiles, and with the freeze movement on the rise, how could he remain in the ivory tower? He genuinely loved teaching, and Ramapo had treated him well. But his radical pedagogy was not producing social revolutionaries. Increasingly, his students were concerned with building careers.[37] He longed to put academia behind him—it was clearly no place for someone who wanted to change the world.

FIGURE 10.1 In March 1982 more than 170 Vermont towns endorsed a nuclear arms freeze. That June hundreds of Vermonters traveled to New York to join a massive peace march in Central Park

Rob Swanson. UPI.

The academic Marxists, with their pedantic articles, published in obscure journals, seemed ever more disconnected from actual political movements. Even the Frankfurt School writings were becoming something of an academic industry. To Bookchin's mind, Horkheimer and Adorno's most important contribution had been their "exciting concerns with the domination of nature and [their] attempt to develop an alternative to the rationalization of the world," but their professorial heirs, rather than take up that crucial problem, retreated into "the bloodless tenets of semiology" and generated a mountain of writings laden with "intellectual obscurantism."[38]

Yet the promising new peace, ecology, and feminist movements desperately needed thinkers, a "revolutionary intelligentsia" that was at home in the streets as well as on the printed page. Today's academic intellectuals, Bookchin thought, should stop flinging rarefied quasi-Marxist jargon at each other in scholarly treatises, put an end to their "lengthy refrigeration in the academy," and step out into the public realm, to help clarify movement issues and dilemmas, to raise awareness, to write with passion, and to speak truth to power.[39]

On April 16–18, 1982, the journal *Telos* invited Bookchin to participate in a conference at Carbondale, Illinois, where leftist academics would consider the issue of ecology and the welfare state. At the first session, as the professors were discussing "the logic of the welfare state" and something called "deconstructionist ecology," Bookchin lost patience. He leaped to his feet and began scolding the panelists and listeners alike. *The peace, ecology, and antinuclear movements are out in the streets*, he said. *Why aren't you out there with them, helping those vibrant, genuinely radical movements, instead of walling yourselves off here with your opaque language?*[40]

Paul Piccone, editor of *Telos*, retorted that the movements Bookchin praised were susceptible to "nationalism, crude self-interest, the defense of existing privileges, and . . . integration within the existing logic of domination." Radical consciousness, he explained, must "undergo a whole series of more modest and realistic political mediations." People like Bookchin should refrain from "self-righteous moralizing" and instead "spell out concrete proposals designed to shift the welfare state toward the reconstruction of a public sphere."

Bookchin shot back that social movements had "a social vision that goes beyond the logic of the welfare state." The clash escalated, until finally he snapped, "This is the kind of academic environment that I am getting out of." Someone later observed that his "truculent charisma and explosive rhetoric" were "undoubtedly the high point of the conference." But no love was lost between Bookchin and the academic world—their divorce would be uncontested.

Wayne Hayes would miss stopping by the Brass Rail in Hoboken with Bookchin after work for boilermakers, but he understood that his friend needed to leave Ramapo. The college had created a business school, and in the age of Reagan, it was quite popular, so much so that the environmental studies school was closed down.[41] Ramapo, said Hayes, was becoming "more routinized and regimented, more outcomes-assessment oriented." *More instrumental*, Bookchin might have added.

In March 1982 he announced that he would take early retirement. Two weeks after his sixty-second birthday, on February 1, 1983, he retired to emeritus status.[42] Now he could live fully on the margins, where the action was.

In Burlington, he hit the ground running. In several impassioned essays he warned of looming ecological destruction and possible thermonuclear war. Ours, he said, might be the last generation that can still avert the virtual destruction of humanity, and the complex biosphere upon which it depends.[43]

Burlington itself had no town meeting; chartered as a city, it was governed by a city council known as the board of aldermen. Bookchin's goal was to create citizens' assemblies, like small town meetings, in each of the city's six wards (as outlined in "On Neighborhood Democracy," written for the 1981 Conference of Neighborhoods). The city's mushrooming neighborhood movement and its new socialist mayor made the prospects for creating those assemblies promising. In fact, if Bookchin could inspire citizens in town meetings statewide to broaden their agendas to address the ecological crisis, then they could potentially spark a nationwide movement for assembly democracy that could challenge the social order. The nuclear freeze issue had demonstrated that Vermont towns could have an impact on the whole country.

As it happened, the Sanders administration expressed an interest in democratizing the city. In June 1981 Burlington was designated a Standard Metropolitan Statistical Area—which meant that it was eligible to receive federal block grants for community development projects (CDBGs).[44] So Sanders's staff were busy developing proposals and applying for grants. One requirement was to show that citizens had participated in formulating the proposals through a democratic process.[45]

In the fall of 1981, perhaps inspired by the "On Neighborhood Democracy" resolution, the Sanders administration proposed the creation of Neighborhood Planning Assemblies (NPAs), one in each of the city's six wards.[46] In September 1982 the board of aldermen passed an enabling resolution, mandating that an NPA be organized in every ward. "All voters in a particular ward of the NPA shall become voting members," said the resolution. Mayor Sanders told a reporter, "We are attempting to do something that has not been done in many communities across the country, and that is to involve large numbers of people in day to day affairs of government."[47]

The NPAs were not town meetings, vested with decision-making power: they were merely assemblies of citizens to comment and advise the city government, to "make recommendations with respect to government decisions."[48] Their one concrete power was to elect representatives to a committee to screen projects for CDBG funding. Still, thought Bookchin, they were a good start, and a prodemocracy movement could push to endow them with ever-greater authority.

The city convened the initial meeting for each NPA, and at Bookchin's urging, the local social ecologists attended. "We thought [that] these neighborhood meetings were great," Tokar recalled, but rather than meet just "when there's a block grant process, managed by city hall," they should become "neighborhood assemblies as ongoing entities with decision-making power."[49]

Burlington was also discussing how much growth to allow within its borders. Murray, fresh from New Jersey, warned in June 1983 that while growth would generate jobs, profits, and tax revenues, it would also create pollution, shoddy goods, deadening work, congestion, and citizen passivity. The wrong kind of growth would transform Burlington from a convivial, neighborly community into a crass shopping mall. To avoid this fate, it should foster community-oriented, creative enterprises and technologies, in a clean, healthy environment. He and the dozen or so local social ecologists organized the Burlington Environmental Alliance (BEA) to make the city a beacon of forward-looking ecological self-management.[50]

One of the more sensitive areas of Burlington's environment is a bottomland along the Winooski River called the Intervale. For several years the city had been planning to construct a power plant there that would generate electricity by burning wood chips.[51] When the project was initially proposed in 1977, it had garnered praise as a bold step toward energy self-sufficiency. But implementation had been delayed, and now in 1983 the state Public Service Board was holding hearings on whether to authorize it.

BEA mobilized to oppose the 52-megawatt plant, which would burn some 500,000 to 800,000 tons of wood chips each year. Not only would the project harm Vermont forests, they argued, but the noise, air pollution, and truck traffic would damage the Intervale's complex ecosystem. BEA created a poster captioned "The Wood Chip Plant Is Coming," showing a landscape of tree stumps, and plastered it all over town.[52]

BEA was certain that Mayor Sanders would join them in opposing the plant, but in September, Sanders came out in favor of it. The anarchists who had just voted for him were stunned. Said Alan Kurtz, "I feel ripped off." Then on September 14, after almost a year of hearings, the state Public Service Board approved the plant.[53] The wood chips were coming after all.

The socialist mayor was behaving strangely about the waterfront, too. Instead of calling public hearings on its future, as he'd promised during his campaign, Sanders created a task force to study the question and design an "alternative vision that would reflect the character of our city." The task force turned the issue over to an architect—who proceeded to design a new plan, this one with a twenty-two-story condo, a marina, and a hotel complex with shops.[54] Bookchin and the social ecologists were aghast. This was no alternative at all—it was actually worse than the old Pomerleau plan.

Even more astonishing, Sanders was actually meeting with Antonio Pomerleau, who said he was ready to invest $30 million in the redesigned project. Yes, the waterfront project would bring revenue to the city's coffers, in the form of property tax. But that wasn't the point—Burlington wasn't supposed to be for sale. "In spite of his

Robin Hood rhetoric," the local alt-weekly editorialized in late 1981, "Bernie Sanders quietly developed a rapport with the city's richest developer."[55]

Opponents of the new design pointed out that it would "severely limit" public access to the waterfront. They managed to get a nonbinding referendum on the ballot that called for a thirty-foot public strip along the shore. The citizens of Burlington voted in favor of it. In November 1982 Pomerleau announced that he was pulling out of the project because of all the controversy and delays.[56]

Curiously, once the NPAs were up and running, Mayor Sanders stopped talking much about democracy. Asked on a local talk show whether he favored town meeting government in Burlington, he said it wasn't needed. The mayor seemed to be coming to the conclusion that he himself should have more power. Alarm bells went off in Bookchin's mind: why should Burlington want to centralize power in the mayor's office, he wondered, when the NPAs were now in place, and Burlington was on the brink of being able to demonstrate to the world how to democratically empower urban neighborhoods?[57]

But something peculiar was going on. In the democratic assemblies that Bookchin had in mind, citizens would be the ones to set the meetings' agendas; but for NPA meetings, city officials were setting the agendas—packing them up in advance with their own issues. The discussions that Bookchin had envisioned would take as long as they had to, so that everyone's voice would be heard; but for NPA meetings, city officials nailed the agenda items to strict time allotments, then breezed through them and adjourned punctually. Where was the bottom-up power flow so basic to assembly democracy? Bookchin was wondering by mid-1983. Where were civic vitality, communal liberty, and unmediated self-rule? "When will we begin to take these assemblies seriously?"[58]

The city council's enabling resolution had specified that at the second round of NPA meetings, the assemblies were to elect their own officers and draw up their own bylaws. But inexplicably, those items, so crucial to democracy, never made it to the agendas. Instead, the city planning department drew up its own set of bylaws and procedures and imposed them on the NPAs. It reduced the frequency of NPA meetings to only four a year. It then submitted these procedures to the city council for approval.

When the council met to consider the new procedures, Bookchin took the microphone during the period for public input. Crucial points of their enabling resolution were being overridden, he advised the councilors. It was their duty to correct this usurpation of power and ensure that the assemblies chose their own officers and bylaws, as originally specified.[59]

But the city council took no such action, and as time passed, the NPAs remained as they were—mere extensions, in Bookchin's eyes, of the city government rather than vehicles for neighborhood self-rule. Some were well attended, and others were not, but by no means did any of them become town meetings. Indeed, "as these assemblies increasingly disagreed with Sanders' administration," said another observer in

1986, "they became less favored, and their voice has been diffused by providing minimal staffing that is always hired by and responsible to the administration, not the assemblies."[60]

Meanwhile, Burlington's lively neighborhood movement had quieted down. The election of the socialist mayor had "co-opted a lot of its energy," as Tokar recalled. "Organizers who had worked for neighborhood organizations all went to work for city hall." Where political life in Burlington had only recently been creative and fluid, it was now being siphoned into municipal officeholding—careerism, as it seemed to Bookchin.[61]

Traditionally, Vermont's political culture was wary of professionalism. The state legislature met for only half the year. The legislators drew no salaries and had no offices; only a small proportion of them were lawyers; and they had to go before the voters every two years. Most other states' officeholders had four-year terms, but in Vermont the governor, the legislators, and all the rest had and still have only two-year terms. The provision was inconvenient for career politicians, but in practice it allowed Vermonters to frequently review the performance of their public officials and hold them accountable.

In January 1983 the Senate Operations Committee proposed to amend the state constitution to replace the two-year term with a four-year term. The change would modernize Vermont, the committee said, and make it more efficient, like the other states.

But Bookchin thought that extending terms to four years would wreck the state's uniquely democratic culture. It would professionalize statehouse politicians and foster New Jersey–style bureaucracy and centralization, complete with high-priced lobbyists. With the proposed four-year term, Vermont's style of local democracy was at stake. He organized a group to defend the two-year term, called the Vermont Council for Democracy (VCD; figure 10.2). He recruited, among others, town-meeting expert Frank Bryan, right-wing libertarian John McClaughry, peace activists Robin Lloyd and David Dellinger, and leftist journalist Greg Guma. The VCD issued a poster showing a crown with a slash through it. The integrity of the "citizens' commonwealth" was far more valuable than efficiency and professionalism, especially as defined by other, less democratic states.[62]

Throughout 1983, in a flurry of interviews, articles, and flyers, Bookchin campaigned to save the two-year term. The sitting governor, Madeleine Kunin, defended the four-year term proposal, arguing that "the cost of campaigning and the length of campaigns make [the two-year term] really difficult." Murray responded that officeholders' best campaign tool should be their record in office, and that the two-year term was Vermont's mechanism for keeping statewide politicians accountable.[63]

In November 1982 the *Burlington Free Press* came out in support of the four-year term. After Bookchin saw it, he charged into the office of Dan Costello, editorial page editor, told him why he was wrong—and actually persuaded him to change his mind. "The *Free Press* editorial board now has exercised its prerogative of reversing

FIGURE 10.2 Testifying to a Vermont Senate committee against a proposed constitutional amendment to extend the governor's and other officeholders' terms from two to four years. Montpelier, March 9, 1983
Photographer: Ron MacNeil. UPI.

its position on the issue," Costello editorialized soon thereafter. Vermont has had fine governors, and "none has seemed to be hampered by the 2-year term." The existing system "has served the state well," so "there appears to be little reason for changing it."[64]

The Vermont Council's campaign even succeeded in changing the mind of Senator Bill Doyle, the proposal's prime mover in the state senate. Now he agreed with Bookchin that the four-year term would professionalize government and said his committee would review the proposal.[65] That was the end of it.

The fate of Burlington's twenty-four-acre waterfront was still up in the air. After developer Pomerleau opted out of the redesigned project, a group of wealthy local investors had begun buying up parcels. They called themselves Alden Corporation but kept their identities secret. In March 1983 they began devising a new plan to develop the lakeshore.

In April the citizens of Burlington were invited to attend their respective NPA meetings to consult on the future of the waterfront. It was the closest the city had yet come to soliciting public input, as Sanders had promised. But as the citizens gathered at the meetings, they found themselves sitting through a slide show in which the Alden promoters presented their vision, which turned out to be extremely vague. Discussion afterward was minimal and tightly managed. The promoters then distributed a questionnaire, asking the citizens what they wanted to see on the waterfront.[66]

The citizens duly filled out the questionnaires, saying they wanted a marina, a fishing pier, a ferry, a restaurant, and a park—and that they did not want hotels, motels, or condos. They wanted public access to the lake; any new buildings should be set back from the shore by even more than the thirty feet now required. In June 1983 Alden hired a prominent Boston architect to incorporate the citizens' recommendations into a plan.[67]

While the architect worked, citizens discussed among themselves. In December 1983 Bookchin's BEA group sponsored a workshop in which participants developed an "ecotopian plan" for the waterfront, with a park, marinas, and natural areas. The Citizens Waterfront Group (CWG) called for an eighty-foot-wide waterfront park for walking, jogging, and bicycling.[68]

In late January 1984, to a packed house at city hall, the Alden group unveiled their preliminary plan, in another slide show, this time accompanied by music. It showed some traces of the much-vaunted "citizen input": a bike path and public access. But the $100 million plan was enormous in scale, dwarfing the plan that Sanders had run against in 1981. It featured a three-to-five-story lakeside hotel, with 150 to 200 rooms; two seven- or eight-story luxury condo towers; 150,000 square feet of office space; a pedestrian boardwalk; a retail pavilion; a waterfront heritage center (to memorialize the beautiful stretch that the plan was despoiling); a marina; parking space; and a small park.[69]

At the public meeting, criticism was immediate and vociferous. The plan didn't provide enough parkland along the shore; the buildings were too densely packed and too close to the water; the design was "like Epcot"; the presentation was "a sales pitch." In February, the various opposition groups joined forces to form the Waterfront Coalition.[70]

Meanwhile, a local attorney and Alden opponent, Rick Sharp, had been doing some research. In 1892 the US Supreme Court had issued the public trust doctrine, according to which all landfill in a state was the inalienable property of the people of that state and must be used in the public interest. And the waterfront was indeed landfill: in the nineteenth century the state legislature had permitted the Central Vermont Railway to fill in the first four hundred feet along the shore. The railroad was permitted to use the filled-in land as an expanded railyard; once that specific use was over, the land was to revert to public use. Burlington's waterfront therefore belonged to the people of Vermont, and they were the only ones with the legal power to decide its fate.[71]

Opponents of luxury waterfront condos now had legal muscle to flex.

* * *

In the spring of 1983, the European peace movement reached its peak of intensity. In Britain eighty thousand campaigners linked arms in a spectacular fourteen-mile human chain spanning three key defense installations, from Greenham Common to Aldermaston. In West Germany, thousands of demonstrators blocked entrances to six US military bases.[72]

In Burlington, a painful fact came to light: the progressive-minded city with the socialist mayor was home to an important munitions producer. A General Electric plant in the south end of town happened to be the world's most significant manufacturer of the Gatling gun, a high-speed antipersonnel weapon whose seven rotating barrels could fire up to 4,200 rounds per minute. It was currently being used on helicopter gunships in the civil war in El Salvador, deployed against the Communist FMLN.[73]

On June 18 several hundred activists protested in front of the GE plant, demanding that it convert to peaceful production. The five-hour rally featured speeches by longtime activists David Dellinger and Grace Paley. Two days later several dozen protesters blockaded truck traffic as it was leaving the plant. Police arrested eighty-eight of them on charges of disorderly conduct and took them to the police station, where they were fingerprinted and photographed, then released. Twenty-nine returned to block the gate again: Bookchin "drove to the police station to pick us up and take us back to the plant gate," recalled one of the blockaders. "It was a great day." Bookchin's only criticism was that the protest's focus was too narrow. As in Clamshell, he wanted the peace movement to broaden itself to oppose hierarchy in general and to support feminism and ecology and democracy and community—as the huge, multi-issue peace movement in Europe was doing.[74]

But surprisingly, Mayor Sanders—who had often solidarized with the Left against US intervention in Central America—refused to endorse the protest. The GE plant was the second-largest employer in the state, he said. Those workers made a decent living and had a right to their jobs. The protesters were "point[ing] the finger of guilt at working people," but "not everybody has the luxury of choosing where they are going to work." Sanders even endorsed the arrests of the demonstrators.

Jaws dropped all over Burlington.[75] Overall, the socialist mayor's track record was showing an unmistakable shift to the right. He had supported the wood-chip plant in the Intervale and luxury condos on the waterfront. His administration had vitiated the NPAs. Now he supported high-powered arms production, on behalf of the factory workers. In a pamphlet called "Workers and the Peace Movement," Bookchin reprised arguments from "Listen, Marxist!" and contended that people who work in factories are not to be reduced to "workers"—they are concerned, like everyone else, about nuclear catastrophe, about peace, about environmental destruction, and about their children's future: "Their human locus is the community in which they live, not the factory in which they work." Radical movements should address them not as proletarians but as community members, as neighbors. "The color of radicalism today is no longer red," he wrote. "It is green. . . . the politics we must pursue is grassroots, fertilized by the ecological, feminist, communitarian, and antiwar movements."[76]

Still, Marxist reverence for the proletariat continued to have a strong grip on the minds of American leftists, not least in Burlington. In 1979 in Nicaragua, the Sandinista National Liberation Front, a coalition of left-wing revolutionary forces that included Marxists, had overthrown a dynastic regime in a bloody revolution. Once in power, the Sandinistas restructured agriculture, redistributed land to the peasantry, and instituted a laudable literacy program; their revolution was welcomed and applauded by the international Left. Their government soon came under attack by reactionary forces, however, as the American president Reagan first tried to isolate and destabilize the regime, then organized and funded the Contras to overthrow it. During the war that continued for the rest of the 1980s, Burlington's Progressives identified strongly with the FSLN, to the point of earning for themselves the nickname "Sanderistas."

Bookchin, however, was exasperated that once again he had to explain the problems with that ideology that he had been contesting for decades. It came as no surprise to him that the Sandinistas brutalized Nicaragua's Miskito Indians, burning villages and forcing them to leave their homelands. Led by Brooklyn Rivera, the Miskitos' Misurasata movement took up arms against the Sandinistas, without allying with the Contras—they were a third force. Bookchin, together with his friend the Seneca John Mohawk, solidarized with the Miskitos and sharply attacked the Sandinistas.[77] Leftists, notably the Marxist Joel Kovel, took umbrage and charged Bookchin with de facto allying with the Contras. He denied it—"my enemy's enemy is not my friend," as I heard him say, around this time, on a radio broadcast over WBAI-FM in New York.

Bookchin could sometimes work effectively as an organizer, as he did on this action. He took pride in his ability to polemicize, seeing a clear, consistent argument as essential for clarifying ideas. "Usually I found his polemical salvos instructive," recalled Howie Hawkins, who knew him from Clamshell days. "He believed that the clash of ideas would sharpen our understanding." Such clarifications were indeed necessary: "There were plenty of would-be Lefties," Hawkins notes, "who deserved a withering critique and an audience of young would-be radicals that desperately needed to be educated on the differences between a liberal accommodation to an oppressive, irrational, and destructive society and a radical approach to social freedom and ecological rationality."[78]

But especially in the heat of argument, when Bookchin felt something important was at stake, his delivery could become harsh, peremptory, and dismissive, and his polemical rigor could slip over into scalding acrimony. To take just one example, Dimitri Roussopoulos recalled Bookchin's speech to an audience in Halifax, Nova Scotia, in the late 1970s. "He was in fine form," Roussopoulos said, "the turnout was good, and the audience was enthralled by his oratory. But there was one Trotskyist, who asked him something about Kronstadt," the 1921 sailors' revolt of 1921; this audience member defended Trotsky's role in suppressing it. "Out comes the ideological Sherman tank," recalled Roussopoulos, "and Bookchin just rolled over that poor person who dared to take the mike. Nothing but bones were left afterward." Murray's point was correct—Trotsky's brutality was indefensible—and furthermore, given the international Left's difficulty, over generations, in weaning itself from Marxism, his interventions on that score were generally salutary. But "the way he did it," says Roussopoulos, of that typical episode, "changed the mood of the audience."[79]

Similarly, in the 1980s, when Bookchin criticized Mayor Sanders for compromising with developers, his point was well taken, but his tone was so harsh that it alienated those who might otherwise have sympathized with his view. When the mayor's supporters rose to defend him, Bookchin slammed them vituperatively. He called activist Marty Jezer's defense of Sanders "a Lyndon LaRouche dossier" and "repellent demagoguery."[80] It was way over the top; I wince to read those words today.

Similarly, in these years young idealists would be attracted to Bookchin's writings, inspired by his utopian vision, and move to Burlington to work with him—only to be put off by his spleen. Suzanne Stritzler, a onetime BEA member, noticed that it happened repeatedly, forming a pattern. People would arrive in the city with excitement in their eyes, then gradually the glow would dim, and a few months later they'd slip away. Greg Guma, an erstwhile Burlington collaborator, finally concluded that Bookchin "spent too much time denouncing those whose approach (or occasional tactics) didn't conform to his social ecological theory and vision.... Murray had trouble disagreeing without splitting."[81]

Bookchin's usual response to such criticism was to dismiss it, saying that he didn't care how he came across. If someone listened to him only for his tone, he said, then it was their problem that they missed the content. But people tend to remember tone at least as much as content, and if the tone of an argument is disproportionate to the content, it can undermine even an otherwise solid case. Debate, as Bookchin often pointed out, is healthy and intellectually stimulating, but at times he seemed to forget that venom and gall (as Weber had called it) can be self-defeating, leading to ineffectiveness and political isolation. "The personality issue got in the way of him being able to practice his politics," said Roussopoulos. Karl-Ludwig Schibel agreed that Bookchin's "vitriolic attacks on his colleagues" were "the main reason why he was not and still today is not adequately received in the intellectual discourse."[82]

What was the reason for the harshness? I think it stems from his sense of urgency about building a sophisticated radical movement to address the looming ecological crisis. To that end, he placed an unusually high value on ideas, on theoretical rigor, on a movement's ideological underpinnings. He had grown up in an era of competing ideologies, and more than his younger friends, he continued to believe that theory and ideas were an essential basis for movement building. Only by continually clarifying a movement's aims and its means could it ultimately be kept from selling out or producing harmful unintended consequences, as had happened too often with past leftist movements. Those who advocated misleading or hybridized or compromising ideas in a movement threatened that movement's promise.

I think the issue dates back as well to the YCL, when the older commissars had placed such heavy, excessive responsibility on the YCL-ers' shoulders at an early age: they had essentially taught Bookchin and his comrades, during the ultrarevolutionary third period, that they were responsible for saving civilization. Such responsibility would be overwhelming for an adult to bear, let alone a child. Now in 1983, when he would see a dubious tendency gaining ground, he spoke out against it urgently, as if the fate of the revolution depended on getting his point across.

In any case, Bookchin's polemical harshness was the flip side of his great personal open-heartedness, which was his normal demeanor with those in direct contact with him and who worked closely with him. They found him emotionally and intellectually generous, sharing even his most original ideas, spending countless hours talking to near strangers, trying to cultivate them as he had done ever since his mentorship of Allan Hoffman. Guileless, unfiltered, he wore his heart on his sleeve and was surprised when others did not. He had no vanity, no interest in personal glory or financial gain. He was the opposite of instrumental. He was eager to the point of insistence to teach and inspire, because achieving socialism rather than barbarism, ecology rather than catastrophe, depended on it. He once remarked that some people have to push too hard, so that others would push hard enough. I'll let that be the last word.

After the Institute for Social Ecology lost Cate Farm, the school might well have gone under. But the director, Dan Chodorkoff, held it together organizationally and found ways for it to continue. In 1982 the ISE organized a second urban alternatives conference at El Bohio in Loisaida: it brought together three hundred activists, representing sixty community and environmental groups, from Green Guerrillas to ACORN, who talked about neighborhood economic development, community self-management, and energy and food production.[83]

Faculty members were finding ways to put social ecology into practice. In 1981 instructor Joseph Kiefer initiated a Hunger Education and Food Garden program in Montpelier. "Many of us find it hard to believe," he observed, "that hunger exists in our own communities." The program taught basic food-gardening skills to low-income children. The produce they grew was donated to a food bank that redistributed it to food shelves, day-care centers, and senior citizens.[84]

Starting in 1983, after a two-year hiatus, the ISE resumed its summer sessions, renting space on secondary-school campuses around Vermont.[85] Enrollment was only around forty—less than half of the norm in Cate Farm's heyday—and the session lasted one month instead of three. But Bookchin still lectured each day for three hours at a clip on social ecology (figure 10.3). In the summer of 1984, the school rented quarters at a school in Vershire. "As farm animals grazed on the lawn through the picture window in the background," recounted a journalist, "social theorist Murray Bookchin slowly rocked back and forth in his chair, delivering a two-hour lecture on cultural anthropology, while students clustered around him eagerly jotting notes."[86]

In the late 1970s, when he was drawing a professor's salary, Bookchin had purchased a piece of land near Goddard, intending to build a house there. But after the loss of Cate Farm, he sold the land and in November 1982 used the proceeds to help his ex-wife-turned-friend Beatrice purchase a home in Burlington's south end. He

FIGURE 10.3 Teaching at the Institute for Social Ecology, ca. 1985
Mark Ivins.

moved into this yellow-painted house, renting a room on the second floor that he used both as a study and as sleeping quarters.

The retired professor was eager to keep teaching, to build the democracy movement, so the yellow house's living room became an informal classroom. In 1983–84 he also taught a course on radical social theory and revolutionary history at Burlington College, a small alternative school.

But mostly he wrote, in a geyser of literary production. At Ramapo, formal teaching commitments had kept him from his typewriter; now it all came pouring out, intensified by renewed political purpose. First he set down the historical underpinnings for his movement for municipal democracy. The city's emergence in ancient times had been a historic advance over tribalism, he wrote, allowing people to associate with one another based not on biological kinship but on residential propinquity. Strangers could find a home, intermix, form ties with their neighbors. Common humanity could transcend tribe. He fell in love intellectually with the civic sphere, where all strangers were potentially equals, and with the idea of citizenship as an ethical compact.

Moreover, at least since medieval times, the municipality had been a historic countertendency to centralized authority. The medieval communes of northern Italy had joined together to form the Lombard League, pitted against the Holy Roman Empire. In 1520–22 the cities of Castile had revolted against the royal authority of Carlos V. These *Comunero* rebels had formed neighborhood assemblies that enfranchised even "the lowest ranks of the community." The *Comunero* cities had formed a "confederation of municipalities" that teetered on being "a dramatic alternative to the nation state."[87]

In modern times, "almost every major revolution has involved—indeed, has often been—a conflict between the local community and the centralized state." The Paris Commune of 1871 had been municipal both in location and in spirit, but so, in significant ways, had the American Revolution of the 1770s (initially driven by the Boston Town Meeting) and the French Revolution (energized in some stages by the Parisian sectional assemblies). The Russian and Spanish revolutions had been carried out by peasants newly arrived in urban factories who found one another concentrated in the city and shared their resentments of the unremitting pace of industrial capitalism. Accustomed to the seasonal rhythms of the countryside, they refused to cooperate with its mechanistic rigors—hence Petersburg in 1917; hence Barcelona in 1936. "The more I look into the great revolutionary movements that opened the modern era," he wrote, the more they seemed to have been by nature urban. They had been based in "specific neighborhoods in Paris, Petrograd, and Barcelona, and in small towns and villages," where people took action "not only as economic beings but as communal beings."[88]

Today, he argued, the potential for social freedom still reposes in the municipality. To be sure, existing municipal governments, with their city managers, mayors, and councils, are states-in-miniature, but popular democratic struggle could rework them, transform them into citizens' assemblies. In Bookchin's eyes, the democratized municipality, and the municipal confederation as an alternative to the nation-state, was the last, best redoubt for socialism. He presented these ideas and arguments, which he called libertarian municipalism, in their fullest form in *The Rise of Urbanization and the Decline of Citizenship*, published in 1986.[89]

In other works he explained in practical terms how to undertake this municipal-revolutionary process. Interested individuals should form groups to study the history and politics of assembly democracy and especially to unearth the long-forgotten democratic traditions in their own locality. Those long-forgotten institutions can be unearthed and brought to new life. After all, "the Commune still lies buried in the city council; the [Parisian] sections still lie buried in the neighborhood; the town meeting still lies buried in the township."[90] Once the study group members have educated themselves, they can become a political group and enter the public sphere. They can run candidates for the city council, on a program calling for the creation of popular assemblies. When and if those candidates are elected to office, they pass enabling resolutions or use other means to change their municipal charter, to devolve power from the council to the citizens' assemblies. They gain power precisely in order to renounce it.

The assemblies, once created, municipalize local economic life, producing for use rather than for profit. They institute small-scale manufacturing and renewable energy facilities. They decentralize the city and build green spaces and urban gardens. They make the automobile redundant by improving public transportation.

They create an ethical society, as "the free municipality transforms an ecological ethics from the realm of precept into the realm of politics."[91]

To handle common problems over larger regions, the democratized municipalities elect delegates (mandated, recallable) to coordinating councils, whose functions are purely administrative, executing policy rather than making it. The confederated municipalities act as a brake on centralized state power and ultimately replace it. Will the confrontation be violent? Or will the confederation erode the state machinery gradually? Bookchin left the question open. For now, he condensed his program into a slogan: "Democratize the republic and radicalize the democracy!"[92]

NOTES

1. Between Jan. 11 and 17 he spoke at Frankfurt University, Kassel, Hanover, and Berlin. Wolfgang Haug to author, Aug. 8, 2010. On the political scene, see Horst Mewes, "A Brief History of the German Green Party," in Margit Mayer and John Ely, eds., *The German Greens: A Paradox between Movement and Party* (Philadelphia: Temple University Press, 1998), 31.

2. Jutta Ditfurth, *Das waren die Grünen: Abschied von einer Hoffnung* (Munich: Econ Taschenbuch Verlag, 2000), 54–58, author's translation.

3. Bookchin, "The German Greens," *Gadfly* 2, no. 5 (Nov. 30, 1986); the *Gadfly* is a University of Vermont student newspaper.

4. "Interview with Murray Bookchin" (Sept. 1981), *Open Road* [Vancouver, BC] (Spring 1982). On Zurich's youth movement, see Stefan Kipfer, "Globalization, Hegemony, and Political Conflict: The Case of Local Politics in Zurich, Switzerland," in David A. Smith and Jósef Böröcz, eds., *A New World Order? Global Transformations in the Late Twentieth Century* (Westport, CT: Praeger Publishers, 1995), esp. 188–91. For what Murray might have said to the Swiss anarchist youth, see what he told Italian anarchists around that time: Paolo Finzi, "Intervista a Bookchin," *A: Rivista Anarchica* [Milan], no. 93 (June–July 1981), 16–19.

5. Bookchin, *Open Road* interview; "An Appeal for Social and Ecological Sanity" (1983), in *The Modern Crisis* (Philadelphia: New Society, 1986), 149.

6. He moved into a rental at 118 Hayward Street in Burlington.

7. Wayne Hayes, interview by author, Aug. 25, 2009.

8. Aristotle, *Politics*, I, 5, 1326b25, quoted in Bookchin, "The Myth of City Planning" (1973), in *TES*, 143; Bookchin, "On Neo-Marxism, Bureaucracy, and the Body Politic" (1978), in *TES*, 237–38; "Self-Management and New Technology" (1979), in *TES*, 119. Of course, Athens had excluded women, slaves, and resident aliens from citizenship, as Bookchin consistently pointed out. Modernity corrected that failing, fulfilling the Athenian principle that every adult is competent for citizenship.

9. Bookchin, "Marxism as Bourgeois Sociology" (1979), in *TES*, 199.

10. Bookchin, "Toward a Philosophy of Nature: The Bases for an Ecological Ethics" (1982), in *The Philosophy of Social Ecology: Essays on Dialectical Naturalism*, rev. ed. (Montreal: Black Rose Books, 1995), 44, hereafter *PSE*; Wayne Hayes, interview by author, Aug. 25, 2009.

11. Scott Mackay, "Election Changed Burlington Politics," *BFP*, Mar. 7, 1982; Alexandra Marks, "Waterfront Dreams," *Vermont Vanguard Press* (hereafter *VVP*), Aug. 13–20, 1982.

12. Scott Mackay, "Storm Clouds Move in over City's Waterfront," *BFP*, Apr. 12, 1982; Marks, "Waterfront Dreams"; Rob Eley, "Lake Access Issue on Ballot to Test Sentiment of Voters," *BFP*, Feb. 18, 1981, 3B.

13. Jo Schneiderman, "Neighborhood Power," *VVP*, May 8–15, 1981.

14. "Citizens Party Prepares for Elections," *BFP*, Jan. 11, 1981; "Sanders Tax Plan," *BFP*, Jan. 27, 1981; "Mayoral Candidates," *BFP*, Feb. 22, 1981; Alan Abbey, "Mayoral Candidates Fire Verbal Salvos at Forum," Feb. 19, 1981, 1B; and Debbie Bookchin, "Assessing Sanders," *VVP*, Dec. 31, 1981–Jan. 8, 1982.

15. Mackay, "Election Changed Burlington Politics." After a recount, Sanders's margin was reduced from twelve to ten votes. See *BFP*, Mar. 5, 1981. See also D. Bookchin, "Assessing Sanders."

16. Alan Abbey, "Sanders Stuns Paquette in Close Mayoral Race," *BFP*, Mar. 4, 1981, 1; Greg Guma, "Changing of the Guard," *VVP*, Mar. 13–20, 1981; Alan Abbey, "Sanders Drew People into Process," *BFP*, Mar. 8, 1981; Bookchin, "Democracy versus the Ten Percent Solution," *VVP*, Mar. 27–Apr. 3, 1983.

17. Alan Kurtz, Brett Portman, Brian Tokar, et al., "Dear People," proposal written for Burlington Conference of Neighborhoods on Neighborhood Democracy (written May 13, 1981), Box 12, MBPTL.

18. Schneiderman, "Neighborhood Power"; Alan Abbey, "Conference Planned to Weld Alliance of Neighborhoods," *BFP*, May 11, 1981; Alan Abbey, "100 Residents Draw Up a 'Blueprint' for Burlington," *BFP*, May 17, 1981, 3B.

19. Michael Riordan to author, June 12, 2009.

20. Bookchin, *The Ecology of Freedom: The Emergence and Dissolution of Hierarchy* (Palo Alto, CA: Cheshire Books, 1982), 141.

21. This section summarizes ideas not only from *The Ecology of Freedom* but also from Bookchin, "Toward a Philosophy of Nature: The Bases for an Ecological Ethics" (1982), "Freedom and Necessity in Nature" (1986), and "Thinking Ecologically" (1986), all published in *PSE*. "Rethinking Ethics, Nature and Society" (1985) and "What Is Social Ecology?" (1984) were published in Murray Bookchin, *The Modern Crisis* (Montreal: Black Rose Books, 1987), hereafter *MC*; "Radical Social Ecology: A Basic Overview" (1983) was published in *Harbinger* 3 (Fall 1985). "The Radicalization of Nature" (1983), the Frankfurt speech, was published as a *Comment* special paper (n.d.). He named his philosophical ideas "dialectical naturalism."

22. Bookchin, "Thinking Ecologically," 102.

23. William Trager, *Symbiosis* (New York: Van Nostrand Reinhold, 1970), vii, quoted in Bookchin, "Radical Social Ecology," 34.

24. Bookchin, "What Is Social Ecology?" 65; "Rethinking Ethics," 10n.

25. On Vrba, see Bookchin, "Toward a Philosophy of Nature," 62; *Ecology of Freedom*, 360–61. On "participatory evolution," see Bookchin, "Freedom and Necessity in Nature," 78.

26. Bookchin, "What Is Social Ecology?" 63.

27. Hans Jonas, *The Phenomenon of Life: Toward a Philosophical Biology* (New York: Harper & Row, 1966), 83, quoted in Bookchin, "Toward a Philosophy of Nature," 64.

28. Hegel, *The Philosophy of History*, quoted in Herbert Marcuse, *Reason and Revolution: Hegel and the Rise of Social Theory* (Boston: Beacon Press, 1941), 229.

29. Bookchin, *Ecology of Freedom*, 342. On "free nature," see Bookchin, "Thinking Ecologically," 136. Johann Gottlieb Fichte, *Die Bestimmung des Menschen* (1800), trans. R. N. Chisholm as *The Vocation of Man* (New York: Bobbs-Merrill, 1956), 20. Bookchin discusses Fichte's remark in *Ecology of Freedom*, 315; "Response to Andrew Light's 'Rereading Bookchin

and Marcuse as Environmental Materialists,'" *Capitalism Nature Socialism* 4, no. 2 (June 1993), 105.

30. Brian Tokar, interview by author, Aug. 7, 2009.

31. John Clark, *The Anarchist Moment: Reflections on Culture, Nature and Power* (Montreal: Black Rose Books, 1984), 11.

32. Karen Field, review of *Ecology of Freedom*, in *American Anthropologist* 86 (Spring 1984), 161–62. Andy Price has perceptively noted that Bookchin's biased anthropological evidence is admittedly speculative and intended primarily to show a potentiality. See his *Recovering Bookchin: Social Ecology and the Crisis of Our Time* (Norway: New Compass, 2012).

33. John Fekete, "Ontology and Value: The Ecology of Freedom," *Canadian Journal of Political and Social Theory* 7, no. 3 (Fall 1983), 167–80; Theodore Roszak, "The Obsessive Drive to Dominate the Environment," *San Francisco Chronicle*, May 6, 1982; Stanley Aronowitz, "Past Perfect," *Village Voice*, July 20, 1982; Robin Clarke, "Earth Deserves a Better Fate," *New Scientist*, Dec. 16, 1982.

34. Bookchin, "Reflections of an Old New Yorker," *Raise the Stakes* (1981), 7; *Raise the Stakes* was a publication of the Planet Drum Foundation in San Francisco. The correspondence is Hess to Bookchin, Aug. 5 and Nov. 10, 1980; Bookchin to Hess, Sept. 26, 1981, all in Box 17, MBPTL.

35. Harriet Hyman Alonso, *Peace as a Women's Issue: A History of the U.S. Movement for World Peace and Women's Rights* (Syracuse, NY: Syracuse University Press, 1993); "Thousands Protest Basing of Missiles in West Europe," Associated Press, Dec. 13, 1982.

36. Scott Mackay, "Disarmament Rally Swells to 700,000," *BFP*, June 13, 1982. In August the US House of Representatives would reject the nuclear freeze resolution 204 to 202. In May 1983 the House would pass, 278 to 140, a resolution calling for a "mutual and verifiable" nuclear weapons freeze. It was nonbinding and largely symbolic. Tom Raum, "House Passes Modified Arms Freeze," *BFP*, May 5, 1983, 1.

37. Bookchin, "Intelligentsia and the New Intellectuals," *Alternative Forum*, no. 1 (Fall 1991), 6–7.

38. Bookchin, "Finding the Subject: Notes on Whitebook and Habermas, Ltd.," *Telos*, no. 52 (Summer 1982), 95–96; "The Role of the Intellectual in the 1980s," *Telos*, no. 50 (Winter 1981–82), n.p.; "Introduction" (1979), in *TES*, 12.

39. Bookchin, "Role of Intellectual" (1981).

40. Carbondale *Telos* Group, "Ecology and the Welfare State: A Conference Report," *Telos*, no. 52 (Summer 1982), 131–33. All quotes from the conference come from this report.

41. Michael R. Edelstein, "Sustaining Sustainability: Lessons from Ramapo College," in Peggy F. Barlett and Geoffrey W. Chase, eds., *Sustainability on Campus: Stories and Strategies for Change* (Cambridge, MA: MIT Press, 2004), 274–75.

42. Marilyn Sacchi (Ramapo College personnel assistant) to Bookchin, Sept. 23, 1982, Box 12, MBPTL. Born in Jan. 1921, he retired at sixty-two in Jan. 1983. At this writing, Ramapo still offers a class in social ecology and a graduate program in sustainability studies. See www.ramapo.edu/masters-sustainability/.

43. Bookchin, "On Neo-Marxism, " 245; "An Appeal."

44. Rob Eley, "'SMSA' Status Brings Benefits to Burlington Area," *BFP*, Oct. 4, 1982.

45. "Block Grants 'Buy' New Battlefield," *BFP*, Sept. 23, 1981, 6.

46. Vermont Council for Democracy, "Statement on the Burlington Planning Commission's 'Procedures & Structure' for the Neighborhood Planning Assemblies as Submitted to the Burlington [Vermont] Board of Aldermen" (1983), MBPTL and author's collection; "City

Plans New Neighborhood Approach: Neighborhood Planning Assemblies," *Burlington Citizen* 8, no. 11 (Nov. 1982), Box 12, MBPTL.

47. Quoted in Debbie Bookchin, "A Dynasty Crumbles: A Fresh Start for Sanders," *VVP*, Mar. 6–13, 1981.

48. "Sanders Neighborhood Assemblies Irk City Planners," *VVP*, August 20–27, 1982.

49. Greg Guma, "Changes in City Hall Push Neighborhood Assemblies Towards Puberty," *Burlington City Pulse*, June 1, 1983, Box 12, MBPTL.

50. Bookchin, "Growth: Miracle or Menace?" in *Burlington City Pulse*, June 1, 1983, Box 4, MBPTL; "Burlington Environmental Alliance," n.d., MBPTL and author's collection.

51. During the 1976 OPEC oil crisis, the Burlington Electric Department proposed building an "energy park" in the Intervale. It would feature a $20 million solar photovoltaic plant, a hydroelectric dam on the Winooski River, and a 100-megawatt combination wood and trash-burning station to turn refuse into electricity. It seemed environmentally enlightened at the time. In 1977 the plan was expanded: waste heat from the generating plant would be used to warm a year-round fish hatchery and greenhouse complex. Most of the plan never was realized, and the incinerator and the wood-chip plant were split into two operations. The wood-chip plant's size was halved to 50 megawatts.

52. Brian Tokar and Alan Kurtz (principal authors), "Report of the Burlington Environmental Alliance to the Burlington Board of Aldermen Regarding the Proposed Resource Recovery Facility," May 9, 1983, in Box 4, MBPTL. See also Alan Abbey, "Petition Drive Against Wood Plant Started," *BFP*, Aug. 4, 1981; Peter Freyne, "Wood Chip Power? Questions Grow as Power Plant Decision Nears," *VVP*, July 31–Aug. 7, 1981.

53. Alan Abbey, "Sanders Gives OK to Wood Chip Plant," *BFP*, Sept. 8, 1981, 1B; "Wood-Burning Plant Criticized by New Group," *BFP*, Nov. 19, 1982, 1B.

54. "Waterfront Future Is Still Cloudy," *VVP*, Feb. 26–Mar. 5, 1982; Joe Mahoney, "Pomerleau Envisions Condos Rising 22 Stories," *BFP*, Sept. 25, 1981, 8B.

55. "Small Changes and Broken Promises," editorial, *VVP*, Dec. 31, 1981–Jan. 8, 1982; Scott Mackay, "Sanders and Pomerleau: Burlington's Odd Couple," *BFP*, June 6, 1982, 1.

56. "Spit into a Park," *VVP*, May 29–June 5, 1981; Marks, "Waterfront Dreams"; Scott Mackay, "Pomerleau Puts Waterfront Condo Plan on Hold," *BFP*, Nov. 9, 1982; Scott Mackay, "Waterfront Protection Plan Aired: Control Would Shift to Aldermen," *BFP*, Dec. 2, 1982, 1B.

57. The program was *You Can Quote Me*. See Northern Vermont Greens, "The Waterfront Issue: Is Burlington Being 'Sold Out'?" (Fall 1985), in Laura Cepoi, ed., "The Double Edged Sword: Ecological Theories and Ideologies among the Greens," Dec. 1986, unpublished compilation, in Box 4, MBPTL.

58. Bookchin, "Must We, in Burlington, Follow Fin de Siecle Galveston, Texas?" *BFP*, June 8, 1983. See also Bookchin, "Power to the Planning Assemblies," *VVP*, June 19–26, 1983; and "Socialism in One City? The Bernie Sanders Paradox," *Socialist Review*, no. 90 (Nov.–Dec. 1986), 51–62.

59. Vermont Council for Democracy, "Statement."

60. Aileen Lachs, "The Thriving and Dying NPAs," *VVP*, Apr. 28–May 5, 1985. See also Greg Guma, *The People's Republic: Vermont and the Sanders Revolution* (Shelburne, VT: New England Press, 1989), 71–72; Bryan Higgins, "Dilemmas of Urban Socialism on the West Coast of New England," *Socialist Review*, no. 90 (Nov.–Dec. 1986), 65–66.

61. Tokar interview; Bookchin, "Popular Politics or Party Politics?" Apr. 27, 1984, *River Valley Voice* (1984), republished as Green Program Project paper no. 2, MBPTL.

62. Vermont Council for Democracy, "Save the Two-Year Term in Vermont!" (Dec. 20, 1983), advertisement in *Rutland Herald*.

63. Kunin quoted in Chris Braithwaite, "Off the Record," *VVP*, Jan. 30–Feb. 6, 1983. See also Bookchin, "Two Year Term Keeps Governor Accountable to People," *BFP*, Mar. 17, 1983; "Vermonters Need to Act to Preserve State Constitution," *BFP*, Dec. 15, 1983; "Towards a Vermont Monarchy," *VVP*, Jan. 22–29, 1984. See also Joshua Mamis, "The Spirit of '77: Taking Liberties with the Vermont Constitution," *VVP*, Jan. 15–22, 1984; "Group Fights Four-Year Term," *VVP*, Jan. 30–Feb. 6, 1983.

64. Editorial, "In Retrospect: 2-Year Term for Governor Has Its Merits," *BFP*, Apr. 12, 1983.

65. Mike Connelly, "Group Vigorously Debates Methods of Government," *BFP*, June 23, 1983, 1B.

66. Northern Vermont Greens, "Waterfront Issue"; "Avoiding Marina Del Rey, Vermont," *VVP*, Feb. 5–12, 1984.

67. Scott Mackay, "Residents Describe Waterfront Desires," *BFP*, Apr. 14, 1983; Scott Mackay, "Boston Architect Hired by Investors in Waterfront Plan," *BFP*, June 15, 1983.

68. William H. Braun, "Access to Waterfront Debated," *BFP*, Dec. 12, 1983, 1B; Carol Conragan, "More Ideas on the Waterfront," *VVP*, Dec. 11–18, 1983.

69. Nelson Hockert-Lotz, "Beyond the Waterfront," *VVP*, Jan. 29–Feb. 5, 1984.

70. Don Melvin, "Coalition Forms to Fight Alden Waterfront Plan," *BFP*, Feb. 10, 1984, 1B; Don Melvin, "$100 Million Waterfront Plan Unveiled," *BFP*, Jan. 27, 1984; John Donnelly, "Dreams Flashy, Details Lacking," *BFP*, Jan. 16, 1984, 1A.

71. Rick Sharp, "The Public Trust Doctrine and Its Applicability to the Filled Land in Burlington Harbor," in Cepoi, "Double Edged Sword"; Hollis Hope, "Twenty Questions at the NPA," *VVP*, Feb. 26–Mar 4, 1984.

72. Mark A. Smith, "14 Miles of Humans for Protest Chain Against Arms Race," *BFP*, Apr. 2, 1983, 1.

73. Debbie Bookchin, "Fastest Guns in the West: Behind the Veil of Secrecy at Vermont's Largest Arms Plant," *VVP* July 17–24, 1981; and Cuthbert C. Mann, "Under the Gun," *VVP*, May 22–29, 1983.

74. Mann, "Under the Gun"; Mike Donoghue, "Hundreds March as Mayor Deals with Other Items," *BFP*, June 19 1983, 1B; Neil Davis, "Arms Demonstrators Target GE Property," *BFP*, June 7, 1983, 6B; Mike Donoghue, "Despite 88 Arrests, Protest at GE Plant is Calm, Amicable," *BFP*, June 21, 1983, 1A; John Donnelly and Jodie Peck, "Although Protesting Death, Smiles Were Demonstrated," *BFP*, June 21, 1983, 1A; Bookchin, "The American Peace Movement: A Green Perspective," *VVP*, June 12–19, 1983.

75. Scott Mackay, "GE Union, Sanders Join Forces to Protest Blockade," *BFP*, June 10, 1983, 1B; Scott Mackay, "Peace Activists Upset with Sanders," *BFP*, June 12, 1983, 5B; "Sanders Endorses March Protesting War Policies," *BFP*, June 15, 1983, 3B; Hilary Stout, "General Electric Protesters Agree to Limited Sit-In," *BFP*, June 17, 1983, 1B; Hamilton Davis, "Is Sanders Leaving the Left?" *VVP*, June 19–26, 1983.

76. Bookchin, "Workers and the Peace Movement" (1983), in *MC* (Black Rose edition), 191; Bookchin, "Rethinking Ethics," 45.

77. On the Miskitos, see, e.g., Shirley Christian, "Rebellion Will Continue, Leader of Miskitos Says," *New York Times*, July 29, 1985; John Corry, "On 13, Sandinistas vs. Miskitos," *New York Times*, July 29, 1986. For Bookchin's views on this and other issues of these years,

see "Murray Bookchin Speaks with Dan Higgins About the Status of the Political Left" (video), Channel 17 Town Meeting Television, Sept. 15, 1986, http://bit.ly/1QuvUPp.

78. Howie Hawkins, "Remembering Murray Bookchin," March 18, 2013.

79. Dimitri Roussopoulos, interview by author, July 6, 2009.

80. Bookchin, letter in response to Marty Jezer, *Socialist Review*, nos. 93–94 (May–Aug. 1987), 180–82.

81. Greg Guma, blogpost, Oct. 2, 2007, http://yhoo.it/1OIlyaN.

82. Schibel to author, Jan. 2010.

83. The conference was held on May 8–9, 1982. Bookchin gave the keynote speech, "The New Municipalism." Chino Garcia spoke on neighborhood government; so did Ruth Messinger, a community activist who had won a city council seat by calling "for a decentralization of power and support for citizen action." Michael Monte, a community development specialist from Burlington, spoke on neighborhood organizing. The conference discussed energy and food production, neighborhood economic development, and community self-management. See also Harry C. Boyte, *The Backyard Revolution: Understanding the New Citizen Movement* (Philadelphia: Temple University Press, 1980), 6; "1974," memo, n.d., in ISE Archive, Marshfield, VT.

84. "Gardening, Hunger, and Social Ecology: A Community Action Project," *Institute for Social Ecology Newsletter* (Rochester, VT) 4, no. 4 (Autumn 1985). Later the program would develop into Food Works, a nationally known, independent, not-for-profit organization. See foodworksvermont.org.

85. At the Stowe School in 1983, the ISE held classes in "ecological food production (biological agriculture and aquaculture), social theory, feminism and ecology, holistic health, and alternative technology (solar energy, passive solar building design, wind power)." Faculty and speakers included Murray Bookchin, Dan Chodorkoff, Ynestra King, Grace Paley, Eleanor Ott, Bill Maclay, Joseph Kiefer, David Dellinger, and others. Advertisements for summer program of the Institute for Social Ecology in *BFP*, May 22, 1983, 7B; *VVP*, May 8–15, 1983; and *VVP*, June 5–12, 1983.

86. Dave Goska, "Social Ecology Survives," *VVP*, July 8–15, 1984.

87. Murray was influenced by Manuel Castells, *The City and the Grassroots* (Berkeley: University of California Press, 1983), 5–21.

88. Bookchin, "Rethinking Ethics," 40; introduction to *Limits of the City*, 2nd ed. (Montreal: Black Rose Books, 1986), 7; "The Ghost of Anarchosyndicalism," *Anarchist Studies*, no. 1 (1993), 19.

89. It was later republished as *Urbanization without Cities* (1992) and *From Urbanization to Cities* (1995).

90. Bookchin, "Theses on Libertarian Municipalism" (1984), in *Limits of the City*, 2nd ed. (1986), 178.

91. Bookchin, "Rethinking Ethics," 42.

92. Bookchin, "Democratizing the Republic and Radicalizing the Democracy," *Kick It Over*, Winter 1985–86, 9.

11

Green Politico

THE NEWS IN March 1983 was head-spinning: in West Germany, in federal elections, a group dedicated to ecological politics had garnered enough votes to gain seats in the Bundestag. They arrived on the first day of the new parliamentary session wearing colorful clothing, carrying Planet Earth beach balls and placards on peace and ecology, and dragging felled pine trees, withered from acid rain.[1] They refused to call themselves a party—they announced that they were merely the political arm of the huge extraparliamentary ecology and antinuclear movement that had been bringing Germans into the streets for over a decade.

Bookchin and his friends and students in Burlington read the news reports and rubbed their eyes in disbelief.

In the fall of 1977, it seems, the German antinuclear movement had reached an impasse. The government had cracked down severely on all oppositional groups—it had met attempted nuclear site occupations, for example, with brutality. The movement "was being killed by a state armed to the teeth," recalled the journalist and activist Jutta Ditfurth. Activists debated what to do next. Some advocated going the parliamentary route, entering local and provincial legislatures, on platforms calling for no nukes, wildlife conservation, and greater local self-government. It would mean participating in the hated power-party system, but it might be worth it if they could change some laws. "We had to find a political space where we could continue to work," said Ditfurth. Because of the crackdown, "there weren't many choices."[2]

Political ecologists began running for seats in provincial legislatures. They didn't define themselves ideologically—after all, pollution, acid rain, and nuclear radiation affected everyone. They got a few votes but not enough to gain seats.

In Frankfurt, local radicals met in July 1978 at the home of Ditfurth and her partner, Manfred Zieran, and "some of us in the nondogmatic left decided, after long hard discussions, to risk working in parliament." True, the mainstream political process might end up neutralizing their radical goals, and "some of our friends warned us about the mechanisms of integration.... But we hoped for new political possibilities—like the creation of a radical democratic milieu."[3] They formed the Hessian Green List, to run for seats in the Hessian *Landtag*, or provincial legislature.

In December 1979 the NATO Euromissile announcement fueled the activists' new program of running candidates. In January 1980 a thousand political ecologists met in Karlsruhe and officially united as the Greens (*Die Grünen*). They were environmentalists, antinuclear activists, squatters, radical feminists, Christian pacifists, punks, anthroposophists—and leftists, notably the Hessian Green List and some ecosocialists from Hamburg and West Berlin. Many of these Greens, not just the leftists, were familiar with social ecology—Bookchin's books were available in German translation—and took for granted the inseparability of social and ecological questions. The congress chose to advance, not an ideology, but four "pillars": ecology, social justice, grassroots democracy, and nonviolence.

Right-wingers were also present at the Greens' founding congress in Karlsruhe, like Herbert Gruhl, author of a 1975 best seller called *A Planet Is Plundered*, which argued that an ecological dictatorship was necessary to coerce people to reduce their consumption of natural resources. Even some pro-Nazi groups were interested in joining. Some advocated putting aside old political differences, in the name of saving the planetary biosphere: Rudolf Bahro, an East German dissident who had made his way to the West, called on the Greens to welcome the whole political spectrum, from eco-socialists to Gruhl. Two months after Karlsruhe, the Greens adopted a basic program, laying out an array of progressive social and economic demands, even invoking the potentialities of modern society for utopia.[4] The right-wingers and neo-Nazis couldn't tolerate it and departed. The leftists rightfully prided themselves on having prevented ecology from becoming a radical-right or even neofascist issue.

Thereafter peace activists, feminists, socialists, and conservationists flocked into the Greens. It was a movement of alternative people against bureaucrats, of the grassroots against the repressive state, of visionaries against functionaries. In the next few years the Greens would portray themselves as transcending political divisions, being "beyond Left and Right"; their popular slogan was "we are neither Right nor Left but up front." But their program of 1980 was truly radical Left, and distinctly social-ecological.

Distrust of parliamentary politics ran deep among the West German Greens. Once Green representatives entered a legislature, everyone knew, they risked becoming professional politicians. Even the most idealistic person could become a compromising deal-maker—it was the nature of the system. But they considered themselves principled and their opposition to nuclear weapons and environmental destruction as nonnegotiable—the very future of the planet depended on their ability to maintain this position unwaveringly, in the face of all challenges and temptations. So the Greens developed a set of rules to keep future officeholders accountable to the grassroots and thereby bound to principle.

Green policy decisions would be made only by rank-and-file members, in local and provincial assemblies or congresses. Any Green legislators would be required to advance these policies (imperative mandate). No Green legislator could at the same time also hold a party position (separation of mandate and office). After two years, Green legislators would have to give up their seats to other Greens (rotation). Green legislators and party officials would have to be 50 percent women (gender parity). And Green legislators had to turn over most of their salaries to the party for projects and initiatives. These democratic procedures were essential to the Greens' political self-definition. To distinguish themselves from conventional parties, they didn't even call themselves "the Green Party," just "the Greens." One spokesperson, Petra Kelly, called them an "anti-party party."

In March 1981 the Hessian Green List in Frankfurt won more than 5 percent of the vote—enough to gain seats in the city council.[5] The six councilors who arrived at city hall that first day (Ditfurth was one) wore gas masks and smocks and carried a sign that read "Green Anti-Catastrophe Service: Parliamentary and Extraparliamentary." The other councilors, appalled, wouldn't let them in. "We stood before the city hall doors, which were barricaded from inside," recalled Ditfurth. But finally those inside had to admit them.

Even as elected councilors, the six considered themselves merely an extension of the extraparliamentary movements. "We made spectacular opposition politics in the city hall and beyond," recalled Ditfurth. "We demonstrated in the streets against air and water pollution. We organized mass assemblies against Startbahn-West [the airport runway extension] and against the nuclear installations in Biblis and Hanau.... We forced debates on toxic wastes from Hoechst."[6] Meanwhile they tried to expose the city council's failings, to make "local citizens aware of the inadequacies of representative democracy." One day, they hoped, the citizens would "no longer allow themselves to be administered, but rather increasingly represent their interests themselves."[7]

In March 1983, turbocharged by the peace movement, Greens all over West Germany entered the federal elections, demanding a shutdown of nuclear power plants and the scrapping of the Euromissiles. On election day, they surprised even themselves by garnering enough votes to earn twenty-seven seats in the Bundestag. Newspapers all over the world, including those in Vermont, ran stories about their eye-popping entry into the Bundestag with dead tree branches.

In Vermont, Bookchin and his *compañeros* watched closely as the German Greens went on to advance ecology, feminism, and disarmament in the Bundestag. Especially in contrast to the dismal, Reagan-dominated American scene, recalled Brian Tokar, "the visionary policies and electoral victories of *die Grünen* in Germany appeared to be nothing short of a political miracle."[8]

Bookchin knew well, from the experience with the MCM in Montreal, that a party's political success can attract opportunists who wish to lead it away from the movement that spawned it.[9] Even in Burlington, Bernard Sanders seemed to be turning into a conventional politician. But street demonstrations and human chains were no option either—they were fleeting, transient (as Murray had pointed out in "Spring Offensives" in 1971). A movement had to enter the political sphere if it wished to make lasting change. It was a dilemma, but libertarian municipalism, Bookchin thought, would solve it. A movement could enter the electoral arena and gain power—and by renouncing power in favor of permanent citizens' assemblies, it could resist becoming normalized.

The German Greens were fascinating because they were grappling with the same Scylla-and-Charybdis problem in their own way: demonstrations expressed popular opposition but were impermanent, but traditional parties led to bureaucracy, following the "iron law of oligarchy." With their remarkable structures of accountability, the Greens seemed to have found their own third way, to have transcended the dilemma. Their political focus, like Murray's, was primarily local: they were attempting to create popular power by entering municipal politics.[10] But if they were to stay grassroots democratic, Bookchin thought, the Greens would need to call not merely for environmental reforms but for a devolution of power to citizens' assemblies in each of those places. He could teach them how to do it.

Solidarizing with the Germans, Bookchin and his *compañeros* renamed themselves the Burlington Greens. In the summer of 1983, one of the Germans' founding members actually visited Burlington: Rudolf Bahro, the former East German dissident, who now called himself a social ecologist. He was just passing through—he had come to the United States to visit the ashram of Bhagwan Shree Rajneesh in Oregon. When Bahro arrived in Burlington, "all dressed in red"—the Rajneeshis' color—"it freaked us out!"[11]

Still, they admired Bahro and mined him for information about the German Greens. In October 1983 Bookchin and Bahro lectured at Dartmouth College to some 150 people on building a Green movement in the United States. The German Greens, Bookchin said, were attempting to create a new politics that was decentralized, ecological, and ethical. Whether that attempt would be successful "we have yet to ascertain." The Greens might very well go the way of the now-conventional MCM. To avoid that fate, they would need sophisticated consciousness, a libertarian focus, and intense commitment. Meanwhile in the United States, "we can learn as much from their errors as from their successes." For "a Green network we certainly require."[12]

That fall the German peace movement reached its apogee, in anticipation of the Bundestag's upcoming vote on whether to allow the deployment of the Pershing II and Cruise missiles. On October 22 a million and a half Germans demonstrated against deployment. Peace vigils and fasts, peace marches, caravans, and sit-ins surrounded missile sites and military bases—one protest followed another. As promised, the German Greens kept themselves anchored in the peace movement. Then, on

238   Ecology or Catastrophe

November 22, the Bundestag voted on deployment. The twenty-seven Green parliamentarians were among the 226 who voted no—but 286 voted in favor. With breathtaking speed, only a few weeks later, the missiles were in place.

The Greens—the peace movement's parliamentary arm—had failed. Entering politics had produced nothing. Defeat hung in the air. What should the idealistic German Greens do now?

By this time, Black Rose Books was publishing whatever Bookchin wrote and even giving his earlier, out-of-print works new editions. "I put my heart and soul into it because I really believed in the ideas," Roussopoulos told me. The feeling was mutual: Murray's collaboration with Dimitri was "a source of great support, both personally and politically, for nearly two decades."[13] Whenever Bookchin visited Montreal, he stayed at Dimitri and Lucia's graystone rowhouse, in the Milton-Park housing cooperative that they had helped organize (figure 11.1).[14]

Roussopoulos agreed with Bookchin that libertarian municipalism, a form of anarchism, was "the most convincing form of radical politics."[15] Both felt that anarchism generally was mired in its past, too focused on the nineteenth century, a veritable mausoleum. Anarchists' main political tool, direct action, was about protest,

FIGURE 11.1 With Dimitri Roussopoulos in Montreal, ca. 1988
Janet Biehl.

not about building lasting "forms of freedom." In order to become meaningful for the late twentieth century, anarchism had to overcome its aversion to political participation. Libertarian municipalism offered them the way to do that: it was a nonhierarchical, democratic politics for citizen empowerment. They were sure that anarchists would recognize the value of the idea and go for it.

It shouldn't be a great leap for them. After all, communal empowerment had once been part of the anarchist tradition. The canonical nineteenth-century thinkers, especially after the 1871 Paris Commune, had all supported a communalist strain of anarchism that encompassed "a sense of civic virtue and commitment." Proudhon had said that local communes and associations should group together in confederations in which "the delegates of townships" would have authority. Bakunin had said that after the revolution "the basic unit of all political organization" would have to be "the completely autonomous commune." These communes would form "federations of autonomous communes" at the provincial level; they would then federate nationally and internationally, replacing the nation-state. Kropotkin had agreed that in the postrevolutionary order, "independent communes, towns, and villages" would "break down the state and substitute free federation for parliamentary rule."[16]

Later anarchists had taken other routes, eclipsing this municipalism in favor of anarcho-syndicalism, but now that times had changed, the working class had proved to be nonrevolutionary, the age of ecology was at hand, and new social movements were on the rise, Bookchin and Roussopoulos wanted to revive that old tradition of the revolutionary commune. Libertarian municipalism, as a new iteration of anarcho-communalism, expressed that revival.

In the next years, puzzled anarchists would query Murray about his new program. "So, Murray," they would ask, as Canadian anarchist Ron Hayley did in 1985, "are you saying that anarchists should run for city governments?"

"I'm saying," Bookchin would reply, "that city government, as you call it, has to be restructured at the grassroots level."[17]

But how? was the common objection.

Bookchin would answer, by creating citizens' assemblies. We did it in Burlington by using grassroots pressure, and the city council did create them, but then once they were in place, the mayor's office neutralized their radical potential. What does the logic of that dilemma tell you? It tells you that we eco-anarchists—or anarcho-communalists—have to run for office and take power ourselves so that at that moment they won't do what Burlington's mayor did. They will instead choose to devolve power to the neighborhoods and vest it structurally in those citizens' assemblies.

But Murray, Hayley objected, "hasn't city government become really statified in the last ten years?" Or as others would later phrase it: weren't municipal governments actually just miniature nation-states, and city councils miniature parliaments?

To which Bookchin would reply by distinguishing between politics and statecraft. *Statecraft* included parliamentarism and hierarchy and domination and all the associated trade-offs and deal-making that he despised just as much as any anarchist; statecraft was endemic to state, provincial, and national governments.

But the local level, the neighborhood and community level, was qualitatively different: it was the realm of *politics*, where community self-management was possible: "When I use the word politics, I go back to the original Hellenic meaning [of] ... an active citizen body managing its own affairs." Politics need not degenerate into statecraft "if it is confined to the civic level" and also if it "is consciously posed against the State." For once the grassroots democracy was created, it would municipalize the economy, confederate with other such assemblies of eco-communities, and thereby form an alternative or dual power to "counteract the centralization of power."[18]

But Murray, was the invariable rejoinder, *how can you hope to achieve this goal by participating in electoral politics? Lots of well-intentioned radicals have thought they could change the system that way, but the same thing happens every time—the system changes them, and they become part of the problem. Wouldn't that happen to anarchist city councilors, too?*

Of course, Bookchin admitted, like anyone else, anarchist councilors could be co-opted. But we are all too aware of that sorry history now, aware of how that kind of absorption into the system has historically torpedoed previous radical movements. Future anarchist councilors could prevent themselves from being co-opted by studying the whole problem, and developing a sophisticated understanding. So when the time came for them to make the fateful choice, they would do so not naively but with an acute consciousness of what was at stake: that the very preservation of the biosphere depended on the choice that they and others, in other liberated municipalities, would make. And so they would choose with moral probity. There was no substitute, Bookchin often said, for consciousness, "for a conscious reconstruction of our relationship to each other and the natural world."[19]

Persuading anarchists, Bookchin began to realize, might prove to be a challenge after all—a rejection of politics was deeply ingrained not only in their ideology but in their political identity. But he was confident of his powers of persuasion, and that the historical logic of this approach would bear him out. After all, did anarchists really want to remain irrelevant to the course of events? Did they want their ideology to remain a museum piece? As Roussopoulos mused years later, if you want to take over cities but you don't want to run in elections, "then what are you going to do? Storm city hall with rifles?"[20]

One day in the early 1980s, Murray was in Dimitri's study, leafing through Sam Dolgoff's anthology of Bakunin's writings (he had helped Dolgoff get it published long ago), when he came across a passage that made him gasp and shout. Dimitri rushed in. Murray pointed to a passage where Bakunin said that municipal politics was qualitatively different from politics at the provincial and national levels: "The people ... have a healthy practical common sense when it comes to communal affairs. They are fairly well informed and know how to select from their midst the most capable officials. Under such circumstances, effective control is quite possible, because the public business is conducted under the watchful eyes of the citizens and vitally and directly concerns their daily lives. This is why municipal elections always best reflect the real attitude and will of the people. Provincial and county governments, even when the latter are directly elected, are already less representative of the people."[21]

In other words, the hallowed anarchist forefather had considered the municipality to be a valid terrain of anarchist activity. What greater legacy could an anarchist call upon than Bakunin! Bookchin exulted. The passage established the anarchist legitimacy of libertarian municipalism, Roussopoulos told me. "It was no longer his invention—it had a pedigree."[22]

Roussopoulos was encouraging anarchist publishers in Europe to translate Bookchin's work: Antistato in Italy, Atelier Création Libertaire in France, Diethnis Bibliotheke in Greece, and Trotzdem Verlag in West Germany. As a result, European anarchists were becoming curious about this new politics, and Roussopoulos and Bookchin were eager to explain it. To give them a venue, the Europeans organized an international anarchist gathering that would "create a new anarchism." The *incontro*, as it was known, opened in Venice on September 24, 1984. "Ciao! Anarchici!" read the banner under which more than three thousand *compañeros* streamed during the week, young and old, anarcho-syndicalists and punks, straight and gay, from as far away as Malaysia and New Zealand. Two hundred German and Swiss autonomous youth arrived, with iridescent chopped hair, many of them squatters. They were "a bit wild and woolly," complained the aging English anarcho-syndicalist Albert Meltzer, who preferred to fraternize with the aging CNT veteran Luis Edo. They spoke a babel of languages, but when they couldn't comprehend each other's words, they gestured and danced.[23]

Bookchin's speech was eagerly anticipated. He took the podium, in dark green work clothes, a row of mechanical pencils arrayed in the breast pocket, as always, and launched into it. Capitalism is simplifying the planet, he told them. It is commodifying both people and the natural world. The historical subject is not the worker but the citizen. The feminist, ecology, and community movements must create decentralized human communities, tailored to their ecosystems. They must democratize towns and cities and confederate them, and create a dual power against the state. When he mentioned that anarchists should run in elections, the old *cenetista* Edo let out a shocked cry: "Murray!" But most suspended judgment out of respect, listening closely.

After the plenary sessions, Bookchin continued talking about it to the *compañeros* as they walked along Venice's canals and *campi*, well into the night. Bob D'Attilio recalled that Murray was "a remarkable presence in addressing a group, crowd, street gatherings. I think he amazed the comrades with his passion, his spontaneity, his energy. They loved him." Murray was "outstanding," Roussopoulos recalled. "At every occasion he would engage in long discussions and debates."[24]

One afternoon while talking to a group of twenty or thirty anarcho-syndicalists in a piazza, Bookchin explained the distinction between workplace and community, between worker and citizen. "There was a lot of hesitation," D'Attilio told me, "perhaps yes, perhaps no." Bookchin repeated that anarchists should run in municipal elections, when an old *cenetista* woman stood up—she had been among the Spanish exiles in France. "We voted in municipal elections!" she beamed, smiling broadly.[25] Later, Luis Edo made his way over to Murray and embraced him and kissed him. "That's the kind of situation that he created," Roussopoulos said.

Some German anarcho-punks complained to Albert Meltzer, the old-school anarchosyndicalist, that the translators were doing a poor job. "It almost sounded as if they were talking of a political party to fight elections on a 'green' basis," they said in disgust. "I had to tell them," Meltzer said, that "their understanding and the translations were perfect and what was inadequate was the discrimination of the organisers." But Bookchin paid no attention to such grousing and buttonholed some of the older German anarchists, urging them to get involved with the German Greens and bring this new politics with them—the Greens needed it badly.[26]

All in all, Bookchin found the Venice *incontro* "a rare gathering . . . committed to searching discussion." The Italian publisher Paolo Finzi told me in retrospect that it was "a magic moment." Bookchin represented "a sound anchorage with anarchist 'tradition' and at the same time tried to explore new realms of thought and action. Murray was a living patrimony for a movement that long had a too-narrow relationship to its own great ideas."[27] As a coming-out party for libertarian municipalism, it was promising.

Afterward, Bookchin traveled around Italy for a few weeks, speaking in small cities and towns where he'd been invited. In Carrara, one of the older anarchists told him a fascinating story.[28] Starting in the 1880s, anarcho-syndicalists had dominated politics in the town. They had led strikes and introduced the six-hour workday—and were immensely popular. From 1945 until the 1960s, Carrara was all but in their hands—if they had chosen to run for office, they could have had a majority on the city council. But the anarchists did not run for office. Instead, the Communists and Socialists did—and now, in 1984, those parties controlled Carrara. Walls around town bore the hammer and sickle; the union of marble workers, once anarchist, was now part of the Communist labor federation. The anarchists were reduced to a handful of elderly people in a broken-down reading room.

What had happened? The problem, this man told Bookchin, was that the anarchists were absolute political abstentionists. They regarded even voting, let alone running for office, as blasphemy. The Communists had easily moved into the political vacuum—and occupied it.

In late October, Bookchin took a train up to Frankfurt to catch up with the Greens, on whom he now pinned great, even revolutionary hopes. "Around the time of Venezia," Roussopoulos told me, "he was getting much more involved with the Germans, almost obsessive. He was a passionate Germanophile. Like Lenin and the Bolsheviks, he thought Germany was the key to the future."[29] How would the Greens move forward, now that they had lost the struggle against the Euromissiles?

Bookchin found Jutta Ditfurth (figure 11.2) to be a kindred spirit: she was a charismatic speaker, as he was, driven by a ferocious commitment to upholding a principled revolutionary politics, given to quoting Rosa Luxemburg; militant and tenacious,

FIGURE 11.2 Jutta Ditfurth fights to ensure that the German Greens remain true to their original principles, 1987
Wikimedia Commons.

yet also charming, warm, and infectious in spirit. She and her allies were Bookchin's great hope within the Greens.

In her Frankfurt apartment, Jutta explained to Murray the recent developments.[30] You have to go back to the Hessian *Landtag* elections of September 1982, she might have said. Neither of the two major parties in Hesse got enough votes to form a government on their own. The Greens got 8 percent. The Social Democrats (SPD) realized that if they could persuade the Greens to ally with them, they could form a "red-green" governing coalition.

When they proposed it, the Greens of course said no. Such a coalition was inconceivable to them—it was incompatible with their self-identity as the principled anti-party party of fundamental opposition. And they loathed the SPD as unprincipled deal-makers.

The day after that Hessian election, a local taxi driver named Joschka Fischer was thinking about the situation and got the idea that a red-green coalition would really not be such a bad thing, especially if the Greens could negotiate good terms. Up to that point, Fischer and his anarchist (Sponti) friends had not been Green members, but now they paid their dues and signed up. They created a Realpolitik Working Group to urge the Greens to, well, get real. Fundamental opposition was "the babblings of yesterday," they said, and a fundamental transformation of society was unobtainable.

Those who pretended otherwise were pseudo-radical purists—"fundamentalists," or *fundis*. Fischer and his friends won over several hundred Hessian Greens to their point of view, and in January 1983 a provincial assembly elected him to the Hessian list. Then, on March 6, 1983, Fischer had been one of the twenty-seven Greens to enter the Bundestag in that colorful parade.

In the next provincial election in Hesse, in late 1983, Ditfurth continued, the Greens' vote fell to 5.9 percent. They still had enough to qualify for seats in the *Landtag*, but the percentage was less than before. Fischer and his fellow realpolitikers blamed the *fundis* for this setback. *See?* he said. *Voters don't want fundamental opposition!* So shaken were the Hessian Greens that their grip on the reins of leadership loosened. Fischer easily took those reins for himself.

That was when the local Social Democrats came knocking. *Rule Hesse with us,* they appealed to him, *in a red-green coalition. My Green base won't approve it,* he surely replied. *Okay then,* said the SPD, *suppose you Greens don't form a coalition with us—but you do agree to sit back and "tolerate" our SPD government. You wouldn't have to join a ruling coalition. You'd just have to say you "tolerate" its budget. How about it, Fischer?*

Fischer replied: To get my Green base to agree to that, you'd have to shut down Hesse's two nuclear power plants and cancel the Frankfurt airport runway extension.

No, they said. *We agree to build no more new nuclear plants, but that's all. And we'll accept some of your Green policies on energy and forests.*

Fischer took this proposed "toleration agreement" to a Hessian Green congress and recommended to the assembled movement activists that they should approve it. Symbolic gestures, he told them, like human chains and peace marches, change nothing. It's legislators, not demonstrators, who make policy. To get at least part of what you want from your adversaries, you have to negotiate. Politics is the art of compromise. It's really just horse-trading.

The Frankfurt radicals (they rejected the name *fundi*) were appalled and fired back. Fischer is asking you, they told the congress, to "tolerate" the continued operation of the two nuclear power plants. That's unacceptable. These people are sellouts, power-hungry opportunists, wheeling and dealing. We Greens have to stick to our original goal: fundamental social change.

The Hessian Greens listened to both sides. Then they voted to "tolerate" the SPD minority government.

And here we are, Jutta might have said to Murray in October 1984. The Greens are in a ruling coalition with the SPD in Hesse. Fischer and his buddies are trying to normalize the Greens so that they become just another party in a corrupt system careening toward ecological disaster. But from the start we have been about fundamental opposition, about grassroots democracy and ecology and peace and social justice. We are about creating a sane future. My friends and I are still committed to that, to the movements from which the Greens sprang, and that's what we'll fight for.

Bookchin would fight for it, too. In late October he spoke in Hamburg, West Berlin, and Hanover, accompanied by his friend and translator Karl-Ludwig Schibel (figure 11.3). Among Greens, he met with realists as well as radicals,[31] arguing

FIGURE 11.3 With Karl-Ludwig Schibel, his German translator and friend, 1988
Janet Biehl.

against the toleration agreement, by which the Greens would surrender their very identity. But then, the Greens should not have entered into the legislatures in the first place, he said—a parliamentary party will always demoralize and vitiate the extraparliamentary movement that gave rise to it. What the Greens should do is abandon provincial and federal politics and create citizens' assemblies at the neighborhood level.[32] And when he met with German anarchists, it was to importune them to join the Greens and work with the radicals to strengthen the wing that was committed to fundamental opposition.

As always, his appearances were followed by late-night discussions in cafés. He savored Germany's feisty radical scene and the verve and imagination of its alternative communities. He admired the Greens' "spontaneity and fervor, the sparkling brilliance of their speakers, and above all the wide range of social, political, and moral issues with which they deal." The great debate was creative, he thought. He scrutinized every development in Green politics, because "the soul of the movement hangs excitingly" on whether the Greens moved to the left or made compromises with the establishment.[33]

He shared a platform with Ditfurth in Kassel on October 31 and praised her principled commitment to fundamental opposition. He also lauded the Hamburg eco-socialists Thomas Ebermann and Rainer Trampert, who had evolved from Marxists to eco-socialists. They weren't just putting a green coat over a red identity—"they've largely abandoned all their Leninist principles, and have moved in a highly libertarian direction. . . . That has been terribly encouraging."[34]

The leftist Greens, as individuals, possessed an exhilaratingly high level of theoretical understanding, he thought, and their intense idealism gave them the power of endurance. Not since the 1930s had he witnessed political actors so ready to undertake the laborious, protracted, thankless work of implementing social change. They weren't merely expressing themselves—they were ready to take responsibility (in ways that their parents' generation had not). They were adults.[35]

By comparison, Americans, as political actors, seemed erratic, plagued by "an appetite for episodic 'highs' and 'kicks.'" They entered and left movements "as though they were skating through shopping malls." When he got back to the United States, he resolved, he would not drop his standards to accommodate their immaturity.[36]

Still, the American eco-activists had a great advantage: they had the country's utopian traditions at their backs, its town meetings, its mistrust of the federal government, and its respect for individual rights. Germans, he suggested, could learn from those libertarian traditions, while Americans could learn from Germans' theoretical rigor. If they could ever somehow join forces, the resulting movement would be a formidable force.

He was eager to share his philosophical ideas with the German Green intellectuals, especially his solution to the Frankfurt School's dilemma about reason and ethics. In a lecture at Goethe University, the original home of the Frankfurt School, he explained that one could ground ethics ontologically in nature without raising the specter of Social Darwinism: he told his listeners about the evolutionary continuum between nature and society, its gradations and phases and mediations, and the organic unfolding of latent potentialities for freedom and consciousness.[37] Whether his audience understood that he was trying to answer the Horkheimer-Adorno dilemma, or were even aware of it, is unrecorded.

A few days after Bookchin left Germany, in November 1984, the Hessian economics minister (an SPD member) approved the construction of a new plutonium-processing plant near Frankfurt. That was more than even Joschka Fischer and his *realos* could stomach. On November 19 the Greens withdrew from the toleration agreement, and the SPD's provincial government in Hesse collapsed.[38] Now the radicals had a chance to put the party back on course. In December a Green congress elected Ditfurth as one of three speakers of the BuVo (*Bundesvorstand*, the party executive at the federal level). From that high-profile position, she could speak for fundamental opposition.

Across the Atlantic, the Reagan administration's antienvironmental policies made urgent the formation of an American Green movement. In May 1984 activists met at the North American Bioregional Congress, in the Ozark foothills of Missouri, and formed a Green politics working group. Later that year, a larger group of people met in St. Paul, Minnesota, to continue the discussions; the ISE organized the meeting.

Sixty-two people attended, among them Dan Chodorkoff of the ISE and Chino Garcia of CHARAS, who advanced the position that the US Greens should take a decentralized approach to organizing their movement. They should establish local groups, then federate them in regional alliances, through an openly democratic, bottom-up process. But others at St. Paul rejected this message—they preferred to establish a national Green organization, even a top-down Green party, right away. Among them were Charlene Spretnak, author of a book about the German Greens (whom she viewed through a New Age lens). As a compromise, the St. Paul gathering established a loose Green organization, called the Committees of Correspondence, to begin organizing.

Bookchin was chagrined by the calls for a national party, but the nature of the American electoral system was, ironically, a consolation. In the US winner-take-all system, the candidate with the most votes gets elected, period. By contrast, in the German system of proportional representation, a party needed only 5 percent of the total vote to gain seats. That had catapulted the Greens onto the national level too quickly. The American system, fortunately, would shut Greens out until they could win 51 percent. Americans thus wouldn't have to grapple with the problem of whether to form a conventional party to participate in statecraft, an obstacle that would only be a boon. With parliamentary participation off the table, the American Greens could build on the example of the New England town meeting. And Vermont would play a crucial role, both as the locus classicus of the town meeting and as the place that had launched the nuclear freeze issue onto the national landscape.

But before Bookchin could engage fully with Vermont, Europe beckoned once again. A year after his previous visit, he returned to Germany—Schibel had translated *The Ecology of Freedom*, then found a publisher and organized a book tour. When he arrived in October 1985, the intraparty battle for the soul of the Greens had escalated. Ditfurth, bold, forthright, and militant, had become the most prominent spokesperson for the radicals, pitted against Fischer, the leader of the *realos*. In public debates and at party congresses, each appealed to the base for support on behalf of their respective factions. Unrelentingly, the *realos* gained ground: by March 1985 they called the shots even in the Frankfurt party.[39]

Meanwhile Greens in other parts of Germany were winning seats in legislatures at all levels of government. As Green parliamentarians honed their political skills and cultivated media connections, their ties to the movement loosened, and their caucuses became increasingly independent. The structures of accountability that were intended to prevent normalization were weakening. One high-profile Green, Petra Kelly, refused to yield her seat when party rules required her to rotate—that would disrupt the continuity of experienced personnel, she argued. When Green newcomers were elected to important positions, those who had more experience manipulated and sidelined them. To get around the ban on individuals' holding both party and legislative office simultaneously, a small group of Green politicians alternated positions among themselves for appearances' sake. These erosions of the democratic structures weakened rank-and-file control over the parliamentarians.

At the Frankfurt Book Fair that October, Bookchin stationed himself at his publisher's display, where *Die Ökologie der Freiheit* received considerable attention. But outside in the streets, Frankfurt was in an uproar. Two weeks earlier, on September 28, a thirty-five-year-old antifascist, Günter Sare, had been marching in a demonstration, when a twenty-six-ton water cannon knocked him over with its stream—and then ran over him, killing him. The city's alternative scene convulsed with fury. Black-hooded anarchists ripped into the streets to fight with police. Ditfurth denounced Sare's killing as a "police state in action."

Just at that moment, Hesse was holding yet another provincial election; once again, as in 1983, no single party gained the necessary majority, and once again Greens gained enough seats to make the difference. This time, the SPD wanted the Greens to agree to a real governing coalition, not merely a toleration. The two parties negotiated. Holger Börner, the local SPD headman, handed Fischer his party's terms: the Greens must accept the existence of Hesse's two nuclear facilities. Fischer submitted Green demands, in turn, but Börner rejected them. Meanwhile, the streets were aflame with outrage over Sare's death. When Fischer and his friend Daniel Cohn-Bendit spoke at a Frankfurt University teach-in, the audience threw eggs at them. Ditfurth lambasted the *realo* Hessian Greens for even considering an agreement with the water-cannon-deploying, nuclear-power-plant-permitting, airport-runway-extension-supporting SPD.[40]

Bookchin had to leave Frankfurt for the book tour, but he and Ditfurth agreed to meet up at a later point. Schibel had organized a weekend seminar in a town outside Hamburg on October 11–13. Green intellectuals and politicos from all factions came to hear Murray Bookchin. He tried talking to them about nature philosophy and the Frankfurt School but soon saw their eyes glaze over—the only question that could hold their attention was whether the Hessian Greens should agree to that governing coalition with the SPD.[41]

Murray begged them to forget about all such parliamentary coalitions. Real politics wasn't about mobilizing for the next election—it was about creating citizens' assemblies in neighborhoods and communities. He sat down with Thomas Ebermann, the Hamburg eco-socialist whom he admired as "one of the most gifted voices of the 'socialistic' wing of the Greens."[42] What did Ebermann think of his municipalist program? he asked. Would the leftist Greens consider that route?

Ebermann, who was familiar with Murray's ideas, replied by dismissing Murray's basic premise that local politics was somehow closer to the people than provincial or federal—in Germany, he said, even local legislators got caught up in machine politics. To which Murray responded that the point of electing people to a city council was not to institute whatever environmental reforms the local economic interests would allow, but to create citizens' assemblies and devolve power to them by changing the city charter.

German cities don't have charters, Ebermann objected.

Then create assemblies outside the power structure, Bookchin said, and insist on structurally empowering them! Demand urban land for green spaces, for ecological programs. Demand housing, control over education, free public transportation, and so on, and escalate these demands into points of friction, of contestation. People will flock to the assemblies you call, and you will be on your way to democratizing the towns. Then confederate those democratized towns and cities and demand power from the provincial government, then from the federal government. As each level demands power from the one above it, then gains it, you will restructure Germany into a popular democracy. But first you need a civic movement to create those neighborhood assemblies. Without such a movement, you'll never go beyond wheeling and dealing in the back rooms of the Bundestag.

Ebermann replied that Bookchin's plan was too idealistic. Municipalities would never cooperate—capitalist society creates competition among them. Finally he dismissed Murray's ideas precisely for being ideas, for being a theory. Theories are part of the problem, he said, and only fearful people need them. We're not waiting for some great person to rescue us with a brilliant idea or the perfect formula. The best ideas will come from the streets, where the real fight goes on.[43]

As I listened to the audiotape of this conversation, decades later, while doing research for this book, I thought I heard Murray's heart breaking in the background. He had been so eager and had placed so much hope in these Germans, who he thought shared his love of ideas. Ebermann's brush-off must have hurt.

Still, disciplined agitator that he was, Bookchin continued on his book tour, moving from one city to the next, giving inspiring speeches, sometimes two a day.[44] On October 21 he met up with Ditfurth, who surely gave him the terrible news, if he hadn't heard it already: a few days earlier Joschka Fischer had reached a deal with the Hessian SPD to form that governing coalition. The SPD had rejected almost all the Green demands, but Fischer had accepted the coalition anyway. He had won only one concession from the Social Democrats: the Greens could name *the minister of environment and energy*. That is, the Greens could name one of their own to oversee the two nuclear power plants, plus a new plutonium-processing plant that was in the works.

But the last word lay with the Hessian Green rank and file. They would have to give the proposed coalition agreement a thumbs up or a thumbs down. They would do so at a membership assembly, to be held a few days later at Neu-Isenburg, outside Frankfurt.

In the intervening days, Bookchin and Ditfurth campaigned together for principle and the grassroots and democratic accountability; they appeared jointly at a high school, to a packed auditorium, denouncing the *realos'* opportunism and power grabbing. Parliamentarism corrupts even the best-intentioned people, he told the audience, and once enmeshed in statecraft, they become an encumbrance on any radical movement.[45]

He spoke at Goethe University again, then Heidelberg University, and finally, on October 27, he reached Neu-Isenburg. Instead of the usual two or three hundred members who attended these assemblies, more than a thousand Greens packed the hall for this crucial one. Longtime Frankfurt activists turned up in force, those who had opposed the airport runway extension since the early 1970s, and those who had fought the construction of nuclear power plants since Wyhl, and even those who had worked on the early citizens' initiatives in the 1970s. But the *realos*, too, turned out, all of Fischer's Frankfurt pals: "Everyone who wanted a little office for himself was there," recalled Ditfurth.[46]

Should the Greens accept the coalition with the SPD? Both sides weighed in on the momentous question, alternating *fundis* and *realos* at the microphone. When Ditfurth's turn came, she said the proposed coalition "is not realism, it is the road to integration into the ruling system." Accepting it would actually require the Greens to accept the existence of nuclear power plants! But in truth "the only coalition partners for us are the social movements." They are growing, they are ready for us, and today "we have a chance to move toward the reconstruction of this industrial society into an ecological, social, and democratic society." Or rather, she tried to say these things—all the while the *realos* were heckling her, trying to drown her out. "The atmosphere was heated to the boiling point," she recalled. Anyone who "tried to defend the green program was shouted down." One of Fischer's old cronies cynically "compared those of us who opposed the coalition to members of a Nazi 'tribunal.'"[47]

And then it came time for the Hessian Greens to vote. As Bookchin watched from the floor, they raised their yellow membership cards to register their choice. By a two-to-one margin, they approved the red-green coalition.[48]

No longer were the Hessian Greens an extension of a movement—they were now a conventional party. Fischer, true to form, didn't miss a beat: he declared himself a candidate for minister of environment and energy.

Bookchin made a final appearance the next day, October 28, with Ditfurth, in nearby Siegen, aching from the painful loss. "It's deeply depressing," Ditfurth grieved. "Instead of throwing sand in the engine, [the Greens] are spreading a new and better kind of oil."[49] Bookchin listened to her words with immense respect.[50] And surely wondered: If the anarchists had entered the Greens and fought on the *fundi* side, could they have made a difference?

A few weeks later, on December 12, 1985, Joschka Fischer became Hesse's minister for energy and environment. When he did, the rank-and-file movement-oriented Greens found that they had a dilemma: they could no longer campaign for a shutdown of nuclear power plants. One of their own, after all, was minister in charge of ensuring their continued operation.[51]

By then Bookchin had moved on to a Parisian suburb, where some of his Italian and Swiss anarchist friends were gathered for a conference. He spoke to them with his usual passion about municipalism, but in the year since Venice, they had thought it over and reached their verdict. It was not the one Bookchin had been hoping for.

The municipality could not possibly be a realm of freedom, they insisted to him. Perhaps it could have become a realm of freedom a century ago, one *compañero* said—perhaps—but today's municipalities could not possibly transform themselves into citizens' assemblies. It was out of the question. The nation-state had swallowed the municipality, chewed and digested it, and assimilated it into its now seamlessly monolithic structure. Neighborhood power too was over, another asserted, because even neighborhoods belonged to history.

Nor should we mourn the impossibility of citizens' assemblies, someone interjected, since they perpetuated economic inequality and were therefore potentially regressive.

Bookchin found it depressing. Why were they all so eager to spurn libertarian democratic institutions, rather than think out how to create them? Why were they so indifferent to today's antihierarchical movements—peace, feminism, ecology? Did a movement have to explicitly label itself as anarchist in order to gain the anarchists' seal of approval? The new movements were anarchist in practice, even if they didn't use that precise label: they all expressed an emancipatory moral sensibility and pointed to broad notions of freedom.

Most recently, the radicals in the German Greens didn't call themselves anarchist, but for the past few years they had deeply explored the whole issue of movement-based democratic accountability. The *fundi-realo* fight had been a momentous battle to prevent a grassroots-democratic movement from being absorbed into the state. Yet the anarchists had sat on the sidelines. They should have joined the Greens' left-libertarian wing, he thought, and tried to make it stronger and even more libertarian. But they had not, and now the German battle was lost. Still, Green movements elsewhere were full of promise. Anarchists should enter those movements and advocate assembly democracy there, to prevent a repetition of the German fiasco . . .

As he spoke, his listeners were fuming. Finally one *compañera* could hear no more. *Anarchists—enter the Greens?* she snapped. *But we'd lose our identity as anarchists!*

Murray was flabbergasted. Political labels were a side issue, he said—one could travel many routes to get to a destination. Movements for freedom had a long history—he'd written about them in his books. They weren't necessarily explicitly anarchist, and many of them long antedated anarchist ideology, but that didn't disqualify them from being part of a tradition of freedom. Didn't his listeners care about them, or want to build on them?

Another anarchist demurred: no traditions of freedom existed, only traditions of domination.[52]

Bookchin could see it was hopeless. Afterward he reflected that these Europeans, although highly cultivated, were paralyzed by dogma and perhaps even cynicism. Surprisingly, they were making him homesick. Americans might be naive and impetuous by comparison, prone to seeking kicks, but they could usually be counted on for a can-do spirit.

As he traveled through Britain in the late fall of 1985, he urged anarchists in the peace and ecology movements to get involved with Green politics, to a mixed response. Finally in late November, he spoke at the History Workshop at Leeds Polytechnic. At a postconference workshop on anarchism, he opened his heart to his anarchist listeners.[53] I genuinely need to know what you think, he appealed to them. We have movements today that aren't anarchist in name—peace, antinuclear, feminist, Native American. But they have anarchist qualities. They reject hierarchy. They want to empower people rather than legislators.

How, in your opinion, should anarchists relate to them? Do we check first for an anarchist label, and if it's missing, we stay away? Do we run down our list of anarchist criteria, and if one or two are missing, then we say goodbye? Or do we work with them anyway?

How would you have related to the German Greens? he asked. They were based in a huge antinuclear, prodemocracy, ecology-oriented movement that brought multitudes of people into the streets. De facto anarchists cofounded that movement and cowrote the Greens' program.[54] The Green idea of a "nonparty party" had anarchist roots. The radicals, the *fundis*, stood for democratic accountability, and some even thought assembly democracy was a good idea.

In the *realo-fundi* fight, these de facto anarchistic Greens tried to keep the party from becoming parliamentary and conventional. Was that a fight worth joining? he asked. Most German anarchists didn't solidarize with the *fundis*—in fact, the black-masked ones in Berlin had seemed to prefer to throw bricks. What would you have done?

And what should we do in the United States? he asked further. The German Greens have been inspirational to us. Now in New England, in the name of Greens, we're calling for assemblies and local control and municipal confederations to counteract the nation-state. Should we expect anarchists to solidarize with us? Or will they stay aloof?

Judging from the audiotape, the English anarchists present didn't entirely understand the question, and those who did responded no more favorably than the continental anarchists had.

Bookchin was ready to go home. If municipalism was going to happen anywhere, it would be in the United States, specifically New England. He was finished trying to promote the town meeting to Europeans. Now he would get to work in Vermont.

NOTES

1. Paul Hockenos, *Joschka Fischer and the Making of the Berlin Republic* (New York: Oxford University Press, 2008). For more on the German Greens, see Elim Papadakis, *The Green Movement in West Germany* (London: Croom Helm, 1984); and Werner Hülsberg, *The German Greens: A Social and Political Profile*, trans. Gus Fagan (London; Verso, 1988). See also Brian Tokar, *The Green Alternative: Creating an Ecological Future* (Gabriola Island, BC: New Society, 1994).

2. Jutta Ditfurth, *Das waren die Grünen: Abschied von einer Hoffnung* (Munich: Econ Taschenbuch Verlag, 2000), 79.

3. Ibid., 63–65.

4. Ibid., chap. 4. See also Ditfurth, *Entspannt in die Barbarei: Esoterik, (Öko-)Faschismus und Biozentrismus* (Hamburg: Konkret Literatur Verlag, 1996); Raimund Hethey and Peter Kratz,

eds., *In bester Gesellschaft: Antifa-Recherche zwischen Konservatismus und Neo-Faschismus* (Göttingen: Verlag die Werkstatt, 1991); Janet Biehl, "'Ecology' and the Modernization of Fascism in the German Ultra-Right," in *Ecofascism: Lessons from the German Experience* (San Francisco: A.K. Press, 1995), 32–73; Rudolf Bahro, *From Red to Green* (London: Verso, 1984); Die Grünen, *Programme of the German Green Party* (London: Heretic Books, 1985).

5. Ditfurth, *Das waren die Grünen*, 103. See also Thomas Scharf, "Red-Green Coalitions at Local Level in Hesse," in Eva Kolinsky, ed., *The Greens in West Germany: Organisation and Policy Making* (New York: Berg, 1989).

6. Ditfurth, *Das waren die Grünen*, 102–4.

7. Römerfrakion quoted in Thomas Scharf, *The German Greens: Challenging the Consensus* (Oxford, UK: Berg, 1984), 120.

8. See Brian Tokar, "The Greens as a Social Movement: Values and Conflicts," in Frank Zelko and Carolin Brinkmann, eds., *Green Parties: Reflections on the First Three Decades* (Washington, DC: Heinrich Böll Foundation of North America, 2006).

9. Bookchin, "Popular Politics vs. Party Politics," *River Valley Voice* [Turner Falls, MA], 1984.

10. Bookchin, "The German Greens," *Gadfly* (University of Vermont) 2, no. 5 (Nov. 31, 1986).

11. Brian Tokar, interview by author, Aug. 7, 2009. In the next year or two, the Burlingtonians would devour Rudolf Bahro's books: *From Red to Green* (1984) and *Building the Green Movement* (1986).

12. Bookchin, "An Appeal for Social and Ecological Sanity" (1983), in *MC*, 154. On his appearance with Bahro, see Charlotte Gerstein, "Peace Speakers Talk in Center," *Dartmouth*, Oct. 3, 1983; and Mathias Dubilier, "Bahro, Bookchin and the Greening of New England," *VVP*, Oct. 9–16, 1983.

13. Bookchin, introduction to *The Limits of City*, 2nd ed. (Montreal: Black Rose Books, 1986), 27.

14. Claire Helman, *The Milton-Park Affair: Canada's Largest Citizen-Developer Confrontation* (Montreal: Véhicule Press, 1987).

15. Dimitri Roussopoulos, interview by author, July 6, 2009.

16. Bookchin, "The Ghost of Anarchosyndicalism," *Anarchist Studies*, no. 1 (1993), 8; Pierre-Joseph Proudhon, *The Principle of Federation* (1863), trans. Richard Vernon (Toronto: University of Toronto Press, 1969), 41, 45, 48; Michael Bakunin, "Revolutionary Catechism of the International Revolutionary Society or Brotherhood" (1865), in Daniel Guerin, ed., *No Gods No Masters: An Anthology of Anarchism*, trans. Paul Sharkey (San Francisco: A.K. Press, 1998), book 1, 142, 132; Kropotkin, "Modern Science and Anarchism" (1913), in Roger Baldwin, ed., *Kropotkin's Revolutionary Pamphlets* (1927; reprinted New York: Dover, 1970), 163, 185–86.

17. Bookchin, "Democratizing the Republic and Radicalizing the Democracy," *Kick It Over*, Winter 1985–86, 9.

18. Ibid., 9. See also Bookchin, "Theses on Libertarian Municipalism" (1984), in *Limits of the City*; and Bookchin, *The Rise of Urbanization and the Decline of Citizenship* (San Francisco: Sierra Club Books, 1986).

19. Bookchin, "The Meaning of Confederalism," *Left Green Perspectives*, no. 20 (1990), 5.

20. Roussopoulos interview.

21. Michael Bakunin, "Representative Government and Universal Suffrage" (1870), in Sam Dolgoff, ed., *Bakunin on Anarchy* (New York: Knopf, 1972), 218–24, esp. 223. "I would want to restate his formulation to mean that municipal elections can more

accurately reflect the popular will than parliamentary ones," Bookchin wrote in "Deep Ecology, Anarchosyndicalism, and the Future of Anarchist Thought," in *Deep Ecology and Anarchism: A Polemic* (London: Freedom Press, 1993), 55.

22. Roussopoulos interview.

23. "Introduction: Venice 1984," *Black Rose*, no. 11 (Winter–Spring 1985), 4–12, http://bit.ly/1bibHtl; Albert Meltzer, *I Couldn't Paint Golden Angels* (San Francisco: A.K. Press, 1996); Roussopoulos interview. For photographs of the *incontro*, see Agnaldo S. Maciel et al., *Ciao Anarchici Venezia 1984*, International Co-edition (Milan: Antistato; Geneva: Noir; Lyon: Libertaire; Montreal: Black Rose, 1986).

24. Roussopoulos's account in *Black Rose*, no. 11 (Winter–Spring 1985), 24–25, http://bit.ly/1bibHtl; Bob D'Attilio to author, July 16, 2009; Roussopoulos interview.

25. Roussopoulos interview.

26. Meltzer, *I Couldn't Paint Golden Angels*; Wolfgang Haug to author, Dec. 8, 2009.

27. Bookchin's account in *Black Rose*, no. 11 (Winter–Spring 1985), 21–23, http://bit.ly/1bibHtl; Paolo Finzi to author, Mar. 30, 2010.

28. He lectured at Carrara, the speech transcribed and translated as "La crisi ecologica: Le sue radici nella società: Problemi e soluzioni," published as a brochure by Circolo culturale anarchico di Carrara.

29. Roussopoulos interview.

30. Ditfurth, *Das waren die Grünen*, chap. 5.

31. Schibel to author, Jan. 2010. Murray met Eberhard Walde (Greens' national secretary and a party founder), Helga Trüpel (a member of the Bremen Greens' regional executive), Claudia Pilz (the Hamburg Greens' provincial secretary), Antje Vollmer (spokesperson for the first Green caucus in the Bundestag), Ralf Fücks, and Hannegret Hönes (1985–87 in the Bundestag).

32. Bookchin, "Parteipolitik oder populistische Politik: Anmerkungen eines in Deutschland reisenden Amerikaners," *Kommune* 1 (Jan. 18, 1985).

33. Bookchin, "New Social Movements: The Anarchic Dimension," in David Goodway, ed., *For Anarchism: History, Theory and Practice* (London: Routledge, 1989), 266.

34. Bookchin, "Democratizing the Republic," 10.

35. Bookchin, "Party Politics vs. Popular Politics"; "German Greens."

36. Bookchin, "German Greens"; Bookchin to Beatrice Bookchin, Nov. 15, 1984, Box 17, MBPTL.

37. Bookchin, seminar lecture at the University of Frankfurt, Nov. 12, 1984, published as "The Radicalization of Nature," *Comment* special paper in *Comment* (ca. 1984).

38. Ditfurth, *Das waren die Grünen*; see also Hockenos, *Joschka Fischer*, 204.

39. Ditfurth, *Das waren die Grünen*, 118–19.

40. Ibid., 120–21.

41. Bookchin to John Clark, Oct. 18, 1985, Box 17, MBPTL.

42. Bookchin, "German Greens."

43. Bookchin, conversation with Thomas Ebermann, audiotape, October 1985, copy in author's collection.

44. Bookchin to Clark, Oct. 18, 1985.

45. Bookchin, "Parteipolitik oder populistische Politik."

46. Bookchin to Everybody, Oct. 29, 1985, Box 17, MBPTL; Ditfurth, *Das waren die Grünen*, 121.

47. Jutta Ditfurth, "Die Grünen als Teil des herrschenden Systems?" (Oct. 10, 1985), in Ditfurth, *Träumen, Kämpfen, Verwirklichen: Politische Texte bis 1987* (Cologne: Kiepenheuer & Witsch, 1988), 194; Ditfurth, "Der radikale Weg verspricht Erfolg" (1985), in *Träumen, Kämpfen, Verwirklichen*, 233; Ditfurth, *Das waren die Grünen*, 121.

48. Bookchin to Everybody, Oct. 29, 1985.

49. Ditfurth, "Grüne Wende rechtswärts" (Oct. 1983), in *Träumen, Kämpfen, Verwirklichen*, 236, 237.

50. Bookchin to Everybody, Oct. 29, 1985.

51. Hülsberg, *German Greens*; Ditfurth, *Das waren die Grünen*.

52. Bookchin to Clark, Nov. 5 and 8, 1985.

53. Audiotape courtesy David Goodway.

54. He was referring to Ditfurth's partner Manfred Zieran, who had helped write the Greens' radical, social-ecological program of 1980.

12

Assembly Democrat

AS INSPIRATIONAL AS Bookchin was as a speaker, he knew he was no organizer. To translate theory into practice, he found a skilled *compañero* in thirty-three-year-old Howard Hawkins, whom he had known since Clamshell days. Murray "convinced me to get involved with the Green movement," Hawkins has said. At first he was "really skeptical" because Green politics seemed like "liberal environmentalism [with] no anticapitalist perspective." But the German Greens "got the economic program right," so he began a collaboration with Murray to build a Green movement in New England. Instead of a party, they would work to create the New England Greens as a movement for neighborhood assemblies.[1]

Hawkins lived a few hours from Burlington, on the other side of the state, so when it came to consolidating the Burlington Greens, Bookchin was on his own. He had to attract new people, then educate them. As before, he preferred a small group of committed, educated people to a large group of uninformed people. After all, only "highly informed" members would be able to prevent the "inevitable opportunists" from exploiting the Greens as "a stepping-stone to national leadership."[2]

A small group began meeting at the yellow house on Friday nights. For the first half of each meeting (as in the old YCL), they'd do internal education, reading a work aloud and discussing it, like Bookchin's *Rise of Urbanization* or Polanyi's *Great Transformation*. For the second half, they'd discuss projects, like how to keep luxury condos off the Burlington waterfront.

In 1985, despite the NPAs' expressed preferences for no condos or hotels on the waterfront, the Alden Plan was still in the works: a six-story hotel and 150 expensive condos, boutiques, and yachts loomed in the shoreline's future. The plan had the backing of the mayor, the city council, and the banks. The state, the city, the railroad, and Alden had agreed that a small area of the filled-in land could become a park, alongside the condo development.[3]

Bookchin spoke against Alden "passionately at public meetings," recalled attorney Rick Sharp, a prominent fellow opponent.[4] Then, in the fall of 1985, the city proposed a $6 million bond issue for certain public amenities—a promenade between the shoreline and the luxury enclave, some parks, a boathouse, and a bike path at the lake's edge—to be interwoven with the Alden development. Most of the councilors supported it, as did Mayor Sanders. It was expected to pass easily.[5]

Opponents of Alden—Sharp's group, the Greens, and others—gathered at Greg Guma's Maverick Bookstore to strategize. About twenty-five strong, they condemned the bond issue and the narrow promenade it would construct—ordinary citizens would feel uncomfortable walking along that constricted twenty-five-foot passage. They joked ironically about Sanders's 1981 campaign slogan, about the waterfront not being for sale, and adopted the slogan "Our waterfront is *still* not for sale!" In the next three weeks, they held press conferences, wrote letters to newspapers, conducted educational forums at NPA meetings, and led nature walks along the shoreline.[6]

The Alden supporters, for their part, mounted an advertising blitz in favor of the bond issue. Sanders campaigned hard for it; "working with him," reported the local newspaper, "are developers, stockbrokers, and corporate presidents, the kind of people with whom Sanders has clashed frequently and sometimes explosively." The Greens accused Sanders of "collusion with business interests" and called on "genuine progressives" to break with him.[7] It was the first time since his 1981 election "that Mayor Radical was publicly opposed at the ballot box by some prominent liberal leftists," observed one sardonic commentator.[8]

On December 10 the citizens of Burlington went to the polls. That evening the opponents, including Bookchin, gathered in the city clerk's office, holding their breath as they awaited the results. The bond issue needed 66.6 percent to pass—and it got only 53.4 percent. Sanders's working-class base (the "sans-culotte wards," as Murray was wont to call them) had rejected it. Only one ward, a relatively affluent one, had approved it.[9]

The vote killed the Alden Plan—the investors had said that if the bond was voted down, they would pull out.[10] Twenty-five people had defeated a multimillion-dollar project. Jubilant, the Greens held an impromptu press conference. Bookchin proclaimed "a victory for the people" and recommended that Sanders, the onetime filmmaker, check out his own 1979 film about the venerable socialist and industrial unionist Eugene V. Debs. Debs "didn't go around making all kinds of unsavory deals with some of the very people who did union busting in the strike!"[11]

Afterward the Greens met to discuss how to build a real democratic Left in Burlington. Mayor Sanders, by centralizing power and by privileging business elites, had clearly become an obstacle. He was, to Bookchin's mind, the city's *realo*, its Joschka Fischer, the onetime radical who was shedding his skin and becoming "more committed to accumulating power in the mayor's office than in giving it to the people."[12]

The task of the Burlington Greens would therefore be threefold: locally, to create that democratic Left; regionally, to epitomize New England assembly democracy as a model for Greens in the rest of the country; and nationally, to spark a decentralist, bottom-up Green movement that could spread through the country. For Bookchin, it all came down to Burlington.

Howie Hawkins visited Bookchin often to learn from him. "I got far more from Murray about history and social theory than from all the Dartmouth professors I had combined," he told me. "He was a living connection to the revolutionary tradition and made its history and ideas come alive with detailed historical accounts."[13] Then he'd crisscross the state, calling on his old Clamshell contacts to join the New England Greens. His beat-up car became a moving office, the backseat stacked with file boxes stuffed with programs and statements (some written by Murray) and mailing lists. In June 1985 the local groups came together to form the New England Greens.

Hawkins having done the spadework, Bookchin contributed the inspiration and perspective. On March 15–16, 1986, he told a New England Greens meeting in Keene the story of the German Greens. Unlike so many American Greens, the Germans weren't New Agers; they had no interest in an eco-spirituality or achieving oneness with nature. Rather, they were hardheaded, no-nonsense political activists who had gone into the trenches and fought for and built a nationwide, democratically structured nonparty party. It was a stunning achievement—but they had then become fixated on entering legislatures and on overcoming the 5 percent hurdle, and then they had joined a governing coalition. Learn from their mistakes, he urged the New Englanders. Don't run for statewide office, let alone federal office. Don't participate in statecraft.

Instead, concentrate on your community, on politics in the Hellenic sense. Develop and explain your message. Patiently, slowly, educate your neighbors about Green ideas. Cultivate your ties with them. Hold public forums. Debate people who disagree with you. Organize demonstrations on environmental and social justice issues, and participate in those run by other groups.

When you are sure people know what you stand for, run candidates in elections, on a social ecology platform. Don't do it too soon, or you'll inevitably compromise your positions. Don't make winning your principal goal until you're sure the citizens know what they're voting for. And even then hold fast to your integrity.[14]

In April 26, 1986, a catastrophic nuclear power disaster occurred at Chernobyl in Ukraine. Two weeks later, on June 15–18, the Northern Vermont Greens held a vigil at the gates of Vermont's own reactor, Vermont Yankee, calling for a shutdown. The turnout was large enough that Hawkins and Bookchin formed a Green Mountain Alliance to link environmental, antinuclear, and peace issues. Structured like the Clamshell but truly bottom-up, it comprised affinity groups, regional clusters, and coordinating council; it demanded renewable energy, decentralization, and democratic self-government.[15] Bookchin, indefatigable, churned out manifestos and declarations for this and other groups springing up around New England.

He knew this was the moment for which he had been working all his life. He had a program—social ecology, libertarian municipalism. His thirty years of pioneering work gave him enough stature that his voice would be heard. Now he would show the world how to bring democratic and ecological potentiality into actuality. Confident of the moment and his purpose, he gave some of his finest speeches—at Waterloo PIRG, at Trent University—to call for a new politics, to inspire that long-sought ethical revolt, to build the Green movement in North America. He threw all his oratorical talents into it. Capitalism is undoing the process of organic evolution, he thundered. It is wrecking human networks of mutual aid, reciprocity, complementarity. We must take charge of our own affairs, must "democratize the republic, and radicalize the democracy." By the time he finished a typical speech, audiences (I was in several of them) were on their feet, reeling, stomping, practically dancing, applauding wildly, ready to "be realistic, do the impossible, because otherwise we will have the unthinkable."

In 1986 I attended the summer session of the Institute for Social Ecology. I had been living in New York. Introverted and socially phobic, at thirty-three I was shy and unworldly to the point of dysfunction. Only in books was I comfortable, so I made my living on the printed page, as a copy editor, freelancing for New York publishing houses, and otherwise drifted aimlessly through my anonymous metropolitan world.

I read *The Ecology of Freedom* once, then twice, and then a friend told me about the author's school in Vermont, where people talked about nature philosophy, ecology, and communal politics. I knew and cared nothing about anarchism, but something about that combination intrigued me. I registered for the ISE's two-week session, which was held that summer in a rented ski school in Waitsfield.

When I arrived, I took off my street shoes and felt the cool grass between my toes. The prospect of spending two weeks here, in a close community of several dozen other students, was thrilling. Perhaps here I could even escape a bit from the prison of my own anxiety.

Murray Bookchin was then at his literary apogee. *The Ecology of Freedom* was being acclaimed as magisterial; *Urbanization* had just followed close on its heels; a new book of essays had just come out, *The Modern Crisis*, which, as one reviewer said, made you "want to assume a creator's task—building the ideal society."[16] Yes! Even shy people can have grand dreams. In his classes at the ski school, Bookchin discoursed for hours without flagging, on the history of assembly democracy, the rise of capitalism and the state, cities and citizenship, and Green politics. His personality was larger than life, magnetic, warm.

He seemed to be having some kind of argument with his friend John Clark, who considered the *Tao Te Ching* "one of the great anarchist classics."[17] Bookchin said it was more like a handbook for elitist control of the peasantry. "Taoist moods, Buddhist homilies, and New Age platitudes," he told us, were sentimental poetry. Instead of seeking a cosmic oneness through feelings, he insisted that we apprehend reality through reason, especially through a way of "thinking ecologically" that he called dialectical reason. We have to "ecologize the dialectic," he said, meaning, essentially, to accept responsibility for conscious human stewardship of the earth, in a kind of planetary self-management.[18]

Murray seemed remarkably kind for a man so esteemed; whenever I asked him a question, he took pains to answer me at length. We sat out on the grass one late afternoon and talked about the biologist Lynn Margulis and the role of cooperation in evolution. It all made me so happy that, one evening at dusk, I danced across a grassy field.

But life in this enclosed community, even for two weeks, turned out to be more than the shy girl could handle. One night, unable to govern my too-eager emotions, I unconsciously sabotaged the situation by becoming involved in a group sex episode. When it became known to the other students, it caused an uproar and much distress; the school held several community meetings to process the feelings, and behind closed doors, the staff argued bitterly. I found myself once again the outsider, my failure to connect a crushing disappointment. Then one afternoon I dove into a swimming hole without checking the depth—and hit my head on the bottom, dislodging some neck vertebrae. After the session, when I returned to New York, I got medical treatment and intensive psychotherapy for that reckless act of self-destruction.

But Murray and I had parted on friendly terms—bizarrely, in spite of everything, he actually seemed to like me. In fact, he invited me to come up and visit him in Burlington that fall. Starting in September, he told me, he would be teaching a class on dialectical nature philosophy in his living room. Knowing only that I felt comfortable with him as I did with few others, I flew to Burlington for a few days to visit.

One afternoon we sat down in his favorite coffee shop, Dunkin' Donuts—I was introduced early to his politically incorrect eating habits—and talked about the terrible episode. During the ISE's staff meetings, he told me, a faculty member had criticized me for instigating the scandal, but Murray had leaped to my defense: "How dare you blame that lovely young woman!" he'd said.

"You defended me?" I asked in amazement.
"Yes," he replied.
"But you were wrong—it really was all my fault."
Without skipping a beat, he said he didn't care.

Gradually it sank in: during that fiasco, this kind man had actually gone to bat for me. Words cannot describe how good that felt.

His kindness was all the more remarkable because (as I later learned) the ISE had been in a precarious situation that summer. A benefactor had granted the school $100,000 to construct a permanent home, and Dan Chodorkoff had finally found a suitable site in Rochester, Vermont. A sympathetic architect had come up with a design. The ISE had obtained a building permit. But the Rochester neighbors objected to having a hippie anarchist school in their midst. They filed a complaint and tried to revoke the building permit. Just as the school was struggling to hold on to the site, the sex scandal erupted. No wonder emotions had run so high.[19] But that day as we sat in Dunkin' Donuts, Murray didn't mention any of that to me—surely, as I see in retrospect, to spare my feelings.

To me, he was convivial and expressive and open-hearted. I made three trips to Burlington that fall and soaked up all he had to say, in the living room of the yellow house, about nature philosophy. Philosophy is crucial, he told us. He talked about Logos, cosmos, and *Geist*; about potentiality and actuality and increasing complexity; about nature as a ground for ethics; about polis and *oikos*; Odysseus and the lotus-eaters; the pre-Socratics; citizens and city formation; the blood tie and *humanitas*; Russia's black redistribution and the soviets; the Paris Commune and the Parisian sectional assemblies. One afternoon in his room upstairs, he cleared a place for me to sit, amid ziggurats of books and papers, and showed me his copy of Hegel's *Logic*, with much underlining and notes scrawled in the margins. My mind was boggled by this strange and wonderful philosophy and by the fact that it underlay a politics.

For the nature philosophy and radical history indeed had a practical dimension. *I'm trying to build a movement here*, Murray explained. The Burlington Greens were going to run a candidate for city council next March, on a program of ecology, a moral economy, and grassroots democracy.[20] Again he invited me to move to Vermont. I could study philosophy and history and work with the Greens.

I loved his engagement with books and ideas, and Vermont really did seem extraordinary. I had always been interested in politics, albeit as a spectator. Maybe this utopian place, I thought, really was exceptional; maybe here my life could finally mean something. Since my livelihood (freelance editing) was portable, finding employment there wouldn't be a problem.

At the end of my second visit, in mid-October, he drove me to the airport for my flight back to New York. We arrived a little early, so while waiting for the boarding call, we sat in a coffee shop and had coffee and muffins. As we talked, we heard a whoop from the back of the restaurant—the waitstaff and cooks and dishwashers were cheering.

Murray turned to look. "Are they going out on strike?" he wondered, his eyes wide.

I blinked. A strike? It was the fall of 1986, and this was age of Ronald Reagan, destroyer of the air traffic controllers' union.

When the waiter came over, Murray asked what they had been cheering about. "Ah!" said the waiter, grinning. "The Red Sox beat the Mets! It's the World Series."

What kind of world had this man come from, I wondered, that he reflexively assumed that cheering meant a strike? At that moment, I knew I would move to Vermont. I didn't care that his assumption had been mistaken; I wanted to know who could have made that assumption in the first place. I hugged him tight and boarded the plane.

On my third visit, in December, I attended a Burlington Greens meeting, where the members were planning their foray into electoral politics. One night they hosted a fundraising party for their city council candidate in a downtown loft. People were drinking beer, and a few were dancing. Murray sat alone off at one side, hunched over, depressed. "What's wrong?" I asked, joining him.

He sighed. It was the group's candidate, he said. A friend of hers who owned a local advertising agency had made a poster for her, for free. It showed the candidate's grinning face over the slogan: "Bea for Burlington." It didn't mention the word "Green" or any political goals. It was a poster for a personality, he said, not for a movement. Compared to the German Greens' bold, provocative, colorful, issue-oriented posters, it was pathetic.

Murray looked as if he wanted to cry. The poster represented politics as usual—it belied the whole purpose of the campaign for a "new politics," an ethical politics. And it was plastered all over town.

Then he looked up at me and said, "You are my consolation and my reward."

My brain was pretty well exploding.

A few weeks later I gave up my rent-stabilized apartment on the Upper West Side, packed up a secondhand car, and drove north to utopia. I moved into a rental in the north end of Burlington on a bitterly cold night in January 1987.

On Thursdays I attended Murray's philosophy class (he wrote a lecture every week or so, which he read aloud), and on Fridays I attended Burlington Greens meetings. On Saturdays he and I watched movies together, like *Inherit the Wind*, in which Spencer Tracy, wearing suspenders or "galluses," plays Clarence Darrow, defending reason and the Enlightenment against the Bible-thumpers. Murray had taken to wearing galluses too, and calling them that, in homage.

His passion for Vermont was contagious, and I welcomed that pleasant fever. In late 1986, politicians in Montpelier were once again proposing a four-year term for officeholders, and so Murray dusted off his arguments from three years earlier. On January 22 we made the short drive to Montpelier, the state capital, and entered the statehouse, with its golden dome.

Of all the state constitutions, Murray testified to the state senate's Government Operations Committee, Vermont's is the most democratic. It is "a moral compact between free citizens to guarantee themselves liberty, free expression, and a participatory political life." It presupposes the competence of every citizen to actively participate in government and politics. The two-year term is the means by which the citizens hold officials accountable—they simply reelect officials who are performing well. But don't let our state concentrate power, as California and New York have done, he beseeched the senators, by depriving citizens of this tool, so basic to our state's unique political culture.[21] I thought he was wonderful.

By March, we were in love. He told me he'd loved me since the ISE session in Waitsfield. He told me I made him feel "Sur l'eau," the name of an essay by Adorno: "Rien faire comme un bête," to lie on the water and to gaze into the sky, "to be without any further determination and fulfillment."[22]

He'd been afraid he was too old for me, he confessed—he was sixty-six, I was thirty-three. No, it's okay, I said. But it was more than okay—I felt like my life had begun at last.

* * *

The German Greens had found nothing controversial in the idea that the ecological crisis has social causes and that it is closely intertwined with issues of social justice. They presupposed that ecology must be social. But American Greens did not make this connection easily: the earliest environmental efforts in the United States had been for conservation, and even in the late 1980s, certain tendencies in the US ecology movement conceived of pristine wilderness as synonymous with nature and disdained human beings as destructive intruders. Bookchin certainly did not exclude an appreciation of wilderness from his thinking; as far back as 1969, he had urged the ecology movement, among other things, "to guard and expand wilderness areas and domains for wildlife."[23] He had personally always enjoyed the psychological solace of walking in forests—in the Great Smoky Mountains with his *CI* friends, around Ramapo, during his lecture tours, and in Vermont; but he had never considered wilderness preservation to be the primary aim of ecological politics. If anything, social ecology sought to preserve nature—and humanity—by fundamentally changing the social arrangements that gave rise to the destruction of the biosphere.

In the spring of 1987 the eco-radical group Earth First! was gaining attention for its militant direct actions against lumbering and dams out west, under the slogan "No compromise in defense of Mother Earth." Philosophically, Earth First! insisted on what might be called an asocial ecology, valuing nonhuman nature over people. Its cofounder David Foreman, a longtime wilderness activist, said he'd like to see large tracts of the American West set aside and closed off to human intervention altogether. "Wilderness is the real world" ran one of Earth First!'s maxims. To achieve this vision, "a move toward population reduction is a primary step."[24]

In an interview that spring, Foreman proposed to protect the American environment by curtailing Latino immigration. "Letting the USA be an overflow valve for problems in Latin America," he remarked, is "just causing more destruction of our wilderness, more poisoning of water and air." And at a time when Ethiopia was experiencing a severe famine, he recommended an extraordinary way to reduce human numbers: "The best thing would be to just let nature seek it's [sic] own balance, to let the people there just starve." This was misanthropy taken to an extreme.[25]

Strangely, the interviewer expressed no alarm at these outrageous remarks. He was Bill Devall, coauthor (with George Sessions) of the book *Deep Ecology*, named after a newly minted ideology that gave primacy to wilderness and a quasi-romantic experience of solitary forest communion. Its concept of "biospheric egalitarianism" or "biocentrism" could easily be read as saying that human beings have no greater claim to life than nonhuman organisms. Deep ecology condemned any special consideration for human beings, and for their needs, abilities, and outlooks, as shallow and "anthropocentric." Needless to say, it advocated a significant reduction in human population.

Deep ecology might as well have been named "antisocial ecology," so greatly did it contrast with social ecology. Bookchin had always been a humanist as well as an ecologist, arguing that blame for the eco-crisis lies not with people as such but with their social arrangements—capitalism, hierarchy. Giant corporations, not individuals, were guilty of clear-cutting timber and spewing sulfur dioxide and chlorofluorocarbons into the atmosphere. *The Ecology of Freedom* traced how humanity had gone down the wrong path. People, he believed, far from being the primal problem, are the solution: human inventiveness is vital for establishing a responsible human-nature relationship. As for biocentrism, he thought it morally and politically sinister. "If all organisms in the biosphere are 'intrinsically' equally 'worthy' of a 'right' to self-realization," he said, then "human beings have no right" to attempt to eradicate deadly microbes, like the smallpox and polio viruses and HIV, or to expunge disease vectors like mosquitoes. The proposition becomes a writ for cruelty, a justification for human deaths beyond number.[26]

Untroubled by such considerations, Foreman explicitly affirmed that "deep ecology is the philosophy of Earth First! They are pretty much the same thing." In May 1987 the *Earth First!* newspaper (edited by Foreman) published "Population and AIDS," which asserted that the virus was natural as a means of population control; "if the AIDS epidemic didn't exist, radical environmentalists would have to invent one." It blithely theorized that a pathogen "would have to kill 80 percent of the world's human population to end industrialism." (The piece was signed by "Miss Ann Thropy," a pseudonym for Christopher Manes.) In the next issue, Foreman editorialized that Earth First! "enthusiastically embraced" biocentrism. In a war between people and bears, he said, he'd be on the side of the bears. He wished we could go "back to the Pleistocene."[27]

Murray, while he was in Germany, had resolved that when he returned to the United States, he would not lower his standards to accommodate Americans' relative political unsophistication. Least of all would he let such politically dangerous remarks go unchallenged. In May 1987 he wrote a polemical article "Social Ecology vs. 'Deep Ecology,'" dissociating social ecology from biocentrism and drawing a firm line between the two. If deep ecologists want us to accept that all organisms in the biosphere have "equal intrinsic worth," he argued, then they must reckon with the fact that rubric includes not only grizzly bears and tigers but also microbes that are deadly to human beings.

As for the famine in Ethiopia, it was arrant nonsense to speak of nature "seeking its own balance" in a part of the world ravaged by agribusiness, colonialism, and exploitation. Impoverished Ethiopians had no power to damage the environment; it was corporations that were laying waste the biosphere. But deep ecology condemned humanity as a whole, the poor as well as the rich, the downtrodden as well as the privileged, the powerless as well as the powerful.

Murray went on to label deep ecology with various colorful adjectives: "vague, formless, . . . invertebrate," and "an ideological toxic dump." He called Foreman an "eco-brutalist" and a "patently antihumanist and macho mountain-man." He called Earth First! leaders "barely disguised racists, survivalists, macho Daniel Boones, and outright social reactionaries." And fresh from his West German experience, he pointed out that "Hitler, in the name of 'population control,' with a racial orientation, fashioned theories of blood and soil that led to the transport of millions of people to camps like Auschwitz."[28]

* * *

By mid-1987, Greens in the United States had established more than one hundred locals. On July 2 many of their members converged on Hampshire College to attend the first gathering of the American Greens. Murray had been invited to give the keynote. As he mounted the rostrum, the nearly two thousand activists welcomed him enthusiastically, as the ecology movement's grand old man.[29] He stood before the crowd in his usual work clothes and Clarence Darrow galluses and took a deep breath.

First he warned his listeners that he had once been a union organizer and begged for their patience with his blunt manner—when you're talking to industrial workers, he said, a certain amount of gruffness was expected. The audience tittered nervously.

Today, he continued, the Green movement is potentially the most important force for persuading people to finally face the ecological crisis. To do so, Greens must be clear in their own minds about the cause of ecological destruction. They have to understand and then be able to communicate what keeps driving our biosphere further and further toward ruin.

Contrary to what some have argued, he said, it's not human beings as such—it's not overpopulation that's the problem. Nor is it some foulness that's inherent in our DNA, something in our genes that makes us wicked. Nor is it even the various technologies that we have developed over the millennia, although that explanation was certainly popular.

No, he said, the cause is capitalism, a system that compels people to produce and overproduce in order to create ever more profits for owners. It compels firms to grow and then grow some more, in order to outcompete other firms. It is out of control, and in this insane quest for wealth it is ripping up the biosphere.

In the face of this basic reality, he said, the Green movement can't be concerned mainly with, say, conserving wilderness. It must be a social movement as well, addressing social issues, such as the nature of the economic system.

It should not, in the name of preserving the natural world, become misanthropic. It should not become Malthusian. It should not welcome things that kill people, like Ethiopian famine and AIDS. On the contrary, the Greens must be humanists, because people are part of nature too, and while people have wrought destruction, they are also our only possible hope of undoing the damage.

That's why Greens should not get mixed up in New Age religion or eco-spirituality, he said: we should be concerned with nature, not with supernatural beings that don't exist. People imagined those beings centuries ago, before the scientific revolution began to explain natural phenomena, and they're irrelevant now. Rather than mislead people by asking them to believe in fairy tales, Greens should offer them nature on its own terms, which is replete with magnificence and beauty.

His time at the podium was almost up. Greens have to be able to explain to people, he said in a rush, that all ecological problems have their roots in social problems. We have to talk about social oppression and exploitation as well as environmental problems, and we have to support struggles for social justice as well as environmental protection. If we don't, we'll be considered a bunch of privileged elitists who just want to be able to go for a hike in a nice forest or paddle a canoe in a clean stream.[30]

The speech was splendid, and the audience gave Murray the rousing applause and cheers he deserved.

Over the next week, at workshops and plenaries, the conference hashed out the political question: Should the Greens go the route of assembly democracy, as social ecology advocated, or should they form a conventional party? Charlene Spretnak emerged as the spokesperson for Murray's opponents. Not only did she advocate a conventional Green party to run candidates for national office, she had written a book on eco-spirituality—"sustainable religion"—which she thought should have a central place in Green politics.[31]

Bookchin and Spretnak clashed over these issues throughout the week. That bothered many of the conference participants, who regarded conflict as "old paradigm"; they were devoted to harmony and oneness as part of the ecological "new paradigm."[32] They insisted on holding a mediation session, where the two could reconcile their differences. Murray was reluctant—he had been trying to make crucial, substantive distinctions and had no interest in papering them over. But he went along. As might have been predicted, the mediation session got nowhere. The facilitators were stymied.

Fatigued by the banality of it all, Murray found a way out. "Charlene," he said, "you see that big box over there?" He pointed into the air. "Let's put our disagreements inside and shut the lid. Then we'll seal it up and bury it in a hole in the ground. Okay?" She winced but reluctantly agreed. The mediators, delighted with this content-free "reconciliation," happily mimed sealing a box and burying it.

At the closing plenum, by popular demand, Murray and Charlene appeared on stage together and shook hands and embraced, albeit stiffly. The audience cheered at the amazing power of Green politics to overcome differences and achieve peace and harmony. Then Charlene made some parting comments about spirituality and the need for a party, which led to "a standing ovation by some 700 enthusiastic people in the audience."[33] Murray made some parting comments about assembly democracy, and humanism, and secularism—and the audience stood and clapped equally passionately. "It was the same people!" he exclaimed to me afterward, exasperated. "The same people clapped for diametrically opposite ideas!" Didn't they take differing views seriously?

I was spending so much time at the yellow house that Murray invited me to move into the attic, and I did. Only a few months earlier I had been frightened of life; now as I watched him fearlessly challenge whole roomfuls of people, I was learning something about courage.

One afternoon in his room, he showed me a stack of letters, from people who had heard him lecture out on the West Coast somewhere. They all eagerly asked how they could help build the Green movement that he had been talking about. But as I looked closely, I saw that the letters were three years old. And they had gone unanswered! Clearly Murray needed an assistant. I stepped into the vacuum and became his corresponding secretary, a role I would play for the next two decades.

The year before I arrived, he and Howie Hawkins had started *Green Perspectives*, a "theoretical newsletter" that mainly served as a vehicle to disseminate Murray's essays. Now that Hawkins was busy as an organizer, he had no time to continue it. So I took over and became *GP*'s coeditor (with Murray), publisher, contributing writer, and subscription and distribution manager—roles I would play as we published more than three dozen issues until we closed it down in 2001.

So that I could catch up on the Thursday-night philosophy lectures he'd given before my arrival, he showed me the manuscript from which he'd read them, *The Politics of Cosmology*. As an experienced editor, I saw that the prose needed work. His low-budget movement publishers, however, tended to skip editing and publish whatever he sent them just as it was. For the next sixteen years, as Murray wrote almost continually, I would be his copyeditor, editor, and researcher.

For the first time in my life, I felt needed. Murray often said that political action strengthens a person's sense of self; working with him definitely had that effect on me. My social ecology apprenticeship focused me psychologically (much as Murray's apprenticeship with Josef Weber had done for him) and allowed me to develop my own abilities. Gradually the hyperanxious, underachieving drifter-through-life gave way to a competent woman. Within two years of my arrival in Burlington, I was not only his secretary, newsletter publisher, and editor; I was writing my own articles on social ecology themes.[34]

In the Burlington Greens, moreover, I learned how to be a political activist. In the summer of 1987, a local developer proposed to build a 320-slip "world class marina" on an unspoiled wetland in Burlington, where the Winooski River empties into Lake Champlain. The developer, Northshore, wanted to dredge the shallows to allow forty-foot yachts to moor; it proposed to pave over the wetland and build a commercial center and parking lot. And once again, to the Greens' never-ending astonishment, the Sanders administration approved the plan.[35] Once again, the socialist mayor was colluding with wealthy business interests; once again, Bernard Sanders played the Joschka Fischer *realo* role to perfection.

We visited the wetland site and found that thirty low-income families lived nearby, making their living by fishing. I took pictures, wrote a script, and created a half-hour slide show, which we presented at the local library. It criticized both the developer and the mayor with the recurring slogan "Burlington is *still* not for sale!" A year and a half later, the state's environmental board rejected the proposed marina because it would have "undue adverse effect upon the scenic and natural beauty and aesthetics of the area."[36] The victory energized all of us.

I accompanied Murray on his lecture trips—to Seattle and New Orleans and Knoxville; to Quebec City and Guelph, Ontario; and to Italy, twice (figure 12.1). In Montreal, he gave a speech at McGill University that was taped and widely distributed.[37] His fiery lectures, especially abroad, were greeted with raucous ovations.

But the number of lecture invitations he got from places in the United States was declining. The reason was clear: his criticism of deep ecology.

FIGURE 12.1 With the author in Palermo, 1989
Janet Biehl.

In November 1987 *Earth First!* hit back, responding to "Social Ecology vs. 'Deep Ecology.'" Editor Dave Foreman said his embrace of deep ecology remained "enthusiastic," and he made no apologies for his Ethiopian famine remark. In fact, he reiterated it: "Human suffering resulting from drought and famine in Ethiopia is unfortunate, yes, but the destruction of other creatures and habitat there is even more unfortunate." Human overpopulation is the greatest problem for the biosphere, and agreement on this point is "an absolute litmus test for Earth First!" Nor did some portions of humanity bear more responsibility than others: "It is invalid to pardon someone because of the rung they occupy on the economic ladder." Least of all were social issues a matter of concern to Earth First!: issues affecting humans "pale into insignificance" compared to the "most important issue," which was "the preservation of wildness and native diversity." Foreman reaffirmed his adherence to biocentrism: "An individual human life has no more intrinsic value than an individual Grizzly Bear life (indeed, some of us would argue that an individual Grizzly Bear life is more important than an individual human life because there are far fewer Grizzly bears)." As for deadly microbes, "'disease' (malaria) and 'pests' (mosquitoes) are not evil manifestations to be overcome and destroyed but rather are vital and necessary components of a complex and vibrant biosphere." (Did their ability to kill humans enhance their intrinsic value? we wondered.)[38]

Other articles in the *Earth First!* newspaper objected to Murray's tone. "What is worth noticing in the fusillades by Bookchin" is "not the words but the mood": the mood of a combative leftist. A new drumbeat began. Mike Roselle of Earth First! disparaged social ecology as "nothing but recycled leftist drivel." Bill Devall noted that "confrontation, diatribe, denouncing comrades and factionalism are characteristics of the leftist movements." He warned that "if the 'Green' movement in the United States becomes just an expression of old line leftist politics, then it will be added to the junk heap of leftist movements which have ignored the inherent worth of Nature." Chim Blea suggested that Bookchin's speech at Amherst "may have been part of a coordinated attempt by American Redgreens to . . . engineer a coup in the emerging Green Party in the United States." After the "redgreen putsch," Blea said, Bookchin planned to carry out a "redgreen pogrom" against biocentrists.[39]

Murray wrote a reply, but Foreman refused to publish it in *Earth First!*, so we published it ourselves in *Green Perspectives*. He made no apologies for his humanism or for his confrontational style. As for putsches and pogroms, "I'm much too close to seventy to be worried about my ideological 'turf,' my status, or my influence in a movement that threatens to degenerate into an environmental version of the Wild Bunch."[40]

And yes, he explained, he was a leftist. But not a Marxist—an anarchist. And however objectionable the deep ecologists might find the label "Left," the problem that was driving the ecological crisis was not the Christian religion, or technology, or civilization as such, or overpopulation, or "old paradigm" linear thinking, or confrontational personal styles; it was, in fact, the same phenomenon with which the old Marxist Left had long been engaged: capitalism, an economic and social system that was inherently antiecological. Its drives toward unlimited growth, regardless of consequences, were irreconcilable with the underpinnings of life on earth. Already the scale and rapidity of the devastation it had wrought were appalling: the greenhouse effect, the thinning ozone layer, deforestation, desertification, extinctions, rising sea levels, obliteration of rain forests, pollution. Should we permit it to continue to tear up the biosphere, it must sooner or later plunge the world "into an ecological apocalypse."[41]

Given the urgency and gravity of the situation, he said, the ecology movement must not advocate some absurd and undesirable "long march backward into the Pleistocene." That wasn't addressing the problem—it was fleeing from it. Nor must it try to escape into nature mysticism or the abject worship of wilderness. Some deep ecologists advocated that people learn to "think like a mountain," but that was a prescription for mindless inertia, Bookchin pointed out—mountains don't think at all. And rather than ask people to "fall down on all fours and bay at the moon," the ecology movement must face the problem square on and appeal to the reason of sensible people who want to take concrete steps to avoid the looming catastrophe.[42]

It must abandon misanthropy and finally acknowledge that human beings are "distinctive, self-conscious, and conceptually unique evolutionary products." They can and have intervened in nature destructively, but they can also intervene in it creatively. The Green movement must explain, in sane and sensible terms, how they may do so, how they may reconstruct society so as to bring it into harmony with the natural world.[43]

And the struggle for an ecological society, he affirmed, is inseparable from the struggles of oppressed peoples. Back in the 1970s in New York, he said, he had visited an exhibit on pollution and the environment at the American Museum of Natural History. At the end of the exhibit was a huge full-length mirror, with a sign overhead that read, "The Most Dangerous Animal on Earth." As Murray stepped over to it, he had seen a black child standing in front of that mirror. A white schoolteacher was trying to explain the message. Murray thought, *That child? Dangerous?* If the museum really wanted to display the most dangerous animal, he wrote, it should have displayed an image of "corporate boards of directors planning to deforest a mountainside or of government officials acting in collusion with them."[44]

In their next wave of counterattacks, deep ecologists doubled down, expressing themselves as willing to scrap basic tenets of Western civilization, like humanism and rationality, in favor of wilderness. The bioregionalist Kirkpatrick Sale wrote that "it does not really matter what the petty political and social arrangements are that have led to our ecological crisis, or even what dire consequences those arrangements have had for certain individuals, types, nations, or races." What matters is reducing population, revering wilderness, opposing human stewardship (an "anthropocentric" concept), privileging intuition and spirituality over reason, and looking to primal peoples (like Native Americans) as ecological models. Given all that, Sale said he could not see why deep ecology "should evoke anger or calumny."[45]

Actually, it would have been strange if those ideas had not provoked outrage. But as the debate wore on, the critics responded less to the content of Murray's argument than to his tone, judging it as harsh and vitriolic and old paradigm. Yes, his language was indeed sometimes intemperate, but "eco-brutalist" seems to me a fine word for the repellent ideas that Foreman and his allies expressed. Howie Hawkins wondered, "How could anybody on the left not feel that vitriolic comments are in order" against those who tolerate AIDS and famine? And *Fourth World Review* editorialized, "The strong feelings are understandable, . . . for what is at stake are issues even greater than . . . their protagonists appear to grasp."[46]

Murray made no apologies for being a writer of polemics, but he was perplexed by the focus on his tone and personality. Instead of addressing his criticisms, the deep ecologists seemed to prefer tarring him in personal terms. "They kept on looking at my demeanor," he said, at the way he expressed his outrage rather than at what was outraging him. Even Murray's well-meaning students would come up to him and assure him that Foreman was really a terrific person. "I have no doubt that personally he may be very lovable," he remarked.[47] But deep ecology was a despicable ideology, and someone had to break through the polite facade and say so. "Old paradigm" he might be, but he took ideas seriously and didn't care whom he alienated in the process of making distinctions between them.

By mid-1988 we were told that Murray's criticism of deep ecology was actually stirring up hostility toward social ecology. Friends who had once admired him and praised his work now disavowed him. Even those who still agreed with him were afraid to say so, given the peer pressure.[48]

In June 1988 Brian Tokar (who had moved to the Bay Area) phoned Murray, asking him to cool it off. You have performed the necessary function of provoking the issue, he advised, but now we need to moderate. And a distinctly chill wind was blowing Murray's way from New Orleans. One night John Clark, his protégé who was also a practicing Taoist, called Murray to tell him that he profoundly objected to his criticism of deep ecology. No idea is entirely wrong, he said; deep ecology had merit that must be acknowledged. In fact, he said, social ecology is part of deep ecology. Soon afterward he sent Murray a quotation from *The Phenomenology of Spirit* that was intended to demonstrate that Hegel himself had actually been a near-Taoist.

Murray's incoming mail slowed to a trickle, and lecture invitations all but dried up. That summer he was even challenged at the ISE, where a faculty member cautioned him about the use of the word "Left." It could scare people away from the Green movement, they said. Why? Murray asked. Was it because the "Reaganization of America" had shifted the whole political spectrum to the right? The German Greens had a slogan identifying themselves as "neither Left nor Right, but up front," but the person who coined it was a reactionary. As long as there was a Right, he often insisted, there had to be a Left. Another old friend accused Murray of "demonizing" deep ecology.[49]

The social-versus-deep dispute continued, in leftist and environmentalist periodicals, into 1989. Murray thought it at least had the salutary effect of bringing differences out into the open. "Real growth occurs exactly when people have different views" and when those views are articulated and discussed, even through "harsh disputes." Far worse would have been "a bleak silence that ultimately turns bland 'ideas' into rigid dogmas." Deep ecology's advocates had previously obscured its true nature with a haze of nature-spiritual-cosmic-oneness fog. But Earth First! had performed the useful service of drawing out its "logical conclusions."[50] And unlike academic deep ecologists who deployed reason to promote unreason, "at least Foreman is logical and consistent."[51]

Still, among Greens who valued "wholeness," the impulse to reconcile social ecology and deep ecology remained strong. Finally, David Levine of the Learning Alliance and Steve Chase of South End Press brought Murray and Foreman together for a high-profile debate in New York on November 5, 1989, in hopes that they could find areas of agreement. Appearing together onstage, they were civil and superficially cordial, and whenever they managed to find some small commonality, the audience applauded. But both protagonists knew that their views were irreconcilable. "Bookchin is entirely correct," Foreman (under his Chim Blea pseudonym) had written back in November 1987, "there is a great gulf between his 'social ecology' and deep ecology."[52] On that point the two men could agree wholeheartedly.

The great debate brought about a certain closure, not because it solved anything but simply because it happened. In their later writings on the subject, neither Bookchin nor Foreman showed even a trace of a change in attitude; the volume was simply turned down.

* * *

In 1988 opinion within the American Greens was drifting toward favoring the formation of a national party, while eco-mysticism and deep ecology and neo-Malthusianism were getting stronger as well. But Hawkins agreed with Murray that "that Green politics should . . . build an anticapitalist and antiauthoritarian political movement."[53]

They envisioned a Left Green Network (LGN) as an organized caucus within the broader US Green Committees of Correspondence. The LGN would call for social ecology, libertarian municipalism, and solidarity with social justice struggles. For internal decision-making, it would use simple majority voting rather than consensus. Far from calling for a "redgreen pogrom" or "putsch," it would function transparently, attempting to persuade not by coercion or intimidation but by honest argument, meanwhile abiding by all democratic processes.

The "Call for a Left Green Network," drafted by Murray and signed by eighty-three people recruited by Howie, was issued January 1989. The founding conference was held in Ames, Iowa, on April 21–23. Murray gave the keynote, saying that Left Greens had the potential to constitute an authentic Left, but warning that the American *realos* were well organized; Left Greens must learn from the Germans and organize themselves, lest they, too, be marginalized.[54]

No sooner was the LGN born than deep-ecology-oriented Greens began bashing it. Lorna Salzman asked, "Is the Left Green Network Really Green?" "The LGN has not abjured the traditional Left process," she pronounced, providing no evidence whatsoever for this statement. Tom Athanasiou responded, "Socialism-as-state-domination is not what the LGN is all about, and after going back and checking the LGN principles, I can't see how anyone could honestly think that it is." The LGN, I wrote, is leftist in that it identifies with the New Left's commitment to participatory democracy. Its antecedents were "left communitarian and anarchist traditions . . . the democratic Parisian neighborhood assemblies of 1790–94, Proudhon's emphasis on mutual aid, the Paris Commune of 1871, Kropotkin's cooperativism, the Spanish anarchist collectives of 1936."[55] I had learned quickly.

As for American anarchists, Murray argued passionately, once again, that they should join the Left Greens. Strengthening the left-libertarian wing was the only way to prevent the US Greens from becoming a conventional party. He implored them not to sit on the sidelines this time. To their objections about the state, he patiently explained that politics (at the local level) is crucially different from statecraft (at state, provincial, and national levels), and that participating in assembly democracy—community self-management—is not the same thing as voting for officeholders in the state machinery.[56]

Burlington, Vermont, nestled in the land of the town meeting, was still the last best prospect for the politics Murray had spent a lifetime constructing. In 1989–90, gratifyingly, new young people were beating a path to his door, arriving starry-eyed at the yellow house to study history and philosophy with Bookchin (in his living room!) and join the Burlington Greens (and be his political comrade!). Murray opened his door and let them in.

Since he considered their level of education to be more important than their numbers, he taught them, mainly about history and philosophy; there was, after all, no substitute for consciousness. On Thursday nights he read aloud to them chapters from the two manuscripts he was writing, alternating *The Politics of Cosmology* (on dialectical nature philosophy) with *The Third Revolution* (on popular movements in Euro-American revolutions). The Burlington Greens not only attended these readings but formed study groups of their own, on philosophy, social theory, and history. At one point, there were as many as ten study groups going on in the city—a mini-university.[57]

We structured the Burlington Greens nonhierarchically and democratically. We made decisions by voting, and voting rights were equal. Prospective members had to attend meetings for six months before requesting to join, and they had to be voted in. Membership reached a peak of twenty-three in September 1989. Most of the members were young and inexperienced, more comfortable with direct action and protest than with electoral politics, but they were bright and eager to learn.

We discussed how Burlington was becoming too expensive for poor and middle-income people, and how city hall was becoming ever more centralized. Business interests were gentrifying the city; expensive restaurants and boutiques proliferated; and ever more homeless people appeared downtown. In the fall of 1988 a local restaurateur offered to give Burlington's homeless people one-way plane tickets to anywhere else in the United States. Outraged, the day after Thanksgiving we set up a one-day soup kitchen outside his upscale restaurant, under a banner that read, "Let the Rich Meet the Poor." We ladled soup and gave away sandwiches. "The real problem facing our community," we said, "is not the homeless or street people; the real problem is a society that tolerates the social and economic conditions that create homelessness in the first place."[58]

Howie Hawkins was now driving around the continent in his beat-up car, organizing the LGN. He stayed closely attuned to developments in the Green Committees of Correspondence. Periodically he'd swoop into Burlington and let us know what was going on. The GCOCs' decision-making process was stymied by a consensus requirement, he explained. Since the consensus process made it so hard to come to agreement on anything, volunteers stepped into the vacuum and ran the show. They tended to be people with enough money and time to travel to meetings. "If the GCOC would just move to electing people to positions of national responsibility" rather than relying on self-appointed volunteers, and move from consensus "to simple majority decisions with protection of minority rights to dissent, it would go a long way toward democratizing the organization," wrote Hawkins.[59]

In late 1988 Bernard Sanders announced that he would step down as Burlington's mayor (so he could run for US Congress). Now was our chance, we thought, to establish a real Green Left in the city. In January the Burlington Greens entered a candidate into the mayoral race, Sandy Baird. A feminist attorney who defended victims of sexual violence and domestic abuse, she'd helped found the local battered women's shelter. Two other Green members ran for the city council (then still known as the board of aldermen). It was a presentable slate.[60]

Our program called for changing the city charter to establish town meetings in the wards; imposing a moratorium on growth; instituting rent control; creating a municipally owned bank; utilizing renewable energy sources; building low-income housing; creating a farmer-consumer network to foster local agriculture; acquiring open land for gardening and parks; and more.[61] We had no expectation of winning on our initial outing—our aim for now was to raise awareness of the issues, present our solutions, and build support. Only when the public fully agreed with what we stood for would we hope for electoral victory.

The writers among us generated a slew of position papers: on democracy under the Sanders administration; on the Alden Plan and the Northshore marina; on the economy.[62] I wrote a seventy-page report on Burlington's environment, covering hazardous wastes, lake pollution, solid waste, air quality, and energy—the first one ever, as far as I knew. Murray, whose opinion I valued most, was greatly pleased and persuaded a local bookstore to carry it.

Baird set the tone for the campaign and defined the issues. The other candidates were forced to come up with their own positions on growth and housing and ecology. Influenced by our group's idea-heavy discussions, she spoke eloquently about democracy. She initiated a serious discussion of downtown traffic. Building new roads and parking garages, she argued, would merely increase congestion—the city needed better public transportation. "Baird is right," opined the local daily. The city must move on developing "trolleys, commuter rail service, light rail links and satellite parking lots with adjacent mass transit service." These are "farsighted approaches."[63]

The Progressives—supporters of Mayor Sanders—criticized the Greens as sectarian splitters and intransigent spoilers.[64] But the Sanders administration's own support for luxury development on the waterfront had created the political opening into which we stepped. Baird called for turning the waterfront land into a public park. "The Burlington Greens' view of Eden on the Waterfront represents an impossible and undesirable extreme, but a useful one nonetheless," declared the newspaper. "Until the Greens arrived, Burlington residents heard only the other extreme in the debate over how to rebuild the waterfront."[65]

On March 7, 1989, Election Day, Baird got only 3.4 percent of the vote. Never mind—we were satisfied that we'd raised the issues, and we'd try again. Repeated attempts were part of our slow and patient strategy.

The following year a Burlington city council seat opened up, and voters in Ward 1 would choose the replacement in an election in March 1990. To keep our momentum going, we wanted to run a candidate. A young Greens member named Steve lived in the relevant ward. A quiet, self-contained man with no political experience, he did not want to run and told us he did not feel capable of it. We assured him that all he would have to do was articulate our program. Several of the older members of the group formed an advisory committee—if he needed help or advice or support, we said, he could turn to them. With great reluctance, Steve agreed. We announced his candidacy and geared up with programs and leaflets and such.

A debate was scheduled for February 28, where Steve and the two other candidates would square off. One of Steve's opponents was Erhard Mahnke, the Progressive president of the city council, running for reelection; the other was a Democrat. We worried that Steve would appear weak in the debate, but much to our relief, he held his own.

Election Day arrived, March 6, and that evening we gathered at the yellow house to await the results. When Paul, a group member, arrived, he looked worried. "I've heard a rumor about collusion," he said. "I'd like to conduct an investigation." He asked each of us questions, and as he proceeded, slowly the picture came into focus. It turned out that the night before the candidates' debate, the Democrat had approached our Steve with a proposal: the two of them should think up softball questions to ask each other at the debate, and tip each other off to them in advance, so they would have an advantage over the Progressive. Steve, perplexed, had turned to his advisory committee for guidance. These wise elders should have forbidden such a thing, but instead, for reasons I will never understand, they gave him the green light. So he went ahead and exchanged questions sub rosa. The next day at the debate, Steve and the Democrat had asked each other the prearranged questions.

Now, as the election results were coming in over the radio, we at the yellow house were just finding out about it. To our horror, Mahnke the Progressive lost—by a margin smaller than the number of votes Steve the Green had received. If he and the Democrat hadn't colluded, Mahnke might well have won. The election was tainted.

Murray and I were aghast that the advisory committee had approved the behind-the-scenes collusion. The Greens were supposed to represent a new politics, an alternative to mainstream wheeling-and-dealing, an ethical challenge to capitalism. It was our raison d'être; if we didn't have the moral high ground, we had nothing.

The Burlington Greens, all two dozen of us, met to discuss what to do. For Murray, me, and another member, Gary Sisco, it was a no-brainer: the Greens had to come clean: write a press release, admit to wrongdoing, and call for a new election. Anything less would render us hypocrites. But to our amazement, all the other Greens objected strenuously. Going public, they said, would be personally too painful, especially for the members of the advisory committee, who had good reputations in town, and we should just let it rest.

For the next four days, most of the Greens put aside everything else and met at the yellow house for protracted sessions of bitter arguing. Over and over, Murray, Sisco, and I explained to the others that we had to go public. Over and over, the others adamantly refused. Some of the group members were only one-third Murray's age—he had placed himself on an equal footing with them organizationally, yet his life's work was on the line. His only recourse was his power of persuasion, which he exercised at length. Voices were raised, nerves frayed, tempers flared, someone overturned a table, exhaustion set in, yet we still could not come to an agreement.

Finally Murray, Gary, and I drafted a press release. "The Burlington Greens find that this breach of public trust and democratic principles is completely unacceptable . . . on the part of the Greens involved" and "publicly apologize to the citizens of Ward One and to Mr. Mahnke." We proposed that the Greens adopt and distribute it. The others said no, it was too shaming—and they were in the majority. Finally, Murray explained that if they did not adopt it, he would resign from the Greens and go public with his own statement. Queasily, they said they might reluctantly adopt it if we inserted some mollifying language ("we understand why they did what they did," or something to that effect). The three of us held our noses and allowed that modification. Then a formal vote was taken, and a majority agreed to issue the press release.

That night we typed it up and photocopied it. The next day we brought it to the yellow house, to divvy up the copies for distribution. But overnight some of the young people had had second thoughts and now tried to stop us. No way—Murray, Gary, and I executed the majority decision. I remember pulling into a parking lot and handing the press release to a local political reporter, who read it and shook his head, muttering, "What a shame." The news story appeared in the press, but unfortunately no new election could legally be held.

Many of us attended the next city council meeting, a few days later. Murray and I sat on one side of the crowded room; across the aisle, the other Greens huddled protectively around the advisory committee members. Loyalty, someone once said, is the enemy of honor. As far as I was concerned, the others could uphold loyalty; I was proud to join Murray in upholding honor. Mahnke, the city council president, sat in his accustomed seat, soon to be surrendered.

During the public comment period, Murray rose and, cane in hand, made the longest walk of his life, up that aisle to the microphone. He stopped and looked Mahnke straight in the eye. *I have worked all my life for a new politics*, I remember him saying, *for an ethical politics. People in my group have behaved unethically, for which I apologize. If such behavior is to be identified with my work, if that is to be its conclusion, then my life has been a waste.*

Mahnke nodded in acknowledgment, and as Murray walked back to his seat, only his cane and his heavy footsteps could be heard.

None of us—including me—understood at the time what had been at stake for Murray with the Burlington Greens. I've come to realize it only as I've worked on this book. None of us understood the decades-long prehistory of his political ideas in the CI group, the Montreal Citizens Movement, the 1970s neighborhood power movement, the thousands of lectures on eco-decentralism and eco-anarchism, the Clamshell Alliance, and the German Greens.

Murray might have been a utopian, but he didn't dodge his real-world responsibilities. He was a leftist, dyed in the wool, but he was no authoritarian manipulator, no putschist. He had opened his doors to inexperienced people one-third his age, and although he was also their teacher, he participated in a political group with them on an equal basis. How many noted political theorists would have exposed themselves in that way?

Most people in political life, when they make a mistake, try to cover it up, or blame someone else, or talk around it, or otherwise dodge responsibility. Murray's public apology was an act of honor, commensurate with his call for an ethical politics. During my research for this book, some of my interviewees told me that Murray didn't walk the talk—that his taste for Twinkies and junk food belied his advocacy of organic farming, for example. But here, in the thing that mattered most to him, an ethical politics—he walked the talk, with integrity.

The Burlington Greens passed out of existence. The young people who had chosen loyalty over honor banded together as the Northern Vermont Greens, met for a few months, then drifted apart. The reputations of the advisory committee members were scarcely damaged, if at all. Erhard Mahnke's career in electoral politics ended.

In the next years Burlington's waterfront was made into a public park. The landfill, owned by the people of Vermont, is today covered with an expansive green lawn and flowering trees; it is home to endless summer festivals, games, and other public events. It is Burlington's great, irreplaceable treasure. The Progressives and their allies would go on to institute many environmental improvements, lacing the city with community farms and gardens; as well, they established or bolstered programs to support the poor, the elderly, battered women, and minorities. Burlington has won numerous accolades for its social and ecological liveliness, and as I write in 2014, it has gained a national reputation for sustainability.[66]

After Bernard Sanders stepped down as mayor of Burlington, he won election to Congress, serving several terms. He is today a superb U.S. senator, serving on committees dedicated both to the environment and to social justice, fighting tenaciously on behalf of the downtrodden. I believe that he is indeed heir to the socialist Eugene V. Debs. I vote for him at every opportunity, but each time I step into the voting booth, I also think of Murray and mentally salute the dream of assembly democracy.

In the fall of 1990, the US Greens gave up their paralyzing commitment to consensus decision-making and shifted to majority voting. They adopted a program that emphasized both social and ecological policies, so much so that Howie Hawkins found it "impressive."[67] The US Greens, he wrote, were on the threshold of becoming "that long-desired organizational framework where environmentalists, antiracists, feminist, and labor, peace, and community activists converge under a common banner into an independent, anti-capitalist, radical-democratic left."[68] Hawkins rechanneled his energy from the LGN into the Greens. Murray and I, too, bade the LGN farewell—it was impossible without its main organizer.

Disconcertingly for Murray, Hawkins also moved away from assembly democracy, from libertarian municipalism. It was acceptable, he publicly began to say, for Greens to run for statewide office, like state governor. Murray insisted that it was not—that libertarian municipalism depends on exacerbating the tension between the city and the state, not blurring it.[69]

Murray was sorry to lose the talented Hawkins—he genuinely liked him. But his transformation presented a problem: when someone who had publicly identified as a social ecologist went off in a different direction, how was Murray to respond? Ideas were terribly important to him—he wanted them to stay intact, their integrity preserved, regardless of individuals' behavior. He felt he had no choice but to publicly dissociate himself, so in 1992 he wrote that if Hawkins wanted to enter conventional politics, and run for statewide office, he should certainly go ahead: he asked only that Hawkins kindly refrain from talking about libertarian municipalism, from using it as a point on some state-government platform. Hawkins, who would go on to run for New York governor and U.S. senator on a Green platform, honored the request.

After the collapse of the Burlington Greens, Murray and I moved out of the yellow house.[70] Our new apartment (at 77 North Winooski Avenue) was humble: it had worn linoleum floors, a raw plywood counter, and an annoying blaster heater. But finally we had our own place. We rented an office downtown, a few blocks away, where we'd walk each morning. He'd do research for his books on his side of the office, while I'd copyedit books for New York publishers on mine.

Murray had spent his adult life trying to take revolutionary socialism beyond Marxism-Leninism to a vision of a rational, eco-decentralist society. At the end of the 1980s, with the normalization of Green parties and the implosion of the Burlington Greens, he understood that he would not be able to fulfill that vision.

Soon after the Greens ended, Murray announced that he was retiring from politics. Murray, Sisco, and I formed a one-evening-a-week study group, along with a few new recent arrivals in search of Bookchin. Murray, I realized, needed to be part of a group, needed to see people's eyes light with hope as he spoke. But it was all hypothetical now, profoundly subjunctive: the revolutionary era was over not only for history; it was over for him, too.

NOTES

1. Maslauskas, "An Interview with New York Gubernatorial Candidate and Rebel Howie Hawkins," Oct. 26, 2010, online at http://bit.ly/oM5iWT.
2. Bookchin, "The German Greens," *Gadfly* 2, no. 5 (Nov. 30, 1986).
3. John Dillon, "Toward a New Waterfront," *VVP*, Oct. 21–28, 1984.
4. Rick Sharp to author, Aug. 17, 2009.
5. Rick Sharp, in *VVP*, Dec. 8–15, 1985; Don Melvin, "Aldermen Clear Way for Waterfront Vote," *BFP*, Nov. 5, 1985.
6. Peter Letzeller Smith, "Waterfront Alignment," *VVP*, Nov. 24–Dec. 1, 1985, 5; Michael Powell, "The Battle for Burlington's Waterfront Heats Up," *BFP*, Nov. 25, 1985; Greg Guma, *The People's Republic: Vermont and the Sanders Revolution* (Shelburne, VT: New England Press, 1989), 113; Howard Hawkins, "Socialist Does Deal with Downtown Developers," *Guardian*, Jan. 3, 1986; Northern Vermont Greens, "The Waterfront Issue: Is Burlington Being 'Sold Out'?" (Fall 1985), in Laura Cepoi, ed., "The Double Edged Sword: Ecological Theories and Ideologies among the Greens," Dec. 1986, unpublished compilation, in Box 4, MBPTL. See also Mark Johnson, "Citizens Group Offers Alternate Waterfront Plan," *BFP*, Dec. 3, 1985, 3B; Candace Page, "Opponents Say City Could Plan Better Public Use of Waterfront," *BFP*, Nov. 19, 1985, 3B.

7. Powell, "Battle for Burlington's Waterfront"; Mark Johnson, "Sanders Unites with Business on Waterfront," *BFP*, Nov. 29, 1985, 1B; the press release is in Cepoi, "Double Edged Sword."

8. Peter Freyne, "Inside Track," *VVP*, Dec. 15–22, 1985.

9. Euan Bear, "Public Distrust Defeats Bond," *VVP*, Dec. 15–22, 1985.

10. Mark Johnson, "Waterfront Plan Goes Down to Defeat," *BFP*, Dec. 11, 1985; Don Melvin, "Alden Corp. Won't Buy Parcel Without Voter Approval," *BFP*, Nov. 20, 1985, 12B.

11. Guma, *People's Republic*, 113; Freyne, "Inside Track." For Sanders's movie about Debs, see http://bit.ly/1jr8PLp.

12. Bookchin, "The Bernie Sanders Paradox: When Socialism Grows Old," *Socialist Review* 90 (Nov.–Dec. 1986), 52.

13. Howie Hawkins, "Remembering Murray Bookchin," March 18, 2013, unpublished, author's collection.

14. Bookchin, speech to New England Greens, Keene, NH, Mar. 15–16, 1986, audiotape, MBPTL.

15. On Vermont Yankee's many technical problems since its construction in 1972, see John A. Dillon, "Yankee's Ingenuity," *VVP*, Feb. 6–12, 1981. Materials on the protest, the Green Mountain Alliance, and the New England Greens generally are collected in Cepoi, "Double Edged Sword," 124–31: "Close Yankee Nuclear for Good—Now! An Appeal for a Vigil on June 14 at Vernon, Vermont," May 1986; "Dear Friends," Green Mountain Alliance, June 30, 1986; Sallie Graziano, "Alliance Startup: Activists Agree on Purpose But Not Structure, of New Group," *Valley News*, July 28, 1986, 2; Eliot Page, "Anti-nuclear Group Outlines Its Plan of Action," *Rutland Daily Herald*, July 28, 1986; Green Mountain Alliance, "Declaration of Nuclear Resistance"; and the constitution of the Green Mountain Alliance (July 15, 1986).

16. "Lamm, Bookchin: Who's the Realist?" *New Options*, July 31, 1986, 7–8.

17. John Clark, *The Anarchist Moment: Reflections on Culture, Nature, and Power* (Montreal: Black Rose Books, 1984), 165.

18. Bookchin, "Thinking Ecologically" (1986), in *PSE*, 98, 97, 119. He called his philosophical ideas (which I sketched in chapter 10 of this book) dialectical naturalism. See Bookchin, *The Philosophy of Social Ecology* (Montreal: Black Rose Books, 1994). For a summary, see Janet Biehl, *Rethinking Ecofeminist Politics* (Boston: South End Press, 1991), chap. 5.

19. Maclay Architects, "Institute for Social Ecology, Rochester, Vt.," online at http://bit.ly/n1x8Sr;. Murray went to bat for the ISE, too. See Bookchin, "Freedom at Stake in School Siting," *Barre-Montpelier Times Argus*, Oct. 9, 1986. In October the state environmental commission would deny the ISE permission to build. But later Dan Chodorkoff would get a job teaching at Goddard, and the ISE would use the college's classroom and dormitory facilities in Plainfield. In the 1990s the institute would finally purchase land in nearby Marshfield.

20. The program called for addressing the problems of growth and pollution and wastes and highways, with proposals for recycling, renewable energy, and "the formulation of a regional plan to share our local energy resources and solve intercommunity waste problems cooperatively"; a moral economy, human-scale participatory institutions, a municipal bank, affordable housing, networks between farmers and consumers; greenhouses and community gardening; small-scale, citizen-controlled enterprises that would produce quality goods; monthly ward assemblies to mandate city councilors; confederation of towns; conversion of weapons enterprises to peacetime uses.

21. Bookchin testified to the Vermont Senate Government Operations Committee on Jan. 22, 1987, Vermont State Archives and Records Administration, Senate, Box LC139, tape 262,

side B, at 923-1621. This quote is from Bookchin, "Can Democracy Survive in Vermont?" *Vermont Affairs*, Winter 1988, 17. See also Bookchin, "The Vermont Constitution: A Legacy Worth Saving," brochure (1987); and "Governor's Two-Year Term Fundamental to Vermont Democracy," *BFP*, Nov. 23, 1986.

22. Theodor Adorno, "Sur l'eau," *Minima Moralia: Reflections from Damaged Life* (1951; London: Verso, 1996).

23. Bookchin, "The Power to Create, the Power to Destroy" (1970; rev. 1979), in *TES*, 44.

24. "Around the Campfire," *Earth First!*, June 21, 1987, 2.

25. Dave Foreman interviewed by Bill Devall, "A Spanner in the Woods: Dave Foreman talks with *Simply Living*," *Simply Living* [Australia] 12 (ca. 1986). Around the same time, Foreman's friend the novelist Edward Abbey railed against immigration from Mexico, on the grounds that the United States "is a product of northern European civilization. If we allow our country—our country—to become Latinized, in whole or in part, we will see it tend toward a culture more and more like that of Mexico." Edward Abbey, "To the Editor," *Bloomsbury Review*, Apr.–May 1986.

26. Bookchin, "Thinking Ecologically," 137.

27. Foreman interviewed by Devall, "Spanner in Woods"; Christopher Manes (as Miss Ann Thropy), "Population and AIDS," *Earth First!*, May 1, 1987, 32; "Around the Campfire," 2; Dave Foreman, in Steve Chase, ed., *Defending the Earth: A Dialogue Between Murray Bookchin and Dave Foreman* (Boston: South End Press, 1991), 107; "Around the Campfire," 2.

28. Bookchin, "Social Ecology versus 'Deep Ecology': A Challenge for the Ecology Movement" (written June 1987), *Green Perspectives*, nos. 4-5 (Summer 1987), 3-5; online at http://bit.ly/1ppDzoZ.

29. Bookchin, "New Social Movements: The Anarchic Dimension," in David Goodway, ed., *For Anarchism: History, Theory and Practice* (London: Routledge, 1989), 265.

30. Bookchin, address to first U.S. Greens gathering, Amherst, Mass., June 1987, audiotape, MBPTL.

31. Charlene Spretnak, *The Spiritual Dimension of Green Politics* (Santa Fe., NM: Bear & Co., 1986).

32. For a useful discussion of "old" and "new paradigms," see Stephen Elkins, "The Politics of Mystical Ecology," *Telos*, no. 82 (Winter 1989–90), 52–70.

33. Bookchin, letter to editor, *Resurgence*, no. 127 (ca. 1988), 46.

34. Janet Biehl, "Ecofeminism and Deep Ecology: Unresolvable Conflict?" *Green Perspectives*, no. 3 (1987), 1–11; "Dare to Know," *Our Generation*, Fall 1988; "Goddess Mythology in Ecological Politics," *New Politics: A Journal of Socialist Thought* 2, no. 2 (Winter 1989), 84–105. These and other articles that I wrote during my collaboration with Murray are posted on my page at Academia.edu.

35. Mark Johnson, "Marina Developer Planning Sanctuary," *BFP*, Aug. 18, 1987.

36. Janet Biehl, "Burlington's Wetlands: The Condos Are Coming," script for slide show produced by Burlington Greens, 1987, author's collection. The quotation is from Biehl, "Burlington's Unbridled Growth: A Position Paper of the Burlington Greens," Jan. 1989, MBPTL.

37. "Bookchin Speaks Out," McGill University, 1990, online at http://bit.ly/qPsHpr.

38. Dave Foreman, "Whither Earth First!?" *Earth First!*, Nov. 1, 1987, 20.

39. Chim Blea, "Why the Venom?" *Earth First!*, Nov. 1, 1987, 19; Mike Roselle, "Ann Arbor Earth Fair Revisited," *Green Letter/Greener Times*, Spring 1988; Bill Devall, in *Green Synthesis*

[San Pedro, CA], no. 28 (Sept. 1988); Devall, "Deep Ecology and Its Critics," *Trumpeter* 5, no. 2 (Spring 1988), 55.

40. Bookchin, "Yes!—Whither Earth First?" *Green Perspectives*, no. 10 (Sept. 1988), 7.

41. Bookchin, "Which Way for the U.S. Greens?" *New Politics* 2, no. 2 (new series) (Winter 1989), 71.

42. Bookchin, "Crisis in the Ecology Movement," *Green Perspectives*, no. 6 (May 1988), 6; "Yes!—Whither Earth First?" 5, 6.

43. Bookchin, "Crisis in the Ecology Movement," *Z Magazine*, July–Aug. 1988, 122.

44. Bookchin, "Looking for Common Ground," in Chase, *Defending the Earth*, 31.

45. Kirkpatrick Sale, "Deep Ecology and Its Critics," *Nation*, May 14, 1988, 670–75; Bookchin, "As If People Mattered," *Nation*, Oct. 10, 1988, 294; Kirkpatrick Sale, "Letter from America," *Resurgence*, no. 125 (Nov.–Dec. 1987).

46. Howard Hawkins, "J. Hughes' Green Liberalism: A Left Green Reply," pamphlet, Green Program Project, 1989. Hawkins was responding to J. Hughes, "Bookchinism: A Left Green Response," *DSA Green News* 2, no. 4 (Dec. 1988), 1–4; "The Great Debate," editorial, *Fourth World Review*, no. 31 (1988), 3–5.

47. Bookchin, "The Question of How to Think . . . ," *Whole Earth Review*, Winter 1988, 16. The ad hominem mudslinging became so dominant that it obscured the issues at stake in Bookchin's 1990s debates. Later scholars had to probe to unearth them. An excellent example is Andy Price, *Recovering Bookchin: Social Ecology and the Crisis of Our Time* (Porsgrunn, Norway: New Compass, 2011). Price concluded that substantively, the deep ecologists never laid a glove on him. Brian Morris, *Pioneers of Ecological Humanism* (Brighton, UK: Book Guild, 2012), is another fine account, as is Vincent Gerber, *Murray Bookchin et l'écologie sociale: Une biographie intellectuelle* (Montreal: Ecosociété, 2013).

48. Among those who defended social ecology or Murray or both were Tom Athanasiou, Carl Boggs, and myself in *Green Synthesis*, Oct. 1989; also Biehl, "Ecofeminism and Deep Ecology."

49. Bob Spivey, "Summer '88 Semester in Vermont: A Personal Report on the Institute for Social Ecology," *Green Synthesis* 29 (Dec. 1988), 10.

50. Bookchin, "Crisis in the Ecology Movement," *Green Perspectives* version, 1, 3.

51. Bookchin, "Which Way."

52. Blea, "Why the Venom?" 19.

53. Howard Hawkins, "U.S. Greens on the Threshold," *Left Green Notes: Organizing Bulletin of the Left Green Network*, no. 8 (June–July 1991), 11.

54. See "The Left Green Network (1988–91)" at http://bit.ly/1GiXsyY. Murray drafted the call; Hawkins drafted the principles. Bookchin's speech at Ames was published as "Which Way for the U.S. Greens?"

55. Lorna Salzman, "Is the Left-Green Network Really Green?" *Green Synthesis*, June 1989, 9; Tom Athanasiou, "A Green Orthodoxy?"; Janet Biehl, "Adherence to Principles"; and Carl Boggs, "Why the Left Green Network Is Necessary," all in "Is the Left Green Network Really Green? Responses from LGN Members," *Green Synthesis* (Oct. 1989), 6–14.

56. Bookchin, "Radical Politics in an Era of Advanced Capitalism," *Green Perspectives*, no. 18 (Nov. 1989), 1–6.

57. Chuck Morse, "Study Groups: Education and Political Practice," *Left Green Notes*, no. 6 (Feb.–Mar. 1991), 11.

58. Quoted in Judith Cebula, "Group Unites Rich and Poor for a Day with Soup Kitchen," *BFP*, Nov. 26, 1988; see also "Burlington Greens Confront Local Scrooge," *Left Green Notes*, no. 1 (Jan. 1989), 6.

59. Howard Hawkins, "Can the GCoC Practice the Grassroots Democracy It Preaches?" *Left Green Notes*, no. 4 (Sept.–Oct. 1990), 13ff.; Hawkins, "U.S. Greens on the Threshold," 11; Howard Hawkins, "Minority Rule Expels NEGA from IC," *New England Green Action*, May 1990, box 4, MBPTL.

60. "Burlington Greens to Run Municipal Candidates," *Left Green Notes*, no. 1 (Jan. 1989). The city council candidates were Paul Fleckenstein and Gary Sisco. As I've mentioned, the city council was then called the board of aldermen. For campaign coverage, see Lisa Scagliotti, "Greens Party's Baird Announces Run for Mayor," *BFP*, Jan. 10, 1989; "Greens Field Baird for Mayoral Post," *Rutland Herald*, Jan. 10, 1989; Scagliotti, "Greens Mayoral Candidate Calls for Halt to Growth," *BFP*, Jan. 27, 1989; Scagliotti, "Mayoral Race Is Wide Open in Final 5 Weeks," *BFP*, Jan. 29, 1989; Scagliotti, "City Mayoral Candidates Concur on Mall, Waterfront," *BFP*, Feb. 3, 1989; Scagliotti, "Greens Call for Democracy," *BFP*, Feb. 9, 1989; Scagliotti, "Steppingstone," *BFP*, Feb. 11, 1989; Scagliotti, "Candidates Tout Changes in Style," *BFP*, Feb. 16, 1989; Scagliotti, "Baird, Not Chioffi, Backed by Caucus," *BFP*, Feb. 18, 1989; Scagliotti, "Campaign '89: The Race for Mayor: Sandra Baird," *BFP*, ca. Feb. 1989; Kevin J. Kelley, "Inside the Race for Mayor," *VVP*, Feb. 16–23, 1989; Kelley, "Greens vs. 'Reds,'" *VVP*, ca. Feb. 1989; James N. Baker, "The Greens of Vermont," *Newsweek*, Feb. 27, 1989, 33; "Stop, Look and Listen," *BFP*, Feb. 28, 1989; Mark Johnson, "Mayor's Race Gets Intense as Finish Nears," *Rutland Herald*, Mar. 4, 1989; Scagliotti, "Mayoral Campaigns Gear Up for Final Blitz," *BFP*, Mar. 4, 1989; Scagliotti, "Lively Mayoral Contest Races to Finish," *BFP*, Mar. 5, 1989.

61. "Burlington Greens to Run Municipal Candidates," *Left Green Notes*, no. 1 (Jan. 1989).

62. Janet Biehl (unsigned), "Burlington's Unbridled Growth: A Position Paper of the Burlington Greens" (Jan. 1989) and "Democratizing City Government: A Position Paper of the Burlington Greens" (Jan. 1989); and Paul Fleckenstein (unsigned), "The Burlington Economy: A Green Perspective: A Burlington Greens Position Paper, 1989," author's collection.

63. "Stop, Look and Listen," *BFP*, Feb. 28, 1989.

64. Guma, *People's Republic*, 181.

65. "The Lakefront Seesaw," editorial, *BFP*, Sept. 2, 1989.

66. On Burlington's sustainability, see Rhonda Phillips, Bruce Seifer, and Ed Antczak, *Sustainable Communities: Creating a Durable Local Economy* (New York: Routledge, 2013). See also City of Burlington, "Burlington's Accolades," http://bit.ly/1AmySfI.

67. Howard Hawkins, "Greens National Conference Takes Positive Directions," *Left Green Notes*, no. 5 (Nov.–Dec. 1990), 12–13. See also Brian Tokar, "The Greens as a Social Movement: Values and Conflicts," in Frank Zelko and Carolin Brinkmann, eds., *Green Parties: Reflections on the First Three Decades* (Washington, DC: Heinrich Böll Foundation of North America, 2006), 90–100.

68. Howard Hawkins, "U.S. Greens on the Threshold."

69. For example, Howard Hawkins, "Confederal Municipalism and Green Electoral Policy," *Greens Bulletin* [Kansas City, MO], Apr. 1992. See Bookchin and Janet Biehl, "A Critique of the Draft Program of the Left Green Network" (May 24, 1991), *Green Perspectives*, no. 23 (June 1991); and Bookchin, "The Furtive Politics of Howard Hawkins," *Greens Bulletin*, May 1992.

70. His ex-wife and children would visit often, and Beatrice would often ask Murray for his views on politics or something she was reading. But mostly the conversations centered on personal matters: son Joseph's job teaching film production at Burlington College, and daughter Debbie's employment as a journalist, then as press secretary for Congressman Bernard Sanders, then as a teacher of public relations at Champlain College, as well as her marriage and in 1994 the birth of her daughter Katya.

13

Historian

―――――――――――――――――――――――――――――――――――――――

ON THE LAST day of 1991, all official institutions of the Soviet Union ceased operations. The long-overdue end of that totalitarian state brought universal relief, not least to Murray, who had been an anti-Stalinist since 1935.

Yet he still loved the red flag for the cooperative utopian dream it embodied. While watching it lowered over the Kremlin for the last time, he listened to a recording of "The Internationale" from the soundtrack to the movie *Reds*—it's sung by the Red Army Chorus at the opposite moment, in October 1917, when the Bolsheviks first take power, with high hopes of creating socialism, calling for "all power to the soviets." "We have been naught," went the lyrics, "we shall be all!"

Now, as Russians rejoiced at the end of the totalitarian system that emerged from that revolution, Murray thought this moment could be clarifying—perhaps the Left could finally dissociate itself from Stalinism and reacquaint itself with older, utopian ideas. For a century before the Bolshevik Revolution, socialists of various kinds had imagined ways to create a cooperative society. The Russians had inherited that legacy, then thwarted and suppressed it. Now that the Soviet Union was gone, perhaps it could be revived. Russia, for example, might potentially decide to transform its soviets into the genuinely democratic institutions that they had been in the 1905 and 1917 revolutions, before Lenin warped them into vehicles for top-down rule. They could become "forms of freedom" for a genuine Russian democracy.[1]

But no such renewal took place, and instead of reviving, the centuries-old radical tradition ebbed and faded into history—or rather, onetime radicals started to bury it. Almost as soon as the Soviet Union collapsed, the international Left began folding up its tents, proclaiming the death of socialism. Socialism had been the lovely dream of their youth, they said, when they had romantically tried to storm the heavens, but now they would reconcile themselves to the existing systems and look for ways to improve them gradually, to mitigate their excesses. A French student was quoted in our local newspaper saying, "We had our fun in 1968, and now it's time to grow up." Murray bristled: for him, the alternative to socialism was still barbarism. If he could only bring together everyone who was saying they were disillusioned, he told me, he'd have enough for a mass movement.

Marx had written that if the working class ever accepted capitalism as natural, then all hopes for revolution would be lost. Now Murray read in the leftist *Nation* that capitalism was here to stay, and onetime radicals had to learn to work within its parameters. "Social democracy is the left's horizon," wrote Todd Gitlin.[2]

Even as global warming threatened the planetary biosphere, onetime radical activists accommodated themselves to the system that was producing it. European Green parties followed the Germans in becoming parliamentary.[3] Their members got excited about the 1992 Earth Summit in Rio, which was, in Murray's view, about corporate polluters and heads of state haggling over ways to manage the crisis and thereby neutralize it as a potential source of radical social change.

Some of his longtime collaborators, however, bucked the trend and created new local political parties calling for ecology and municipal democracy. In Frankfurt, Jutta Ditfurth, after leaving the German Greens in 1991, founded the Ecological Left (Ökologische Linke), with a municipalist orientation. And in 1990 Dimitri Roussopoulos founded Ecology Montreal, whose program called, once again, for transforming Montreal into a green city and decentralizing power to the neighborhoods.[4]

But most of Murray's anarchist friends continued to spurn libertarian municipalism, even when he pointed out that communalism had historically been "the democratic dimension of anarchism."[5] Anarchists will run candidates in elections "when fish fly," said one of Murray's anarchist publishers. "The [anarchist] the rejection of electoral activity, even at the local level," Murray lamented, "has become "a paralyzing dogma."[6]

He had to acknowledge that all his lectures, writings, and classes had not produced a viable movement to change society. Social ecology seemed a lost cause. He felt that he had outlived his era, that he was on a peninsula that was slowly being covered with water. *It's over, pal,* he'd mutter to himself.

Not only had the political spectrum shifted to the right, but alternative political culture itself had changed. Confrontation was no longer considered acceptable behavior. Back in his Trotskyist days, even when debates had been acerbic and vituperative, everyone had understood not to take it personally—the clashes were about clarifying ideas for the sake of the larger cause. But in the 1990s, people took clashes of opinion very personally—and sought to avoid them. Now, it seemed to Murray, political people were supposed to hold only polite, low-keyed conversations and seek out commonalities. He was repeatedly reproached for his militancy and his sharp debating style—he was too aggressive, he was told. In the spring of 1991, he participated at a conference called "What Is an Oppositional Left?" at the University of Vermont. While the other speakers talked about ways to ameliorate the present system, Murray insisted that the system must be fundamentally changed. People in the audience called out, "The Left is dead!" "Trotsky was Lenin who failed!" "Your style is revolting!"[7]

Was the Left really gone? Murray asked himself over and over in disbelief. Where was idealism? Where was anger? Regardless of the fall of the Soviet Union, capitalism was still clawing away at the biosphere, and all the old issues of domination and exploitation and commodification remained. Were these new pragmatists really prepared to give up all their principles for the sake of accommodating the system? What did they want, careers? It was one thing to suffer a defeat; it was another to surrender and even embrace the enemy. Wouldn't they have preferred to go down fighting, to show some "resolution and character," rather than give in to barbarism?[8]

Sometime in 1991, as Murray and I were having lunch at the Oasis Diner in Burlington, we were reading an article in a leftist newspaper about a recent convention of the now-tiny Communist Party USA. The party leadership had decided to exclude some nine hundred people from the meeting—about a third of the membership. (Some things hadn't changed.) The nine hundred people crossed the street and held their own meeting. Then, back at the main meeting, Herbert Aptheker (one of Murray's old teachers at the Workers School) limped up to the podium with his cane and chastised the party for its dogmatism.

When I finished reading the article aloud to Murray, I expected him to scoff, as he so often had, at all those benighted Marxists. But he was silent for a while, then said he felt sorry for them—they were witnessing the demise of all their ideals, the ruin of their political culture. I thought, *He's talking about himself.* They still clung to the fragments, he said, but no signs of a new Left were discernible. Their hearts, like his, were broken.

"Let the dead bury the dead," he would sigh, quoting Marx's *Eighteenth Brumaire.* But the death of socialism was for him the death of meaning. "I feel very much like a stranger in a strange world. . . . My sense of expectancy today [July 1990] is almost zero."[9] He knew he didn't belong, but a society without a Left—with no ideals or vision—"is not a world in which I would want to live."[10]

He turned down the few lecture invitations that came in. What would he tell them? he wondered. He didn't want to teach anymore, or try to influence anyone—he wanted to withdraw from the world. In the summers he was still expected to travel to Plainfield to lecture at the ISE, but he dreaded it. Young people had always looked to him to provide hope, but he no longer knew how to give it to them. In 1992 some Greek friends offered to organize a trip to Greece for him, and because they were so eager to show him the birthplace of democracy on the Acropolis, and because I wanted to go, he agreed. But as the departure date neared, he despaired. He had no answers for them—how would he fill up those speeches? Just then a medical test revealed a possible health problem, giving him a reason to cancel the trip. His Greek friends were painfully disappointed, but he felt as if he'd dodged a bullet.

By then he dreaded leaving Burlington, even going outside our apartment. I'd push him into the center of town in his wheelchair (his osteoarthritis was making walking impossible), but he'd complain that he felt more claustrophobic outside than he did inside.

He decided he wanted to live out his days within our four walls, as a refugee from the past. He would write about philosophy, but only for his own satisfaction—after all, he thought, no one was paying attention. As a freelancer, I worked at home, so we were together most of the time. The apartment was warm and comfortable, with stacks of books. We'd read next to each other in the evenings.

He listened to Rachmaninov, Mussorgsky, and Borodin. Tchaikovsky made him weep. He favored movies, like John Huston's, in which men with grand ambitions end up noble failures—but are defiant to the end. Even more, he loved movies featuring obsessive, lonely, and difficult but kind-hearted male heroes, like Alec Guinness as George Smiley (in *Tinker, Tailor, Soldier, Spy*), Robert Mitchum as Philip Marlowe, and later, Michael Kitchen as Christopher Foyle, DCS. Over and over, we watched Mike Leigh's *High Hopes*, especially the scene in which a young couple visit Marx's grave and read the inscription: "Up to now philosophers have thought about the world. The time has come to change it." "But how?" wonders the young man.

He was my surrogate father and my mentor. His love remade me psychologically: my lifelong anxiety yielded to self-confidence and even enjoyment of life. By a happy stroke of fortune, I was able to reciprocate by providing him with a safe, secure, loving home in this desolate period. We gave each other daily reassurances and structured our life with small routines and rituals. We were each other's alpine sanatorium.

As the revolutionary libertarian Left fell off the political spectrum, Murray resolved at least to be its placeholder. He changed the name *Green Perspectives* to *Left Green Perspectives*. Even if he had to occupy that space alone, so be it. "My sole meaning in life, outside of any personal life, is to recreate ... and to embody the ideal of a Left."[11] If he could not do it in practical terms, by creating a movement, he would do it in theoretical terms. He identified the most valuable principles and arranged them into a composite, which he called "the Left that was": it was democratic, rational, secular, nonhierarchical, libertarian, and ecological. It was internationalist and antimilitarist. It was theoretically coherent. It was humane and ethical.[12]

Above all, it was socialist. If capitalism was irrational, then socialism was rational, the "fulfillment ... of humanity's potentialities for freedom, self-consciousness, and cooperation."[13] If socialism could not exist in reality, it could exist as a moral standpoint, the *what should be*, the ethical standard by which one might judge capitalism. The British socialist George Lichtheim once observed that "the Hegelian system provides a transcendental resting-place for ideals not realized in actuality. It holds out to men the promise not of freedom, but of the idea of freedom."[14] That was what it provided for Murray, the beached whale, in his old age.

* * *

More and more, the American ecology movement seemed to prefer deep ecology and biocentrism. Despite his criticisms, ecology appeared to be becoming less a social movement than a wilderness cult. Parts of it were even going neoprimitive, rejecting civilization as antiecological, and advocating some kind of return to prehistory.

Back in the 1970s and 1980s, Murray had extolled organic society as communal, organic, egalitarian, and harmonious with nature. In writing *The Ecology of Freedom*, he had looked back at prehistory not out of some perverse longing to return but rather to understand the principles that underlay egalitarian communal societies, and to explain how the forces of hierarchy and domination, capitalism and instrumentalism had emerged. By no means was he a primitivist, or believe that prehistoric tribal societies fulfilled humanity's potentialities. His aim, rather, had been to rescue their socially ethical principles and merge them with the positive advances of civilization.

But some of his readers did not grasp his dialectical treatment of domination and freedom and seemed to have misinterpreted *Ecology of Freedom* as a brief for primitivism. In the early 1990s, certain strands of anarchism joined the deep ecologists and Earth First! in calling for a return to the technological, economic, and perhaps population levels of prehistory, and to go "back to the Pleistocene." In late 1992, David Watson, editor of the anarchist newspaper *Fifth Estate*, called for seeing the world through the "wolf's point of view." One night Murray telephoned him to ask if he really thought wolves had a point of view. Watson replied by asking how Murray knew they didn't. Holding the phone to his ear, he shut his eyes and sagged in weariness.[15]

But fanciful ideas about wolves were the least of it. In the name of social ecology, Watson was writing diatribes denouncing Western civilization as "destructive in its essence to nature and humanity," and as "a maladaptation of the species." It was to be rejected "in bulk," along with the idea of progress, which was a mere cover-up for its sins. Prehistoric society, in his telling, had been a world of dancing, singing, celebrating, and dreaming—a countercultural be-in. Aboriginals were natural anarchists—they "refused power, refused property." Free of material desires, they had been able to "do what they like when the notion occurs to them." And even back in the Pleistocene, in their ecological wisdom, they had "deliberately underproduc[ed]."[16]

People today must learn from that Pleistocene wisdom, Watson wrote, reject reason, and "learn to listen" to trees. After all, trees "talk to each other, and they'll talk to you if you'll listen." We must "dismantle mass technics." We must replace cities with forests and small plow-based farms; we must eliminate roads, hospitals, medications, telephones, radios, and computers—and "pull the plug on the communications system." We must oppose all mass generation of electricity—even renewable energy. Not even the wheel merits saving: "The wheel is not an extension of the foot, but a simulation which destroys the original."[17]

Had *Ecology of Freedom* contributed to such mad rejections of civilization and reason? Murray wondered. Around this time, too, postmodernism was gaining popularity on college campuses, attacking Enlightenment reason as imperialistic and totalitarian. In 1991 his students were telling him that anarchists were embracing postmodernism as their philosophy. The counter-Enlightenment—irrationalism, antihumanism—seemed to be running rampant through the culture as a whole.

"I regret," Murray wrote in 1994, "my excessive criticism of the Enlightenment."[18] He had learned to criticize it from the Frankfurt School, especially from *Dialectic of Enlightenment*. The Frankfurts' distinction between instrumental reason and dialectical reason had once been crucial for him, but now that reason as such was under attack, distinction seemed less important than the defense of reason itself. With the rise of antirational, primitivistic anarchism, Murray's four-decades-long intellectual waltz with the Institute for Social Research was over. Their once-beloved writings now seemed to him "farragoes of convoluted neo-Nietzschean verbiage, often brilliant, colorful, and excitingly informative, but often confused, rather dehumanizing and ... irrational." He no longer saw their work as dialectical but as "lack[ing] any spirit of transcendence."[19]

In those years of withdrawal and an evaporating Left, he found a new purpose: to defend the Enlightenment, to defend the power of reason to improve the human condition. Over the course of history, he now affirmed, in "History, Civilization, and Progress," humanity has made concrete advances, material as well as cultural and psychological, toward greater degrees of freedom, self-consciousness, cooperation, and rationality.

I remember Murray explaining all this to me one night, as we were driving along. "Defending the Enlightenment" seemed a huge task, and I hoped we would not be alone, but I jumped into the fray with him regardless. I would take on a newly popular idea in the ecology movement that worship of an earth-goddess could instill in people a love of nature and attitudes of cooperation, and that if enough people worshipped the goddess, it could save the planet. I wrote several articles criticizing this tendency, which I saw as a symptom of the counter-Enlightenment, cultural feminist style.[20]

Unfortunately, one of the sources for goddess worship seemed to be none other than ecofeminism, the doctrine that had emerged in the late 1970s at the ISE, under much different circumstances. But it wasn't the goddess worship alone that bothered me. Ecofeminism promoted the patriarchal stereotype that women are "closer to nature" than men—their supposed deficit in rationality made them more attuned to the earth and its needs and therefore more qualified than men to defend nature. That struck me as not only preposterous but regressive—it perpetuated demeaning patriarchal stereotypes about women. When I told Murray I thought ecofeminism was a poor concept, he didn't hesitate—he encouraged me to write the book I had in mind.

Rethinking Ecofeminist Politics was published in 1991.[21] Angry ecofeminists answered back in blistering reviews of my book, in periodicals that I cared about. The reviews tightened my chest and made me gasp—the old anxieties were coming back. I couldn't bear to read them—Murray had to read them to me aloud, one phrase at a time. I was really too thin-skinned for the rough-and-tumble world of politics, it seemed.

Then, in October 1992, out of the blue, a German writer called Ulrike Heider published a long criticism of Murray in her book *Die Narren der Freiheit* (Fools of Freedom). Knowing the language, I typed out a translation of the main section for

Murray—and could not believe my eyes. It was a hatchet job. Heider called him an authoritarian, a nationalist, a fascist, a Bolshevik, a worker-hating booster of capitalism, and a religious nut—a whole range of extremes, all contradictory. Because he admired the institutions of ancient Athenian democracy, he must support slavery, the domination of women, and war. Because the New England town meeting had roots in Puritanism, he must be a devotee of Christian morality and Puritan patriarchalism.

Its vitriol was incomprehensible. She'd attacked me too, by association; Murray said he felt guilty, that allying with him had been costly for me. But this time, I needed no spoon-feeding. I went back to Murray's original works and read them against Heider's diatribe and identified her technique of distortion. I saw how, time after time, she had yanked Murray's words out of context and sometimes even reversed his meaning. I wrote it all down in a response, coauthored with Murray.[22] It taught me that, when attacked in print, one need not despair—just get to work and write a rebuttal. The more I did it, the more confident I became. I would get a lot of practice in the next years.

In January 1992 the Tulane philosophy professor Michael Zimmerman was in the process of compiling a major anthology on environmental philosophy, and he had charged John Clark with compiling the section on social ecology. Clark selected (in addition to pieces by Murray, me, and himself) an article by David Watson, the anarcho-primitivist admirer of wolf consciousness. Murray was shocked. Clark was widely considered his collaborator, but here he was taking social ecology into the very counter-Enlightenment direction that Murray had now made it his singular aim to oppose. What was happening with Clark? Later that year he wrote, "Let the next Gathering of the Greens conduct all its business in poetry."[23] Replace politics with poetry? On several occasions during the winter of 1992–93, Bookchin and Clark talked about their differences but could not come to agreement.

That fall, meanwhile, the ISE's curriculum committee had decided to drop Clark as a visiting lecturer—his lectures were not well received, and the committee wanted to bring in more women speakers.[24] Clark apparently proceeded to tell people, as he traveled the academic eco-philosophy conference circuit, that he had been purged, and Murray had engineered it. It was not even close to the truth—Murray was not a member of the committee and had never advocated dropping Clark from the school—but thanks to the deep ecologists' earlier accusations about redgreen putsches, the gossip mills started churning, and the smear spread quickly.

Clark's betrayal hurt, but Murray's despondency deepened for other reasons as well. Global capitalism had turned out to be too formidable to comprehend, invading and permeating everything. The young people he met were strangely passive—they would listen to him in silence, mute. Why don't they speak up? he wondered. Since they didn't communicate, they were opaque. His body was betraying him as well—his arthritis-induced hip and spinal pain made sleep difficult. The thought of going anywhere or being involved in anything made him shudder.

Around this time the Greek economist Takis Fotopoulos invited Murray and me to be on the international board of his new periodical *Society and Nature*. It seemed promising, we stayed for a few years, and the magazine published our articles. But we eventually came to believe that Fotopoulos's aim wasn't to promote social ecology but rather to mine it for material for own ideology ("inclusive democracy"), so we resigned. Such encounters made Murray wary of forming new friendships—somewhere down the road, he feared, he would just have to break with that person, which would only arouse further acrimony.

If Fotopoulos had borrowed Murray's ideas, others were borrowing the name *social ecology* and attaching different ideas to that label. Rudolf Bahro held a chair in "social ecology" at Humboldt University in Berlin—but he used the label to refer to his spiritualistic and authoritarian views: he was calling for an "ecological dictatorship" to run a "salvation government"—even a "green Adolf."[25] When we were in Berlin in 1990, Murray spoke at Bahro's class and said he thought the phrase "ecological dictatorship" was a contradiction in terms. That was the end of their once-friendly connection.[26]

In 1992 the business management guru Peter F. Drucker, a notable member of the corporate world, chose to call himself a social ecologist.[27] That same year, the academic ecofeminist Carolyn Merchant published a book that used *social ecology* as a category that encompassed Marxist as well as anarchist ideas.[28] The University of California at Irvine had initiated a program in social ecology. An Institute for Social Ecological Research was formed in Frankfurt, Germany—connected to the Social Democratic Party. None of these social ecologies had anything to do with Murray's work.

It was irritating, but he had no claim on the label—he himself had picked it up from E. A. Gutkind. Whatever became of the name, he understood, was out of his control.

But in the case of people who had learned social ecology from him, he had reason to trust that they would respect the ideas that he had advanced under that label and that if they ever reached the point where they disagreed with him, then they would have the courtesy to not use the label. His old collaborator Howie Hawkins, who had left Murray's fold to join the New York State Greens and run for statewide office, modeled the best approach: he had honorably let go of the labels and never again identified with social ecology or libertarian municipalism. But John Clark not only hung on to the name social ecology, as he moved elsewhere along the political spectrum—he tried to redefine social ecology in his own terms, as a mystical eco-communitarianism grounded in Taoism.

In the summer of 1995, onetime ISE student Mike Small organized a conference to found an international social ecology network, held in Dunoon, Scotland. At Small's invitation, John Clark was the keynote speaker. In his speech, Clark not only ripped into Murray's libertarian municipalism—he rejected the very concepts of citizenship and democracy (asserting, for example, that "it would be a mistake to associate democracy with any form of decision-making"). Social ecology, he said, should not be a political concept at all; social ecologists should instead direct their efforts to "the familial group." Social ecology—depoliticized? Conference participants who were expecting to hear something about Bookchin's ideas were shocked.[29] And when the conference formed a committee to draft principles and aims of the International Social Ecology Network, Clark participated, objecting to any mention of libertarian municipalism and successfully barring it. Instead of a democratic politics, the draft advocated eco-spirituality.

Back in Burlington, we read the document in amazement. Why had the conference permitted Clark to help draft principles of social ecology? How had he been able to bulldoze those who knew better?[30]

Murray, stunned and hurt, wrote a commentary objecting to the choice to invite Clark to keynote "when he had been in the process of shedding social ecology for quite some time." He responded to Clark's speech in detail and laid out his substantive disagreements with him on democracy, politics, history, civilization, progress, religion, and so on.[31]

Certainly Murray didn't own the name *social ecology*, or claim to, but he had given it meaning, built a school around it, and put it on the map. After decades of work advocating social ecology, he hoped that others would at least respect the integrity of his ideas, their coherence. People were free to accept or reject social ecology, but at a minimum he hoped its definition would be clear. As part of the retreat of the Left, radicals-turned-reformists were now developing a bias against political labels and ideologies, as if they were synonymous with dogma and sectarianism.[32] He decided he didn't mind when others criticized him in those terms—it meant he stood for thought and clarity and principle in an era of fuzzy thinking and blurred boundaries. He no longer cared what anyone thought of him—he just wanted his legacy intact. Perhaps future generations would find meaning in his work.

By the mid-1990s, fantasies of going "back to the Pleistocene" and wilderness worship were increasingly popular among American anarchists, as reflected in their periodicals.[33] Least of all did they show any interest in the kind of movement organizing and institution building that Murray advocated. Once upon a time, anarchism had been oriented toward social change—Kropotkin had even said it had a "socialist core." But now such "social anarchism" (to use the retronym) was becoming a dim memory, judging from popular periodicals like *Anarchy: Journal of Desire Armed*. Discussions of the Spanish anarchists and their collectivizations made today's anarchists yawn; instead of building "forms of freedom," they wanted to write on walls with spray cans, throw bricks at police, and break windows. Anarchism, Murray thought, was deteriorating into chaotic dadaism and nihilistic rejectionism—"a playground for juvenile antics," as he called it.[34]

Marx had long ago criticized anarchists as "petty bourgeois individualists." Murray had long disagreed but now thought perhaps he was right, especially in 1993 when the anarchist L. Susan Brown came out with a book endorsing Margaret Thatcher's statement, "There is no such thing as society. There are individual men and women." Anarchism's theoretical level had always been, in Murray's eyes, singularly low. Even

Kropotkin, the best of the lot, had based his notion of "mutual aid" on the purported existence of a "social instinct" in both humans and animals. (Murray rejected this idea because animal behavior is genetically programmed, while human societies are mutable, and people, with their advanced consciousness, can choose to change them.)[35] Compared to today's anarchists, Kropotkin seemed like a genius. Couldn't they think straight?

In the summer of 1994, Ulrike Heider's despicable book was published in English translation. What had happened, Murray wondered, to the anarchism that he had once championed as an ethical movement? Today too many anarchists "lie, distort and edit ideas with moral standards comparable to those of junk bond dealers and corporate raiders."[36] In another sign of anarchism's deterioration, Jason McQuinn, the editor of *Anarchy: A Journal of Desire Armed*, came out as a historical revisionist—the Third Reich killed only a few hundred thousand Jews, he said, and the Holocaust has been magnified "into a larger-than-life tale of historical racial persecution." The extreme Luddite John Zerzan took to denouncing not only technology but even language as a form of oppression.[37] Anarchism had turned into a madhouse. If this was anarchism, Murray seethed, he was no anarchist—he was anarchism's sworn enemy.

He threw down the gauntlet in a withering attack. If anarchism continued in this vein, he wrote in "Social Anarchism or Lifestyle Anarchism: An Unbridgeable Chasm," and if it were to shed its "socialist core and collectivist goal" in favor of individualism, irrationalism, and primitivism, then he would not only break with it, he would oppose it with every fiber of his being.[38] In June 1995 he finished the manuscript and sent it off to the publisher, A.K. Press. We awaited the response.

How acutely he missed the old political culture of the 1930s, when great movements had been in motion! Yes, Lenin had been a tyrant, and Trotsky was dogmatic and imperious, but nobody today compared with them for boldness and brains. In the 1930s people argued about ideas, not motives or personalities: their political goals were high-minded and socially generous; they were fired by passion and moral outrage and imbued with a collective spirit. But in the 1990s, most people who might have been rebellious seemed to prefer to withdraw into private life.[39]

Still, human beings do have the potential to innovate socially, to achieve a free and cooperative society.[40] The rational, ecological society that he advocated was another name for ethical socialism, which remained for Murray the highest ideal of social organization. Those who dared to try to fulfill it were the militants who made the great revolutions, those who fought ferocious social wars in the struggle for socialism, communism, and anarchism. Spanish anarchists stopped Franco's tanks by blowing themselves up with their grenades. Their valor, he told me, must never be forgotten (figure 13.1).

To pay tribute to them all, he went back to his unfinished manuscript for *The Third Revolution*, on the history of popular movements in revolutions—he'd started it back in the late 1980s. The phrase "the third revolution" had been used in the later stages of both the French and Russian revolutions. The first revolution was the bourgeois

FIGURE 13.1 Two historians of the Spanish Revolution: with Vernon Richards in London, 1992
Janet Biehl.

revolution, the one that overthrew the ancien régime; the second was the revolution of the authoritarian Left against the bourgeoisie, resulting in dictators like Robespierre and Stalin who aborted the popular upsurge; and the third was the popular, democratic revolution against the dictators, to claim social and political freedom.

So began, in 1992, his great late-life project: writing the history of the movements he loved. Even as the present grew stranger by the minute, he transported himself to the past, to consort with his familiar heroes and the "forms of freedom" that they created. As he researched the European peasant wars, and the English and American and French revolutions, he reveled in the dialectic of revolutionary social change, the immanent self-transformation of society. I helped with research and organization and edited the manuscript. Cassell agreed to publish the book in two volumes, and we expected to finish in a couple of years.[41]

By the spring of 1995, his osteoarthritis had worsened, so we moved into a more comfortable apartment in downtown Burlington, with handicap accessibility. Pooling our resources, we purchased a three-wheeled electric scooter, which made it easy for him to go into town and sit in coffee shops—and imagine the Parisian cafés of 1789 or 1848 or 1871, where people argued about how to bring their social dreams into reality. Or he'd scoot up to the university library. He finally gave up his trusty Selectric for one of those new word processors, and he'd tap out his history for ten hours straight. At age seventy-five, his stamina was incredible, but then, working on *The Third Revolution*, he was having a ball.

The year 1996 was a time of researching and writing peacefully. He exulted in the study of revolutionary movements and institutions, their potentialities, and the causes for their failures: the 1848 revolution in France, the First International, the Paris Commune of 1871. In our pillow talk, he analyzed revolutionary strategy. What stood out, he admitted, was that even in periods of upsurge, most people had been concerned not with building utopia but with remedying their immediate problems. When they went further, it was only because visionary leadership had pushed them.

On July 1, volume one, published, arrived in the mail, but he'd already moved on to researching the Russian Revolution of 1905, then 1917. Immersing himself finally in the Russian Revolution, he felt as if he'd come home. Reexamining it from the perspective of a seventy-five-year-old, he came to the conclusion that the Left Socialist Revolutionaries (and not the Bolsheviks) had had the best program for Russia: distribute the land to the peasant communes, socialize the little industry that existed, and form a strong peasant-worker alliance against the forces of reaction.

While I awaited his draft of volume two to edit, I compiled *The Murray Bookchin Reader*—Cassell had accepted my proposal in February 1996. That summer I also wrote a short primer on libertarian municipalism, called *The Politics of Social Ecology*, for publication by Black Rose. "It was probably the sweetest, kindest year I've yet had with Murray," I wrote in my journal.

The *Third Revolution* project was not only the antidote to his despair, it gave him plenty to say to young people after all (figure 13.2). Even though the revolutionary era was over, he could keep the revolutionary tradition alive just by talking about it—even if only in that room—and thereby refuse to participate in the general capitulation. Writing *The Third Revolution* thereby was not only a review of scholarship, a reexamination of his life, and an escape from present troubles—it was a kind of direct action in its own right.

Don't let yourself be trivialized, he told students, as he recounted episodes from revolutions past, and don't be afraid to stand alone. Don't mistake the way people are today for the way they could be, in a society that allowed them to fully express their reason, ethics, and creativity. Let yourself be guided by principles, he said, and derive your principles from our potentiality for a rational, ecological society. We will always have the latent potential for ethical socialism. In times like the present, when ethical socialism is not even on the agenda, and no movement advocates it, it still has a role: it is the standard, the *what should be*, by which we judge the *what is*. And sometime in the future, he knew, when ethical socialism returned to the agenda, *The Third Revolution* would be available to help people turn *what should be* into *what is*.

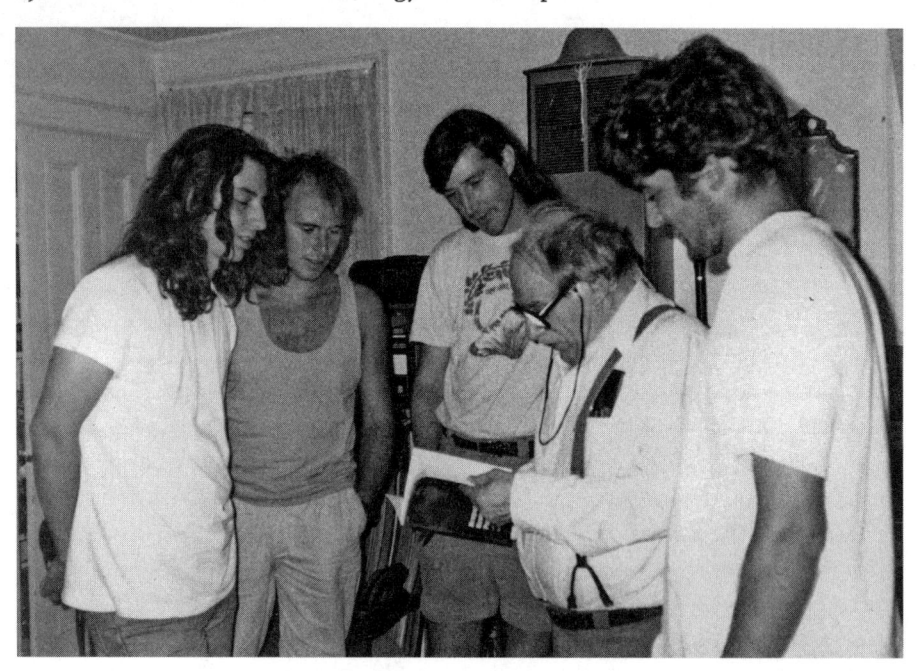

FIGURE 13.2 Murray kept the apartment door open to visiting students. Here with students from Greece, Norway, and Chile in Burlington, 1994.
Janet Biehl.

That summer, plenty of admiring visitors came around. Like Scheherazade, he kept his revolutionary reality alive by talking it into existence, by seeing it light up his visitors' eyes with hope. Among them was Eirik Eiglad, a bright young man from Norway (figure 13.3). A onetime anarchist, Eiglad had been conducting a social ecology study group in his native Porsgrunn since 1992. He and his *compañeros* had read all our books and articles, translated many of them, and distributed the material in their infoshop. He had attended the Dunoon conference and fought John Clark's machinations. In the next years he would organize groups in southern Norway to advocate communalism, which in his country, he told us, was a better name for libertarian municipalism. In 1998 in Telemark, he and his friends founded Democratic Alternative (DA), a communalist movement organization. Communalism "can give us a solid base for the creation of a rational social alternative to the nation-state and market society," read its manifesto: "the empowerment of the people through municipal self-management, confederalism and direct democracy." DA would develop a local and regional structure, including a secretariat, hold congresses, and conduct internal and external education. At present, "unfortunately Communalism is not widely recognized," the manifesto said, "but we are going to change this."[42]

For Murray, Eirik and his friends were another antidote to despair, and he did not disappoint. We stayed in close touch by telephone and through his periodic extended

FIGURE 13.3 With Eirik Eiglad, ca. 1997
Janet Biehl.

visits. In 2001 I found that, due to the work on *The Third Revolution*, I could no longer publish *Left Green Perspectives*, so after forty-one issues, I closed it down. Eirik stepped up to replace it and founded *Communalism*, an online journal to publish, among other things, Murray's writings as he produced them.[43]

In November, Murray finished the chapters on the Russian Revolution. He seemed to have no off switch, writing past exhaustion. By the end of the year, I had finished both my primer on libertarian municipalism and the *Bookchin Reader* and sent them off to their respective publishers; Murray approved both and called the *Reader* in particular a masterpiece, the best introduction to his work. Then I took up editing volume two of *The Third Revolution*. I wrote in my diary on January 18, 1997: "This must be one of the happiest days of my life. I've finished two books. I've been with Murray ten years. I am healthy. What more could I ask?"

I'm so glad we had that peaceful year.

Although John Clark had largely abandoned politics for mysticism, his spiritual plane was not so elevated as to override his bitter streak. For reasons known only to him, he chose to mount a campaign to discredit his former mentor, starting with a caricature of Murray as a Stalin-like commissar of social ecology. Published in *Anarchy: Journal of Desire Armed*, it portrays Murray as purging the brave dissident Clark for ideological deviation at a Moscow-esque show trial.[44]

Why does Clark hate me so much? Murray wondered. His own personality had indeed been shaped by the Communists of the 1930s—they had raised him and trained him to be a commissar. But the deepest layer of his psyche was emotionally generous, stemming from his idyllic Bronx childhood with the stern but loving grandmother. It had made him ebullient and magnanimous with *compañeros* throughout his life—and trusting that students would, at a minimum, treat him respectfully.

In 1998 Zimmerman's *Environmental Philosophy* anthology was republished in a second edition; the section on social ecology was reduced to one article, by John Clark, advancing mystical eco-communitarianism. Murray phoned the general editor, Michael Zimmerman, to ask why he'd agreed to that, and during the brief conversation Zimmerman referred to Murray as having wronged Clark. Murray had no idea what wrong he had committed, and afterward he asked his friends to tell him if they ever heard, just so he'd know.[45] He never did find out.

In the next years, whenever Murray advocated a specific idea, or took a side on an issue, or even singled out a subject for discussion, Clark would attack him for being partial, and hence sectarian, in contrast to Clark's own "wholeness," his spiritual embrace of everything. Of course, the concept of "wholeness" is so empty and uninformative as to be meaningless. And when Murray, in response to critics, defended his life's work and specific political passions, Clark condemned him as not only partial but "pugilistic."[46]

Anarchist responses to the gauntlet Murray had thrown down ("Social Anarchism or Lifestyle Anarchism") had been arriving, and the time had come to pay attention to them. The anarcho-primitivist David Watson published a book-length diatribe, *Beyond Bookchin*, that was an encyclopedia of Murray's defects, at least as Watson imagined them. Bookchin approved of reason and civilization—that made him an icy technocrat, a hyperrational, lab-coated scientist; social ecology must instead embrace the irrational and "listen to trees," which "talk to each other." Bookchin was a hater of primal societies—social ecology must instead reject civilization and go back (whatever that meant) to the Paleolithic and reject all technology, even the wheel. Bookchin, who claimed to be antihierarchical, was actually as a "protofascist," as the research of Ulrike Heider had so definitively shown. Bookchin, the ex-Marxist who claimed to be an anarchist, was "General Secretary" (i.e., Stalin)—that is, when he was not "Chairman" (Mao). Over the years, Bookchin had changed his positions, which showed that he contradicted himself; in other respects he had stayed the same, which showed that he was an inflexible fanatic. Anyone who dared to defend Bookchin was by definition his "sycophant" or, like me, his "hagiographer."[47]

The level of personal attack escalated. Those who had a grudge against Murray (or perhaps envied his originality and prescience) made a sport, it seemed, of trying to provoke the disconsolate old man into overreacting with yet another angry outburst that would further discredit his character. In the summer of 1997 or 1998, an ISE alumnus—I'll call him Derek—showed up on the school's campus at the opening of the session and sat in on Murray's class. A half-hour into the lecture, as Murray was answering a student's question, Derek commenced to shout at him. No teacher would tolerate such a disruption, and Murray threw him out. As Derek left, he screamed, "Fuck you, Murray!" Around dinnertime, in the school cafeteria, Derek and a friend of his turned up and approached a faculty member who was sitting at a table with her students. The friend remarked that the ISE should really be called the "Institute for Jewish Ecology," while Derek grinned maniacally. "I never realized how malevolent people could be," Murray told me.

In response to *Beyond Bookchin* and to Clark's aspersions, Murray wrote "Whither Anarchism?" in which he repeated his mea culpa for the defects of *Ecology of Freedom*: in previous decades "I myself shared an excessive enthusiasm for certain aspects of aboriginal and organic societies." In his 1982 book, "I gave an overly rosy discussion of them" and "waxed far too enthusiastic about primitive attitudes toward the natural world." Aboriginal lives were not to be romanticized: they were actually short and difficult, plagued with material insecurities, food shortages, diseases, drudgery, rivalries, and warfare, problems that civilization, at least to some extent, had ameliorated. He still regarded the principles of usufruct and the irreducible minimum as "sources of valuable lessons for our own time." But in the 1970s, in his eagerness to criticize the instrumentalism of the Enlightenment, he had suggested that "the animistic qualities of aboriginal subjectivity were something that Westerners could benefit from emulating." He admitted the error: "I was wrong. . . . I regard any belief in the supernatural as regressive."[48]

As for Clark, Murray considered the public rupture to be welcome and overdue, since "our ideas . . . are basically incompatible"; but "I would have hoped that our disassociation could have occurred without the personal hostility, indeed vilification that Clark/Cafard exhibits."[49]

I went into research-assistant overdrive, countering Watson's catalog of Murray's alleged defects, exposing his misstatements as I had done with Heider's book a few years earlier. Murray and I handed the article back and forth, until we finished in March 1998. I collected a number of other essays and interviews and assembled them, along with "Whither Anarchism?" into a book called *Anarchism, Marxism, and the Future of the Left*, published in 1999.

We'd worked fast, but hardly was the ink dry on "Whither Anarchism?" when, much to our amazement, Watson suddenly changed his mind about civilization. Opposition to it "in bulk," he wrote, was "empty theoretical bravado," since "communal and liberatory visions and practices" are also "woven into the sinews of civilization itself." Progress had occurred over the course of history after all: "We've inevitably learned some things along history's way . . . which are probably indispensable to us now." As for his earlier celebrations of the Paleolithic, he admitted that primitivism was a "fool's paradise" that he wanted "less and less to do with."[50]

It often felt like we were living in a foxhole, continually fending off fresh attacks. The smears and slurs had to be answered, but it was tedious and time-consuming, and when an opponent changed his position faster than we could rebut them, it was a waste of time. Ultimately, the mudslingers bored Murray, he told me, because they were symptoms of the times. The absence of radical politics, he told me, disturbed him far more than any of their venom and gall.

Next to take a swing at Murray were the academic neo-Marxists, still trying to exhume Karl Marx for service as an early ecologist. Their efforts evolved into something called eco-Marxism, and the periodical *Capitalism Nature Socialism*, or *CNS*, was founded to advance it.[51]

In the late 1990s, *CNS* turned out three consecutive issues containing harsh criticisms of Murray. "They've got a bug about me—they can't stop," Murray sighed. In a contribution darkly called "Negating Bookchin," Joel Kovel (a psychiatrist as well as a Marxist) diagnosed Murray as having a "messiah complex" and of seeing Marx as a "competitor." ("Oh sure—that's the only reason I would have criticized Marx!" Murray laughed.) Kovel condemned Murray as the unreconstructed offspring of the old Enlightenment Left. ("He considers that a shortcoming?")[52]

In 1995 we learned that the *CNS* crowd was compiling an anthology of critical articles on Murray. Andrew Light was to be the editor of *Social Ecology After Bookchin*, another dark title that I thought suggested Murray was dead. ("If only I were," he told me.) A sympathetic contributor, who was privy to the draft table of contents, passed it along to us. Kovel's "Negating Bookchin" was there, as was Clark's more-dialectical-than-thou "Municipal Dreams." Watson was represented, too—"all my fans," Murray said. Seeing what loomed, he contacted the publisher, Guilford, and asked to be permitted to write a response, to be included at the end of the collection. Guilford said it would forward the request to the editor, Light. Murray never received a reply.

In August 1998, shortly before *Social Ecology After Bookchin* was to be published, editor Andrew Light himself turned up on our doorstep—coincidentally, he was in Burlington for a conference—and apologized to Murray for the whole thing. The eco-Marxist James O'Connor, founder of *CNS*, had originated the project four years earlier to bash Murray, he admitted, but then it had languished. It had finally been assigned to Light, even though he didn't share O'Connor's agenda.

Murray told Light that he had asked to be able to write a reply. What happened with that? Oh yes, Light remembered. He'd polled the contributors, asking whether they'd allow it. Their resounding answer had been no. Light realized, he told us, that they were frightened of Murray. Having some training in psychology, he said, could have helped him on this project, since "there was so much Oedipal stuff going on!"[53]

Work on *The Third Revolution* continued—a life raft in stormy seas. The manuscript for the second volume became too long, so Cassell allowed Murray to go into a third.

More anarchists now weighed in, responding to the gauntlet Murray had thrown down. The vast majority said, in effect, that they preferred primitivism to socialism, and bravado to movement building. They didn't seem to care much even for anarchism's nineteenth-century forefathers. Kingsley Widmer accused Murray of standing "in lonely splendor" on the "ghostly shoulders of Bakunin, Kropotkin, and their descendants such as at the Spanish anarchists of two generations ago." Their "antique left-socialism" belonged to "a different time and place and conditions."[54]

Other responses dismissed libertarian municipalism as outside the boundaries of anarchism. Steve Ash, writing in *Freedom*, called it "entirely non-anarchist," based as it was on "direct democracy and rule by the majority." Insofar as Murray claimed to be an anarchist, he said, he was "a square peg in a round hole." Harold Barclay agreed that libertarian municipalists "are not anarchists," since "democracy by anyone's definition is a form of government." The free cities that libertarian municipalism advocated were "essentially micro nation-states." The anarcho-syndicalist periodical *Black Flag* objected that libertarian municipalism was about "the pursuit of power—and we are not going to be led down that road." *Organise!* said that libertarian municipalism was a process wherein "'libertarians' capture the local state and end up captured by it."[55]

Over and over, we heard the basic anarchist objection: that democracy is tyranny, because majority rule is rule. Observed Peter Marshall, "The majority has no more right to dictate to the minority, even a minority of one, than the minority to the majority."[56] Kropotkin himself had asserted that "majority rule is as defective as any other kind of rule."[57] Good luck organizing a liberatory society on that principle, we thought.

In 1962 the historian George Woodcock had observed that anarchism is by no means "an extreme form of democracy," for "democracy advocates the sovereignty of the people," while "anarchism advocates the sovereignty of the person."[58] Murray had thought he could change that description. He could not. In the four years since the publication of "Social Anarchism or Lifestyle Anarchism," anarchists had sent him an unmistakable message: they were not interested in libertarian municipalism. His twenty-odd-year effort to persuade them to accept it had come to nothing.

But my little book summarizing libertarian municipalism had been translated into several languages and was taking on a life of its own. In March 1997 Dimitri Roussopoulos, who had published it, proposed that he and I organize a two-conference series to try, one last time, to get anarchists interested in libertarian municipalism. Murray told us he thought it would be useless, bound to fail, and he didn't want it. But he saw that Dimitri and I were eager for one last try, so he grudgingly gave his consent, but he declined to participate.

The first conference was held in Lisbon, Portugal, in August 1998. Several dozen social ecologists from Europe and North America converged, as did many people who seemed to want either to observe or to criticize. We heard the usual objections that libertarian municipalism was statist, and that the ancient Athenians were patriarchal and had slaves, forever tainting democracy, it seems. Some Spanish anarcho-syndicalists showed up, denounced the project, and generally created chaos. Those of us who actually wanted to try to build libertarian municipalism were far outnumbered by those who opposed it or were indifferent. All in all, it was a disaster.

Plans for the second conference continued anyway, and it was held as scheduled in August 1999 in Plainfield, Vermont. Once again libertarian municipalism, intended as the subject of the conference, was mostly on the defensive. This time North American anarchists showed up to reject it, in favor of apolitical cooperatives.

I had kept Murray apprised of the proceedings, and on the last day he decided to show up after all. He announced that he was breaking with anarchism as his ideological home.[59]

He laid out his reasons. First, anarchism was fundamentally individualistic. Yes, plenty of anarchists called themselves social and collectivist, but participation in a collective requires sacrificing a measure of autonomy, and anarchists refuse that, giving the individual priority.

Second, anarchists oppose laws and constitutions, but no society, not even a free society, can exist without a rational and orderly way of regulating itself.[60]

Third, anarchists consider power to be a malignant evil and demand its abolition. But power, Murray explained, is neither good nor evil. The pertinent issue is who has it—the elites or the people.

He illustrated his point with one more story from the Spanish Revolution. On July 21, 1936, the workers of Barcelona defeated the Francoist rebels and gained control over Catalonia. The committees they established—for supply and transportation and land collectivization and defense, in factories and neighborhoods—had constituted a de facto revolutionary government, and through it the workers and peasants held the regional power in their hands. For guidance on how to manage that power, they turned to the CNT, the world's largest anarcho-syndicalist union.

On July 23 the CNT convened a plenum in Barcelona to discuss the problem. Some delegates argued that the CNT should approve the collectives and committees and proclaim *comunismo libertario*. But others argued that such a move would constitute a "Bolshevik seizure of power." Instead, they said, the CNT should join all the other antifascist parties—bourgeois liberals, socialists, and even Stalinists—and form a Catalan coalition government.

The CNT plenum chose the second course and, in effect, proceeded to transfer power from the workers and peasants to that coalition government—which really was a state. Thereafter this newly formed Catalan state consolidated its power. It restored the old police forces. It gave the Stalinists a free hand to suppress the workers' and peasants' committees. Within about five months, it had demolished the revolution and arrested its supporters.[61]

Murray's point was that anarchists had been confused about the nature of power then—they should have proclaimed libertarian communism when they had a chance—and they were still confused today. But arguing with them was draining and futile. Opposed as they were to both politics and democracy, they would and could never be a force to change society or improve the human condition.

For years Murray had carried an eerily prophetic quotation from William Morris around in his wallet: "Men fight and lose the battle, and the thing they fought for comes about in spite of their defeat, and when it turns to be not what they meant, other men have to fight for what they meant under another name." Murray had fought for anarchism and lost the battle. "I now find," he would write formally in 2002, "that the term I have used to denote my views must be replaced with Communalism."[62]

In November 1999, anarchists were battling globalization and the World Trade Organization in the streets of Seattle. Some of his students who were in Seattle called him on their new cell phones and begged him to reconsider his break with anarchism. The antiglobalization movement loved him, they said—they considered him their grandfather. And look how they were organized—into affinity groups with spokes!

Murray turned on the television and saw footage of anarchists breaking windows—they could have stepped right out of the pages of "Social Anarchism or Lifestyle Anarchism." He heard anarchists proclaiming the revolution, just as he'd heard overexcited students do back in the 1960s. The vaunted affinity groups, it turned out, were actually small groups of people networking over the Internet, and they weren't trying to build permanent "forms of freedom," citizens' assemblies; they were just organizing the protest. He'd seen protests, much bigger than this one, come and go, and while he admired the participants' rebelliousness, no protest, not even this one, could redeem anarchism in his eyes.

No movement for communalism existed at present: it "lies dormant as a prospect for a new politics."[63] But at least he had set down his ideas clearly on paper. If future generations should ever want to make assembly democracy real, they could find a strategy for doing so by reading his work. He'd done all he could—now it would be up to others.

He taught me a quotation from Hegel: "The Owl of Minerva takes flight at dusk." Minerva was the goddess of wisdom, and the owl her symbol; the sentence means, in effect, "Wisdom comes late, after the fact."

The years of work on *The Third Revolution*, he told me, were among the most illuminating he'd ever had. Amid all the controversies, he had still been working on the manuscript, and in September 1999, after the Plainfield conference, I resumed editing it. The project, once again, helped Murray reflect. As he explored the revolutions of the past—as he studied the most upstanding and competent revolutionary leaders, the alternatives they had faced, the choices they had made, and the forces arrayed against them—he concluded that "we never had a chance." Lenin had tried, through his own sheer will, to move the clock-hands of history forward. No one before or after him had made so enormous an effort. He was history's greatest revolutionary, Murray said—but even his revolution had failed, since it had created a new tyranny.

He thought a lot about one particular time and place where the revolutionary people might have had a chance. In Berlin in the fall of 1918, just after the defeat of the kaiser and the collapse of the Reich, a Social Democratic government had been established in Germany. The various leftist groups, energized by the Russian Revolution—the Spartacus League (Communists), the Revolutionary Shop Stewards, and the Independent Social Democrats—were champing at the bit to replace it with a Communist social order. The German proletariat was well schooled in revolution and

Marxism, better than the proletariat in any other country. If ever there had been a moment for socialist revolution, this was it.

In January 1919 the SPD government dismissed Berlin's police chief, Emil Eichhorn, who was a radical. Outraged, leaders of the leftist groups called a general strike. On January 5 half a million workers materialized in Berlin's streets and squares, armed with rifles and machine guns, ready to fight for Eichhorn's reinstatement and to replace the Social Democratic government with a "Workers' and Soldiers' Council Republic." Potentially a proletarian army, they waited in the streets for the leaders to give them the signal.

That day, in a room high above them, behind closed doors, the leaders debated among themselves. Some wanted an insurgency; some wanted to negotiate with the SPD government. As evening approached, they had made no decision. The crowd of armed proletarians, receiving no signal, drifted away, hungry and disappointed. The next day they reappeared in full force, once again armed and ready. But still the leaders could not decide on a course of action. The crowd of workers waited all day in a cold fog and steady rain, then once again, having received no signal, they dispersed. They never returned.

The German government at this time had only a weak military force. Had the strike leaders given a prompt signal, the workers might well have taken over Berlin. But their failure to do so was a gift to the SPD government—a window of time in which to muster the right-wing paramilitary, the Freikorps. A few days later, the Spartacus group finally did call an uprising, but by then the government was able to deploy the Freikorps against the workers who came out, as well as their leaders. That was when they murdered Rosa Luxemburg. Disorganization and indecisiveness had thus squandered a revolutionary possibility of historic proportions.

The story obsessed Murray in his last years. Had today's anarchists been on the scene in 1919, he mused, what would they have done? Shunning organization per se as authoritarian, they might have thrown a few rocks. Perhaps in 1936–37 they would even have opposed the Spanish anarchists' collectivizations.

But now the revolutionary era was over, and for young people today capitalism had become a natural fact, like air and water. Instead of trying to change society, it seemed to him, people in the 1990s were mainly thinking about how to live comfortably. Their lives seemed trivial, the flow of events meaningless, the level of consciousness spiraling downward. Life was a bubble that grew, then burst.

His sessions at the word processor lengthened, generating page after page after page. He'd finish a chapter, and I'd start editing. His sentences were getting longer and more sprawling and running on; my challenge was to preserve their music while bringing them under control. But as I worked, he'd type out more pages, mostly repeating what he'd said before, sometimes bringing in new material, but out of sequence with the story; or he'd tell the same story in a new way, with new insights. I was trying to edit and research and rewrite so that the narrative would at least be coherent. But he kept generating new pages. He couldn't stop, and in retrospect, I don't think he wanted ever to stop writing that book—finishing it once and for all would mean finally lowering the revolutionary tradition into its coffin, and he couldn't allow that to happen.

The situation became onerous for me. For many years I had been attending to his complex needs, both editorial and household, nearly every evening after my workday and every weekend. I had no close friends besides him, and no social life at all apart from his circle of friends. Whenever he saw that work on the book was stressing me, he'd go out on his scooter and bring me a yellow rose with a lovely note attached.[64] I was touched every time, but the strain was becoming overwhelming.

Volume three was still too long—once again Cassell let us split it in two. Volume four was to cover the (aborted) German revolution, and finally Spain—two crucial events that he had been pondering for most of his adult life. He wrote and rewrote, pouring decades of stories, ruminations, conclusions, and speculations into those pages. Finally, after eleven years of this project, and endless rewrites and additions, I declared closure. The manuscript of volume four, in the condition it was in on May 21, 2003, became the final version of that book. We took it over to Kinko's, packed it up, and sent it off to Cassell in London.

Since 1987 I had devoted every spare moment to social ecology, all the while earning my livelihood. I had written books and articles on its behalf and edited nearly every word that Murray had written for publication during this very productive period of his life. Since our home was also my workplace, we had been together for almost twenty-four hours every day since 1987. His world had become my world.

A few days after we dropped off the manuscript at Kinko's, I told him that I was proud of all the work we had done together, but that I needed to do something besides social ecology now. I'd been an artist in my twenties, and I was thinking about picking up my paintbrushes again. He immediately understood and encouraged me. He pointed to one of my old watercolors hanging on the wall and exclaimed, "Yes—you could do that again. In fact, you could draw political cartoons, like Daumier!"

I took some classes in printmaking and quickly developed a bit of skill. He encouraged me wholeheartedly, offering comments on every print I made. He even took to reading books on art, so he could talk to me about it. When my prints were hung on display at a local craft shop, he said he wanted to go see them there. I told him I was afraid that, on his electric scooter, he'd knock over some of the shop's fragile glass objects. *No!* he insisted—*I'm going.* He did go. He saw my work on the wall and didn't knock over a thing.

He ruminated on the period between the two world wars. In those crucial "interwar years," as he called them, he had been told that capitalism was moribund. But it had not been moribund at all. He had been told that it was decaying, putrefying—but actually it had just been revving up. He went over this problem in his mind again and again, as if he were trying to figure out where his path had become decoupled from social reality.

We'd long ago stopped driving to Plainfield every summer for the ISE session—due to his arthritis, he could barely sit up without pain, let alone ride in a car for an hour.

So now the students traveled up to our apartment in Burlington and sat on the living room floor. A few times each summer, he'd fall into his blue La-Z-Boy, pull the lever until he was practically lying down, and talk to them about whatever was on his mind.

In these reactionary times, he told them, your noblest role is to defend the Enlightenment, and to defend the use of reason in public affairs, against the dark forces—irrationality, nihilism, and barbarism—that stand at the gates of civilization. My generation fought Stalinism, he said. Your generation must, at the very least, oppose the dumbing down of the human mind, the growing, appalling ignorance of even the recent past, and the new gospel of self-absorption and narcissism.

What had sustained him for eighty years, he told them, was his devotion to ideas, study, books, and theory; his belief in human potentialities; and his conviction of the need for a rational society as an alternative to barbarism. Don't let yourselves be trivialized, he would tell them, or commodified. Build a movement for an ecological society, and keep it anticapitalist. Don't let irrationalism drown out reason. Remember that human beings can remake society, and that the preservation of the biosphere depends on them doing it.[65]

I did not want to trivialize myself, but I had to admit that ever since the failure of the libertarian municipalism conferences in 1999–2000, my doubts about antistatism had been growing. Despite the many failings of the nation-state, it seemed to me that it had historically also played a crucial role in rectifying social injustices and ensuring legal rights in the here and now. An ecological, rational society would be excellent, but in the here and now, some localities, like many in the American South, are reactionary, and the civil rights legislation of the 1960s had been an indisputable good. The social safety net set up by the New Deal and the Great Society programs had prevented millions of people from suffering. As for environmental action, many local places in the United States opposed it and undertook it only when federal agencies forced them to. Anarchism seemed to provide an incomplete picture of the nation-state, focusing solely on its injustices and abuses and ignoring its progressive aspects. And although the nation-state was too locked in with wealthy corporations, it also seemed to be far more likely to constrain capitalism and mitigate global warming than would a decentralized, stateless society.

One evening after supper I told Murray gently but honestly, that while I loved and cherished social ecology as a beautiful idea, I could no longer act as its public advocate. Murray himself had often said that the revolutionary era was over—that socialist revolution was no longer on the political spectrum. He was content to be its placeholder even if it meant standing alone against the world, but at the age of fifty that was not adequate for me. I told him I was reverting to the political identity I'd had before I met him, which was a liberal Democrat—what he, like the revolutionaries of his generation, disparagingly called a social Democrat.

When I finished, he leaned back in his chair, inhaled, and opened his mouth. For a moment, I thought he was going to remind me that the German Social Democrats had murdered Rosa Luxemburg—had committed that primal, unforgivable blood-crime that had woven in and out of his life. Or perhaps I would hear a denunciation of comfortable, trivial baby boomers who knew nothing of heroic struggles. I braced myself.

Instead, after a moment, his look softened, his muscles relaxed, and he said, "I love you anyway."

And I felt the same for him. But as age made him ever frailer physically, our thirty-three-year age difference became difficult for me to cope with. It wasn't just the strain of my protracted celibacy, or his increasingly complex needs for physical care. In these last years his disappointment in the world was acute. He'd been saying for a decade that he couldn't bear to live in a world without a Left. Intensely depressed, feeling his life was meaningless, he said frequently that he yearned for death.[66]

But I was on a different trajectory: as his mentality became bleaker, I was looking outward, in ways I never had before. The shy and timid person was long gone; his extraordinary affection had transformed me into a self-confident, creative person, ready to spread my wings as never before. Now the outside world seemed newly complex and fascinating, a place in which I might come to feel at home after all. When I was Murray's co-polemicist, I had had no patience for people who didn't share his ideology; now I wanted to listen to them and hear what they thought.

If only I could get more help with the caregiving, I thought, it would give me more space. But how? Having no outside friends to help me figure it out, I turned to a psychotherapist, who suggested, sensibly, that the caregiving could be shared by others. Murray had two adult children living in Burlington—one of them could take him in. Or he could move to Section 8 (senior) housing and be cared for by all his family members and me in tandem. Or we could apply for assistance for home care.

One morning in January 2005 I told Murray that I loved him as much as always, but I needed to change the caregiving arrangements. Perhaps one of his children could take him in. I would share the caregiving load with them, but the burden would no longer fall wholly on my shoulders.

To my surprise and dismay, he interpreted the proposal as me breaking up the relationship. *No, it's not that, not at all,* I said, *I just need help so that my needs can be met too.* But he couldn't seem to understand it, accused me of betraying him, and became distraught and angry. His daughter Debbie seemed to understand what I was asking, and being a resourceful person, she took on the task of applying for assistance from various agencies.[67]

Meanwhile emotions escalated, voices were raised. One morning anger and fear were running so high that he and I couldn't seem to hear each other. I went outside to walk but could not calm myself. Frantic, I realized I was frightened to return to the apartment. He kept several handguns there. A decade earlier, in 1992, a prominent elderly German Green, a man of sixty-nine, had shot his twenty-four-years-younger partner in the head, then used the weapon on himself. Could such a thing happen to me? Terrified, I phoned my therapist, who suggested I call the police and request that an officer accompany me into the apartment. I did so. The policeman went in first, checked the situation, came out and said everything was fine. When I entered, Murray assured me that he would never hurt me.[68]

That incident was a turning point, as it began to defuse the situation. I realized that if he was going to interpret a shift in caregiving as a breakup, then I would have to rethink. Meanwhile Debbie succeeded in securing home care for Murray through Medicaid. Soon a caregiver from the Visiting Nurses' Association was coming over regularly and took on many tasks. It made all the difference: not only was less required of me, but the presence of a neutral outside party reduced the pressure, dispelled fears, and quelled anger, allowing the essential affection between Murray and me to rise to the surface again.

In April 2005 Murray's cardiologist diagnosed him with aortic valve stenosis—he had at most five years to live. The only treatment was to replace the valve surgically. He declined the surgery, preferring to let the stenosis run its course.[69]

Instantly a primal fervor swept me—to make his last days on earth as beautiful and tender as I could.[70] I will always be grateful for the fifteen months that followed the diagnosis. Emotions again escalated, but this time to an acme of mutual affection and compassion. I learned what it meant to give freely, with no strings, no conditions, no guilt, no roles, no barriers. One late Sunday afternoon we sat quietly holding hands in the Dunkin' Donuts on Main Street, near where the Fresh Ground Coffee House once stood. As the sun came in through the plate-glass window, he leaned back on his scooter, closed his eyes in contentment, and pressed my fingers. Together we were *sur l'eau*.

In April 2006 his congestive heart failure led to tachycardia and a moderate heart attack. He was hospitalized, and after treatment the hospital was gearing up to send him to rehab, then hospice.[71] Once again his daughter, Debbie, intrepidly organized matters so that Murray could skip rehab and go directly to hospice care, not in an anonymous institution but in his own home, our home.

His son Joseph arranged for a hospital bed to be set up in our living room. The Visiting Nurses set up an excellent home hospice program. With their help, those three months of caregiving were for me not a burden but a privilege.[72]

I read aloud to him from books he'd loved in his youth—*Martin Eden* by Jack London, and *Pudd'nhead Wilson*. We watched our favorite movies together—*The Name of the Rose*, *The Late Show* (figure 13.4). We sat holding hands for long periods. One day in June he looked up at me and asked, "Am I boring you?"

A procession of friends came to bid farewell—I remember especially Dan Chodorkoff, Brian Tokar, and Chaia Heller from the ISE. Murray's half-brother Bob wished to visit but was unable to travel, having recently suffered a broken shoulder.[73]

One evening in late July I sensed the end was near. Quietly I bade him farewell, assuring him that he had always been perfect for me in every way. For the last time we marveled at the unbelievably fortunate accident that we had met at all. "What are you going to do?" he asked. I told him I'd write his biography, and a radiant smile lit up his face.

FIGURE 13.4 Reading on his deathbed (etching), June 2006
Janet Biehl.

A day later, as his breathing became labored, I read aloud to him, from Will Durant's *History of Philosophy*, about the death of Aristotle, then about the death of Voltaire, hoping to give him courage. When I got to the passage about the engraving on the philosophe's tomb, his eyes flew open, and he stammered it back to me—"Here—lies—Voltaire!" Then he whispered his last words: "I am you and you are me."[74]

Two days later, on July 30, 2006, he set sail on the infinite sea.

NOTES

1. IPS Symposium, Amsterdam, November 1990, 49, 131, transcript at MBPTL and author's collection.

2. Norman Rush, "What Was Socialism . . . and Why We All Miss It So Much," *Nation*, Jan. 24, 1994; "A Symposium: The Left 40 Years Later," *Dissent*, Jan. 1994.

3. Janet Biehl, "Western European Greens: Movement or Parliamentary Party?" *Green Perspectives*, no. 19 (Feb. 1990), 1–7.

4. The Ecological Left (ÖkoLi; later ÖkoLinX-Antirassistische Liste) sent members (including Ditfurth) to the Frankfurt municipal assembly (*kommunale Parlament*) in 2001, 2006, and 2011. See http://www.oekolinx-arl.de. Ecology Montreal, which Roussopoulos

founded in 1990, ran twenty-one candidates in 1990 and received 24 percent of the vote. It existed until 1994. Dimitri Roussopoulos, interview by author, July 6, 2009.

5. Bookchin, "What Is Communalism? The Democratic Dimension of Anarchism," *Green Perspectives*, no. 31 (Oct. 1994), 1–6.

6. Bookchin, letter to the editor, *A: Rivista Anarchica* [Milan], no. 185 (Sept.–Oct. 1991), 40–41.

7. "What Is an Oppositional Left" conference, University of Vermont, Mar. 8, 1991. I was present and heard these shouts.

8. Bookchin, "Intelligentsia and the New Intellectuals," *Alternative Forum*, no. 1 (Fall 1991). This article was originally a speech to the Youth Greens at Goddard College, July 27, 1990. See also Bookchin, "The Question of How to Think . . . ," *Whole Earth Review*, Winter 1988; Bookchin (as Robert Keller), "When Socialists Discovered They Were Liberals," *Green Perspectives*, no. 30 (July 1994), 5–8.

9. Bookchin, "Intelligentsia and New Intellectuals.".

10. Bookchin, "The Left That Was: A Personal Reflection," *Green Perspectives*, no. 22 (May 1991), 8–9.

11. Bookchin, "Intelligentsia and New Intellectuals," 2.

12. Bookchin, "Left That Was."

13. Bookchin, "Reply to John Moore," *Social Anarchism*, no. 20 (1993), 36. See also *PSE*.

14. George Lichtheim, *Marxism: An Historical and Critical Study* (1961; reprint New York: Columbia University Press, 1982), 6.

15. Bookchin, "Twenty Years Later . . . Seeking a Balanced Viewpoint," introduction to *The Ecology of Freedom*, rev. ed. (Montreal: Black Rose Books, 1991); "Whither Anarchism? A Reply to Recent Critics" (1998) in *Anarchism, Marxism, and the Future of the Left: Interviews and Essays, 1993–1998* (San Francisco: A.K. Press, 1998), 206.

16. David Watson, "Civilization in Bulk," *Fifth Estate* 26, no. 1 (Spring 1991), 10, 11; Watson, *Beyond Bookchin* (New York: Autonomedia, 1996), 52, 240.

17. Watson, *Beyond Bookchin*, 68–72, 154; David Watson (as George Bradford), "Bhopal and the Prospects for Anarchy," *Fifth Estate* 20, no. 1 (Spring 1985), 26; David Watson (unsigned), "Notes on Soft Tech," *Fifth Estate* 18, no. 1 (Spring 1983), 4; David Watson (as George Bradford), "Media: Capital's Global Village," *Fifth Estate* 19, no. 3 (Fall 1984), 11.

18. Bookchin, preface to Turkish and Greek translations of *TES*, Feb. 6, 1994. See also Bookchin, *Re-enchanting Humanity: A Defense of the Human Spirit against Anti-humanism, Misanthropy, Mysticism, and Primitivism* (London: Cassell, 1995).

19. Bookchin, "History, Civilization, and Progress: Outline for a Criticism of Modern Relativism" (1994), in *PSE*, 175. The Frankfurt Schoolers used the terms *objective reason* and *subjective reason*, which I find confusing and have avoided here.

20. Janet Biehl, "Women and the Democratic Tradition," *Green Perspectives*, nos. 16–17 (June–Aug. 1989), reprinted in *Society & Nature* 1, no. 1 (1992); and "On Feminism and the Retreat from Reason," *New Politics*, Winter 1990, 180–90.

21. Janet Biehl, *Rethinking Ecofeminist Politics* (Boston: South End Press, 1991), published in Canada as *Finding Our Way* by Black Rose Books in Montreal. It was the first book about social ecology (apart from Gutkind's work) that was not written by Murray Bookchin.

22. Bookchin, "From a Fool of Freedom," reply to Ulrike Heider ("written with the assistance of Janet Biehl," Oct. 17, 1992), published as "Von einen Narren der Freiheit" in *Schwarzer Faden*, no. 44 (1992), 48–50.

23. John Clark (as Max Cafard), "The Surre(gion)alist Manifesto," *Fifth Estate* 28, no. 1 (Spring 1993), 18.

24. Bookchin, "Whither Anarchism?" 224.

25. Bahro, interview by Frank Schumann, "Die deutschen Linken und die nationale frage," in *Streitschrift* (1990), 6, quoted in *In bester Gesellschaft: Antifa-Recherche zwischen Konservatismus und Neo-Faschismus*, Raimund Hethey and Peter Kratz, eds. (Göttingen: Verlag die Werkstatt, 1991). See also Jutta Ditfurth, *Das waren die Grünen: Abschied von einer Hoffnung* (Munich: Econ Taschenbuch Verlag, 2000), 82; and Martin A. Lee, *The Beast Reawakens* (Boston: Little, Brown, 1997), 218.

26. A summary of Murray's speech is "Herrschaftsfreiheit statt Ökodiktatur," in Rudolf Bahro, *Ruckkehr: Die In-Welt Krise als Ursprung der Weltzerstörung* (Horizonte, 1991), 305–6. Thanks to Karl-Ludwig Schibel for referring me to this source.

27. Peter Drucker, "Reflections of a Social Ecologist," *Society*, May–June 1992.

28. Carolyn Merchant, *Radical Ecology: The Search for a Livable World* (New York: Routledge, 1992).

29. Janet Biehl, "Two Conferences: In Reply to Clark's 'Comments for the List' (Oct. 13, 1998)," RA [Research on Anarchism] List, Oct.–Nov. 1998 (accessed 1998).

30. Social Ecology Project to Mike Small, Sept. 13, 1995, MBPTL and author's collection.

31. Bookchin, "Comments on the International Social Ecology Network Gathering and the 'Deep Social Ecology' of John Clark" (written Sept. 1995), in *Democracy and Nature* 3, no. 3 (1997), 154–97; online at http://bit.ly/1lJ7jGE. The editor of *Democracy and Nature* asked Clark for permission to publish his article as well but was refused (154).

32. Bookchin, "The Unity of Ideals and Practice" (1997), in *Anarchism, Marxism*, 319.

33. Bookchin, "'Anarchist' Label Abused," *Kick It Over*, no. 26 (Summer 1991).

34. Bookchin, *Social Anarchism or Lifestyle Anarchism: An Unbridgeable Chasm* (San Francisco: A.K. Press, 1995), 10.

35. Bookchin, "Deep Ecology, Anarchosyndicalism, and the Future of Anarchist Thought," in *Deep Ecology and Anarchism: A Polemic* (London: Freedom Press, 1993).

36. Ulrike Heider, *Anarchism: Left, Right and Green* (San Francisco: City Lights, 1994); in response Murray wrote "A Meditation on Anarchist Ethics," *Raven*, no. 28 (Winter 1994), 328–46, esp. 345.

37. Jason McQuinn, "Holocaust or Bust?" *Anarchy: A Journal of Desire Armed*, no. 34 (Fall 1992), 17; John Zerzan, *Future Primitive and Other Essays* (New York: Autonomedia, 1994).

38. Bookchin, *Social Anarchism or Lifestyle Anarchism*, 54.

39. In the 1920s and 1930s, Hamburg had been one of the most radical cities in Germany, but during the 1940s, Hamburg workers became Nazi police and hunted down Jews. This revelation, in Christopher Browning's *Ordinary Men*, about Battalion 101 of the Nazi Order Police, shocked and sickened Murray.

40. Bookchin, "Introduction: A Philosophical Naturalism" (1990), in *PSE*.

41. *The Third Revolution* would turn into four volumes, published by Cassell between 1996 and 2003.

42. Demokratisk Alternativ, "Prinsipper & Vedtekter, Principles & Bylaws," brochure, 2000, author's collection. Murray and I published Eiglad's articles in *Left Green Perspectives*, including "Bases for Communalist Programs," *LGP*, no. 40 (Feb. 1999), and "The Birth of a Democratic Alternative," *LGP*, no. 41 (Jan. 2000).

43. Although published in Norway, *Communalism: International Journal for a Rational Society* appeared in English only. It continued beyond Murray's death to 2008, when Eiglad and his

Norwegian comrades replaced it with the successor magazine and book publisher New Compass; see New-compass.net. In 2007 the social ecology group Demokratisk Alternativ for Oslo mounted a campaign for the city council. The program called for democratizing Oslo, strengthening local self-government, enforcing social justice, and making the city more ecological. "There are several other political initiatives who present demands to create a greener and more equitable Oslo," recalled Sveinung Legard, "but none other who actually demand the transfer of power to the people that inhabit the city. That's why we wanted to emphasize the direct and participatory democracy aspect of our social ecology politics. This is what makes us unique on the Oslo political scene." Legard to author, 2013. For the program, see http://new-compass.net/node/223.

44. John Clark (as Max Cafard), "Confession to Comrade Murray Bookchin, Chairman and General Secretary of the Social Ecologist Party and Founder of Dialectical Naturalism (Dianat)," *Anarchy*, Spring–Summer 1998.

45. Bookchin, "Turning Up the Stones: A Reply to Clark's October 13 Message" (Oct. 1998), online at http://bit.ly/qc1Yw6.

46. John Clark (as Max Cafard), "Bookchin Agonistes," *Fifth Estate* 32, no. 1 (Summer 1997), 20ff.

47. Watson, *Beyond Bookchin*, 68–72, 9, 19, 15, 16, 40, 37. A second book-length response, in a similar vein, was written by Bob Black. As Brian Morris points out, Watson's and Black's criticisms "are full of distortions and complete misrepresentations of social ecology," falsely accusing Bookchin of being "a technocrat and a neo-conservative." Yet they "confirm the substance of Bookchin's own critique," in that they "denigrate rationalism, and embrace instead intuitions and religious mysticism, defend eco-primitivism and make a cult of technophobia, repudiate civilization in its entirety." Morris, *Pioneers of Ecological Humanism* (Bristol: Book Guild, 2012), 177.

48. Bookchin, "Whither Anarchism?" 187–88.

49. Ibid., 218. Cafard was the pseudonym that Clark, a professor, used for his extracurricular screeds.

50. David Watson, "Swamp Fever, Primitivism, & the 'Ideological Vortex': Farewell to All That," *Fifth Estate* 32, no. 2 (Fall 1997), 18, 19.

51. James O'Connor, in *Capitalism Nature Socialism* [CNS] 5, no. 1 (Mar. 1994), 17.

52. Joel Kovel, "Negating Bookchin," *CNS* 8, no. 1 (Mar. 1997), 3–35. Murray's reactions are from my memory.

53. Quoted in Bookchin, "Turning Up the Stones." The book was Andrew Light, ed., *Social Ecology after Bookchin* (New York: Guilford Press, 1998).

54. Kingsley Widmer, "How Broad and Deep Is Anarchism?" *Social Anarchism*, no. 24 (1997), 77–83.

55. Steve Ash, "Murray Bookchin: A Square Peg in a Round Hole?" *Freedom* (London), Sept. 5, 1998; Bookchin, "Murray Bookchin Replies to a Critic," *Freedom*, Oct. 3, 1998; Harold Barclay, "Libertarian Municipalism," *Freedom*, ca. 1998; M. H., "If Voting Changed Anything . . . ," *Black Flag*, no. 207 (ca. 1996); and review of *Social Anarchism or Lifestyle Anarchism*, in *Organise*, no. 43 (Summer 1996), 17. Murray responded in *Organise*, no. 44 (Autumn 1996).

56. Peter Marshall, *Demanding the Impossible: A History of Anarchism* (London: HarperCollins, 1992), 22.

57. Peter Kropotkin, "Anarchist Communism: Its Basis and Principles," in Roger Baldwin, ed., *Kropotkin's Revolutionary Pamphlets* (1927; reprinted New York: Dover, 1970), 68.

58. George Woodcock, *Anarchism: A History of Libertarian Ideas and Movements* (Cleveland: World, 1962), 30.

59. Bookchin, "The Communalist Project," in *Communalism* [Porsgrunn, Norway], no. 2 (Nov. 2002), n. 18, n. 8, reprinted by New Compass at http://bit.ly/1enHufp and in Bookchin, *Social Ecology and Communalism*, ed. Eirik Eiglad (Oakland, Calif.: A.K. Press, 2007). With this article Bookchin repudiated the definition of *communalism* that was implicit in his 1994 essay "What Is Communalism? The Democratic Dimension of Anarchism." Murray now fondly quoted the Paris Communard Gustave Lefrancais, who had styled himself as "a communalist, not an anarchist, please"; see Kropotkin, *Memoirs of a Revolutionist* (1899; New York: Grove Press, 1968), 393.

60. Wrote Kropotkin: "We foresee a state of society where the liberty of the individual will be limited by no laws, no bonds—by nothing but his own social habits and the necessity . . . of finding cooperation, support, and sympathy among his neighbors." Peter Kropotkin, "Anarchist Communism: Its Basis and Principles," in Baldwin, *Kropotkin's Revolutionary Pamphlets*, 63.

61. Bookchin, *The Third Revolution: Popular Movements in the Revolutionary Era* (London: Continuum, 2005), 4:180–92.

62. Bookchin, "Communalist Project," n. 18. For the William Morris quote, see *Ecology of Freedom* (Palo Alto, Calif.: Cheshire Books, 1982), 33.

63. Bookchin, "Reflections: An Overview of the Roots of Social Ecology," *Harbinger: A Journal of Social Ecology* 3, no. 1 (Spring 2003), 11.

64. Bookchin to author, thirty-six florist note cards, 1991–2002.

65. With the proliferation of ecology programs in universities everywhere, the ISE's enrollment waned, and in 2004 it sold the piece of land that it had previously acquired for classes. It continues today to offer short-term intensives and colloquia and offers a joint graduate program with Prescott College. See www.social-ecology.org. Upon the retirement of Dan Chodorkoff, Brian Tokar became director. Author of books on green politics, ecology, biotechnology, and climate justice, Tokar also serves on the board of Vermont 350.org.

66. Bookchin, "A Journal for an End to This Century," handwritten ms., 9 vols., 1990–97.

67. Author to Lucia Kowaluk, July 24, 2005; Biehl, "Caregiving for Murray," unpublished ms., August 9, 2006, author's collection.

68. The Green couple were Gert Bastian and Petra Kelly. See Alice Schwarzer, *Eine tödliche Liebe: Petra Kelly und Gert Bastian* (Cologne: Kiepenheuer und Witsch, 1993), a book I had read from cover to cover. See also "Incident Report," Burlington Police Department, Jan. 13, 2005.

69. Author to Kowaluk, July 24, 2005. Aortic valve stenosis, if untreated surgically, has a prognosis of about two years. See C. Spaccarotella et al., "Pathophysiology of Aortic Stenosis and Approach to Treatment with Percutaneous Valve Implantation," *Circulation Journal* 75, no. 1 (2011), 11–19, doi:10.1253/circj/CJ-10-1105.

70. It was reciprocated. See Bookchin, "As to the Books," Nov. 2005; and "To Janet, Debbie, Joe and Bea," Jan. 10, 2006; and various book inscriptions to me.

71. Author to Dr. Claudia Berger, Apr. 5, 2006; author to Dr. Matthew Watkins, Apr. 5, 2006; author to Inara De Leon and Paula Harrington, June 18, 2006.

72. Author to De Leon and Harrington, June 18, 2006.

73. Bob Bookchin to Debbie Bookchin and Marilyn Bookchin to Debbie Bookchin and both in "Letters," ca. April–June 2006, author's collection. Dr. Robert Bookchin, a research physician, last visited Murray in October 2004; author's calendar.

74. Author, "June–July 2006," bedside steno notebook.

Epilogue

GIVEN HIS DIALECTICAL cast of mind, Bookchin perpetually looked for emancipatory potentialities. When he studied history he looked for the moments when forces for change could have taken a more liberatory turn; and when he surveyed the present, he looked for ways people could create a better future. Then he tried to inspire them to take that direction.

The United States at midcentury was pregnant with potentialities for radical social change. Bookchin saw early on that industrial agriculture, with its chemical-laden monocultures, posed a threat to human health. He argued for biodiverse farming free of chemicals (not yet called organic). But the alarm was not heeded, and in the intervening years, the scale of that threat has only exploded as agribusiness has wrought havoc in ever more ways on people's health for the sake of profit.

He decried the inhumane monoculture of the megalopolis as stressful, congested, routinized, capitalist-driven, nature-alienated, and polluted, and saw a potentiality for a decentralized, ecological society powered by renewable energy. Many Americans shared his abhorrence of gigantic cities, but when they left, it was not to an eco-decentralist society but to suburbia, to auto-dependent sprawl, shopping malls, office parks, and gated communities whose existence depended on fossil fuels.

Had industrialized societies heeded his warnings and followed his prescriptions in the 1950s and 1960s, we would likely not be facing the ruinous prospects of climate change today. The landscape would be adorned with ecological towns, not choked with highways; food production would be local, not controlled by five gigantic corporations; and the political system would be a citizen-controlled democracy.

But the solutions that he advanced in those early years—renewable energy, organic agriculture, and local democracy—are the ones whose validity is far better understood today. Moreover his message that human beings are capable of rearranging their affairs to create an ecological society is a message of hope. His utopian alternative reminds us what a good society could look like and the social generosity of which human beings are capable. The project of creating it could be a process of not only preserving the biosphere but creating meaningful lives and exercising virtuous human agency.

Some environmental authors have grimly placed the blame for the ecological crisis on human population numbers, destructive technologies, and human nature itself. Bookchin countered all these notions, arguing that the capitalist system, not population numbers, is the problem; that technology is indispensable and over millennia has immeasurably improved our well-being; and that people are the solution, not the problem. Far from being irremediably destructive parasites on the planet, we have a unique faculty of reason that is crucial for developing solutions.

The roots of the ecological crisis, he insisted to the end of his days, lie in our social arrangements. Long before anyone else on the Left did, Bookchin argued that the ecology issue represents the limit of capitalism. In that grow-or-die market economy, businesses must compete to undersell each other and thereby to maximize profits. The argument that capitalism is tearing down the planet is actually a hopeful one, since capitalism is a social arrangement, and social arrangements are malleable.

Today some basic assumptions of "social ecology" have become conventional wisdom. The American environmental movement is no longer a wilderness preservation effort. It now understands that saving fragile habitats requires fighting the large social forces that are threatening both people and nature. Global warming itself has ceased to be an issue of ecology; it is an issue of civilization itself.

Bookchin outlived the era that shaped him, the era of revolutionary socialism, and indeed the era of ideology itself, in which people looked for unitary, secular, coherent philosophical worldviews for explanations of reality. But that desire for coherence gave him an optimistic outlook and infectious enthusiasm. In a cynical time he was earnest, exploring the grand themes of history and trying to bring their lessons to bear in the present. Possessed of the idealism and moral imagination to power change, he kept the long-term goals steadfastly in sight, caring more for the big picture than for details. Despite massive countervailing social forces, he maintained a utopian temperament. He understood that while success might not be immediate, we must nonetheless make choices now to avoid catastrophe in the future.

While no substantial social ecology movement emerged in the West, in another part of the world a movement has arisen to put Bookchin's ideas into practice. In the mid-1990s several of his important books were translated into Turkish, and they found an eager reader in Abdullah Öcalan, the leader of the Kurdistan Workers Party (PKK). Öcalan had been arrested, tried, and sentenced to life imprisonment in 1999; while in solitary confinement, he read Bookchin in translation, and in April 2004, through an English-speaking intermediary, he requested a dialogue. He told Murray that he considered himself an excellent student of his and called himself a social ecologist. Bookchin, however, was too weary and ill to carry on a dialogue and declined, sending Öcalan his best regards.

Öcalan went on to recommend Bookchin's works to the Kurdish people and to adapt them to the Middle Eastern scene. Under his leadership, the PKK transformed itself into a movement for "Democratic Confederalism." When Murray died in 2006, its assembly wrote a touching memorial to him and announced its intention to create the first-ever society based on Bookchin's ideas. Kurdish activists went to work constructing institutions of assembly democracy, gender equality, ecology, and a cooperative economy in southeastern Turkey, albeit under conditions of persecution by the Turkish state.

In the summer of 2012, their efforts bore remarkable fruit in northern Syria after Baath control collapsed there, as a result of the Syrian civil war. In a quasi-autonomous setting, in the absence of the dictatorship, Kurds were free to build democratic confederalism in what they call Rojava. They have established communal assemblies in neighborhoods and towns, where men and women alike participate; the assemblies send delegates to confederal councils at the district, city, and cantonal levels. They are committed to power flowing from the bottom up and to fostering cooperation among the area's diverse ethnic and religious groups; "unity in diversity" is one of their watchwords.[1]

I close this biography with a salute to their quest to create a rational, ecological, democratic society. I only wish Murray Bookchin had lived to see it.

NOTE

1. On the relationship between Öcalan's ideas and Bookchin's, see Janet Biehl, "Bookchin, Öcalan, and the Dialectics of Democracy," *New Compass*, Feb. 2009, http://bit.ly/XRk3oh; Ahmet Hamdi Akkaya and Joost Jongerden, "Reassembling the Political: The PKK and the Project of Radical Democracy," *European Journal of Turkish Studies* (2012), http://ejts.revues.org/4615; and Akkaya and Jongerden, "Democratic Confederalism as a Kurdish Spring: The PKK and the Quest for Radical Democracy," in Mohamed M. A. Ahmed and Michael M. Gunter, eds., *The Kurdish Spring: Geopolitical Changes and the Kurds* (Costa Mesa, Calif.: Mazda Publishers, 2013). On the Kurdish freedom movement's efforts in Turkey, see TATORT Kurdistan, *Democratic Autonomy in North Kurdistan*, trans. Janet Biehl (Norway: New Compass, 2013). On the Kurdish revolution in northern Syria, see Anja Flach, Ercan Ayboga, and Michael Knapp, *Revolution in Rojava*, trans. Janet Biehl, Sherko Geylani, and Björn Frölich (Porsgrunn: New Compass, 2015); Janet Biehl, "Impressions of Rojava: A Report from the Revolution," ROAR, Dec. 16, 2014, http://bit.ly/1FsH5E6; and the website "Ecology or Catastrophe," Biehlonbookchin.com.

Index

Abbey, Edward, 280n25
Adams-Morgan, 159–160, 162–163, 166, 169. *See also* neighborhood movement
Adorno, Theodor, 48n28, 76, 145, 215, 262; *Dialectic of Enlightenment* by, 143–144, 208, 289; ethics and, 208, 210, 246
affinity group, 121–122, 123, 141, 164, 196, 258, 303; in Clamshell Alliance, 179, 180, 181, 184–186
African Americans, 23, 31, 112–114, 145; Black Panthers and, 117, 121, 122, 123, 158; CORE and, 91–93
agriculture: alternative, 96, 130n84, 134, 163–167, 187, 194–195, 233n85, 315; aquaculture as, 158–160, 161, 163, 165, 166, 169, 170, 182, 187, 194; collectivized, 20, 109, 121; cooperative, 133, 141; local, 71–72, 159, 169, 274; organic, 141, 158–160, 164, 166–167, 173n23, 175n47, 182, 188, 277, 316; urban, 134, 159, 162–163. *See also* Adams-Morgan; eco-decentralism; Institute for Social Ecology; Lower East Side
agriculture, industrial, ix, 61–62, 70–72, 83, 86, 89, 109, 115, 158, 315

Alliance for Jobs or Income Now, 93–94
Alternate University, 132–133, 134, 136, 151n6
American Revolution, 170–171, 227, 273
anarchism, xi, 3, 5, 60, 83–84, 91, 92, 95–99, 107, 116, 119–120, 144, 156–157, 171, 181, 213, 259–269, 296; Bookchin champions, 115–116, 121, 123–124, 132–135, 138–141, 145–146, 190–191, 196–197, 269, 272; Bookchin rejects, 301–303; in Burlington, 141, 208–209, 212, 217; Clamshell Alliance and, 180–181, 184; communalism in, 239; democracy and, 147–148, 238–241, 285, 298–303; ecology and, 89–90, 144–145, 174; European, 108–111, 119–120, 145–146, 206–207, 241–242, 243, 245, 248–252; Marxism and, 140, 167, 178, 191, 194, 291; in Montreal, 138, 163–164, 187–188; organization and, 19–20, 119m, 121–122, 141, 146, 179–181; power and, 302, 304; primitivism and, 288–289, 292–293, 298–301. *See also* affinity group; antigovernmentalism, American; counterculture; eco-anarchism; eco-decentralism; Spanish anarchism
anarchism, Spanish. *See* Spanish anarchism

319

Anarchos, 114–116, 120, 123–125, 138, 147–148, 193
antiglobalization movement, 303
antigovernmentalism, American, 115–116, 187, 246
antinuclear movement, xi, 166, 191, 208, 215, 234, 252, 258; Clamshell Alliance and, 178–182, 184–187; German, 178–179, 213, 235–236; Ravenswood reactor and, 87, 91, 180; Women for Life on Earth and, 196
Aptheker, Herbert, 286
Aronowitz, Stanley, 213
Aschner, Bernard, 62
Ash, Steve, 301
assembly democracy. *See* democracy, assembly
Asturias general strike (1934), 12
Athanasiou, Tom, 272
Athenian *polis*, 70, 72–74, 77, 90, 115, 207, 261, 290, 301
automobile, 58, 70, 71, 82, 109, 132, 133, 135, 136, 150, 189, 227

Baer, Steve, 165
Bahro, Rudolf, 235, 237, 291
Baird, Sandy, 273–274
Bakunin, Mikhail, 83, 140, 144, 239, 240–241, 300
Baraka, Amiri, 116
Barber, Benjamin, 162
Barclay, Harold, 301
Beck, Julian, 84
Berkeley Ecology Action, 134
Biehl, Janet, 259–317 passim; *The Murray Bookchin Reader* by, 295, 297; *The Politics of Social Ecology* by, 295; *Rethinking Ecofeminist Politics* by, 289
Black Mask, 107, 111, 116
Black Panthers, 117, 121, 122–124, 158
Black Rose Books, 138, 163, 192, 238–239, 295
Blackwell, Russell, 97–99, 109, 121–122, 200n12
Blea, Chim (pseudonym for David Foreman), 269, 271
Block, David, 208
Bob, Murray, 59
Bob, Reni, 54, 59
Boggs, Carl, 191

Bolsheviks, 12, 15, 16, 19, 22, 43, 92, 123; revolution of, 3, 5, 14, 18, 28, 31, 34, 284; Trotskyism and, 29–31, 32
Bookchin, Beatrice (first wife), xiin3, 54–55, 77, 78, 100n20, 127n17, 142, 225, 283n70
Bookchin, Debbie (daughter), 100n5, 283n70, 307, 308
Bookchin, Joseph (son), 100n5, 283n70, 308
Bookchin, Katya (granddaughter), 283n70
Bookchin, Murray, life of: birth and childhood, 4–7; marriages, 54–55, 283n70; death of, 308–309
Bookchin, Murray, influences on. *See* Engels, Friedrich; Frankfurt School; Gutkind, Erwin Anton; Kropotkin, Peter; Luxemburg, Rosa; Marx, Karl; Marcuse, Herbert; Mumford, Lewis; Peirats, José; Weber, Josef
Bookchin, Murray, philosophical and political ideas of. *See* affinity group; communalism; cooperation; democracy, assembly; dialectical naturalism; dialectical philosophy; eco-anarchism; eco-decentralism; ethical revolt; humanism; libertarian municipalism; postscarcity; potentiality; social ecology; socialism
Bookchin, Murray, political organizations belonged to. *See Anarchos*; Burlington Environmental Alliance (BEA); Burlington Greens; Citizens Committee for Radiation Information; Clams for Democracy; Clamshell Alliance; *Contemporary Issues* (*CI*); Congress of Racial Equality (CORE); Decentralist League; Ecology Action East; Green Mountain Alliance; Left Green Network; Libertarian League; Montreal Citizens Movement; New England Anarchist Conference (NEAC); New Left; New York Federation of Anarchists; Socialist Workers Party (SWP); SWP Minority; United Auto Workers; United Electrical Workers; U.S. Greens; Vermont Council for Democracy; Young Communist League (YCL); Young Pioneers of America
Bookchin, Murray, teaching activities of. *See* Alternate University; Burlington College; Goddard College; Institute for Social Ecology; Ramapo College of New Jersey

Bookchin, Murray, writings of: "Arms for Hungary," 75, 77; *Crisis in Our Cities*, ix, 88–89, 146; "Desire and Need," 107–108; "Ecology and Revolutionary Thought," 90, 97, 115, 132, 151n5; *Ecology of Freedom, The*, 142–145, 209–213, 247, 259, 264, 288, 299; "The Forms of Freedom," 115; "History, Civilization, and Progress," 289; "The Legacy of Domination," 95–96; "Limits of the City, The" (1960 article), 73, 74, 79n5, 85, 148; *Limits of the City, The* (1974 book), 79n5, 148, 161; "Listen, Marxist!" 122, 123, 139, 194, 222, 108; *The Modern Crisis*, 259; "A Note on Affinity Groups," 184; "Open Letter to the Ecology Movement," 189; *Our Synthetic Environment*, ix, 83, 85–86, 87, 89, 101n25, 156, 161; *Politics of Cosmology, The*, 267, 273; *Post-scarcity Anarchism*, 132, 138–140, 179, 184; "The Power to Destroy, the Power to Create," 134; "The Problem of Chemicals in Food," 62, 74; "The Radical Decentralist Project," 123, 124; "Revolution in America," 114, 138; *The Rise of Urbanization and the Decline of Citizenship*, 227, 256; "Social Anarchism or Lifestyle Anarchism," 293, 298, 301, 303; "Social Ecology vs. 'Deep Ecology,'" 264, 268; *Spanish Anarchists, The*, 121–122, 142, 180, 184; "Spring Offensives, Summer Vacations," 147–148, 237; "State Capitalism in Russia," 55; "Stop the Bomb," 64, 75, 77; *The Third Revolution*, 273, 293, 295, 297, 300, 303; *Toward an Ecological Society*, 192; "Towards a Liberatory Technology," 95, 97, 115, 126n11, 133, 158; "Whither Anarchism?" 299; "Workers and the Peace Movement," 222

Bookchin, Nathan (father), 3–4, 6, 9, 25n9
Bookchin, Robert (half-brother), 25n9, 308
Bookchin, Rose Kalusky (mother), 2–4, 6, 9, 10, 12, 30, 69, 82
Borkenau, Franz, 19
Bowery Poets Coop, 96–97
Brand, Stewart, 133
Brown, Jerry, 156, 188
Brown, L. Susan, 292
Brownstein, Leon, 59
Brownstein, Sandy, 136, 141
Bryan, Frank, 171, 219

Buber, Martin, 96; *Paths in Utopia* by, 94, 208
Burlington, Vermont, 141–142, 145, 157, 171, 194, 207; Neighborhood Planning Assemblies (NPAs) in, 216–218, 220, 222, 256, 257; study group in, 208; waterfront of, 208–209, 217–218, 220–221, 222, 256–257, 274, 277. *See also* Sanders, Bernard
Burlington College, 226
Burlington Environmental Alliance (BEA), 217, 221, 224
Burlington Greens, 237, 256, 258–259, 261–262, 267, 272–277

Callenbach, Ernst, 157
Cameron, Angus, 83, 142
Cannon, James P., 28, 29, 31–32, 38–39, 40, 43
capitalism, x, 5, 44, 64, 92, 109, 114, 143–144, 156, 158, 161, 164, 183, 187, 191, 249, 290, 304, 306; agriculture and, 61–62, 70–72, 83, 86, 89, 109; commodification of nature and social life by, 57, 59, 62, 84, 145, 199, 241, 286, 306; ecological limits of, 132, 135, 269, 306; history of, 73, 143–144, 227, 259, 288; as irrational society, 58, 59, 76, 88, 94, 223, 287, 289, 293, 306; in Marxism, 8, 9, 10, 12, 15–16, 21–22, 28, 34, 37, 39, 60, 190; natural evolution and, x, 90, 210, 239, 259; in postwar period, 41–42, 55, 70, 79, 92, 95, 198–199, 305; as root of ecology crisis, x, 61, 132, 149, 269, 286; scarcity and, 79, 93–94, 95, 133, 150; technology and, 60–61, 93; Weber on, 42–43, 52–54, 55–57, 59, 74, 79. *See also* economy; ethical revolt
Calvert, Greg, 113
Carson, Rachel, ix, 85–86
Cassidy, Robert, 160–161, 168
Castro, Fidel, 112, 113, 114, 198
Chahroudi, Day, 158, 165
Chaput, Leo, 194–195
CHARAS, 161–162, 165, 169–170, 187, 192–193, 208, 246
Chase, Steve, 271
chemicals in food, ix, 61–65, 76, 82–83, 89–90, 109, 131, 315; coloring agents and preservatives, 62, 71, 72; fertilizers as, 61–62, 71–72, 86, 158–159, 165; herbicides as, 61, 71, 86; pesticides as, ix, 61–62, 71–72, 85–86

Chernyshevsky, Nikolai, 1, 2, 4
Childe, V. Gordon, 142
Chodorkoff, Dan, 157–160, 162–163, 164–165, 169, 197, 225, 246, 260, 308, 313n65; Cate Farm and, 164–165, 169, 197–198
cities, 112, 140, 161, 188–189, 248, 288; communes in, 96–97, 148, 162; decentralization of, 70–73, 83–84, 88–89, 90, 138, 148–149, 161–163, 188, 205, 241; energy and, 87–88, 133, 165; farming in, xi, 71, 134, 141, 149, 159, 161, 163, 169, 171, 227; gigantic, 70–71, 73, 83, 86, 88, 90, 135, 147, 156, 162, 189, 315; green spaces in, 10, 70, 71–72, 136–137, 227, 248; history of, 70–74, 90, 135, 143, 168; homesteading in, 162, 170–171; as locus of rebellion and revolution, 2, 60, 93, 109, 113–114, 116–118, 122, 136–137, 146, 164, 171, 206, 226–227; stress in, 63, 70–71, 83–84, 89, 133, 315. *See also* automobile; eco-decentralism; libertarian municipalism; neighborhood movement; town and country
Citizens Committee for Radiation Information, 87
Citizens Waterfront Group (CWG), 221
civilization, Western, 15, 36, 94, 135, 147, 182, 223, 269, 289, 292, 306, 316; cities and, 71, 74, 90; ecology and, 89, 145, 263; industrialization of, 62, 70, 86, 88, 115, 183; Luxemburg on, 37, 41, 90; rejections of, 107, 118, 270, 288–289, 298–299
Clams for Democracy, 185, 186–187
Clamshell Alliance, 178–181, 184–187, 206, 208, 222, 223, 256, 258, 276
Clark, John, 171, 192, 212; clashes with Bookchin, 259, 271, 290–292, 296, 297–298
Clark, Wilson, 158, 165, 188
Clarke, Robin, 213
Cohen, Linda, 170
Cohn-Bendit, Daniel, 248
Coleman, Jane, 201n25
Committee for Non-Violent Action (CNVA), 85, 91
Commoner, Barry, 191
communalism, 239, 285, 296, 303, 313n59. *See also* libertarian municipalism
commune, as counterculture institution, 107, 133, 134, 141, 142, 145–146, 148, 159, 162, 167, 206, 213; medieval, 70, 226; revolutionary, 94, 96, 99, 117, 118, 209, 227, 239–239, 261, 272, 295; rural, 1, 107, 111, 125, 133, 134, 141; urban, 96–97, 105, 141, 162
Communism, international, 3, 7, 8–9, 10, 12–13, 15–17, 20, 24, 34–35, 55, 58, 119–120, 222, 242, 303–304. *See also* Trotskyism
Communist Party of the United States, ix, xi, 3, 6–24, 59, 92, 106, 298. *See also* Young Communist League (YCL); Young Pioneers of America
Comunero movement (Spain), 226–227
confederalism, 226–227, 228, 239, 240, 241, 249, 252. *See also* libertarian municipalism
Confederación Nacional del Trabajo (CNT), 20, 97, 109–110, 121–122, 180, 186, 241, 302. *See also* Spanish anarchism
Congress of Racial Equality (CORE), 91–93
consensus decision-making, in Clamshell, 179–185; in U.S. Greens, 272, 273, 277
Contemporary Issues (CI), 46, 52–83, 85, 90, 91, 93, 94, 114, 148, 263, 276. *See also* Weber, Josef
cooperation, 94, 107, 146, 280; in natural evolution, 210, 211, 260, 287, 289; as radical social ideal, x, xi, 57, 72, 84, 106, 133, 171, 210–211, 272, 284, 287, 289, 293, 317; in social evolution, 145, 147. *See also* cooperative
cooperative (counterculture institution), 133–134, 146, 148, 162, 170, 188, 189, 238, 296, 302
Costa-Pierce, Barry, 167, 169, 182, 187
counterculture, xi, 99, 106–109, 116, 121, 125–126, 131–133, 138, 140–141, 157

D'Attilio, Bob, 241
Debord, Guy, 109, 111, 134, 190
Debray, Regis, 113, 117
Debs, Eugene V., 3, 257, 277
Decentralist League, 171
decentralization, 140, 161, 184–185, 199, 206, 241, 256, 306; counterculture and, 94, 95, 107, 132, 133; eco-technics and, 90, 96, 133, 156; Greens and, 237, 246, 258, in libertarian municipalism, 147–148, 227; urban, 71–72, 88–89, 138, 162, 164, 173n23, 174n35, 176n47, 188–189, 227, 285. *See also*

Decentralist League; eco-decentralism; neighborhood movement
deep ecology, 263–264, 267–272, 272, 288, 290
Dellinger, David, 147, 219, 222
democracy, assembly, x, xi, 94, 95, 115, 192, 207, 209, 222, 227–228, 277, 287; anarchism and, 147–148, 251–252, 258, 259, 266, 272, 277, 285, 291–292, 301–303. *See also* Adams-Morgan; Athenian *polis*; Burlington Greens; Clamshell Alliance; Confederación Nacional del Trabajo (CNT); *Comunero* movement; German Greens; Kabouters; libertarian municipalism; Montreal Citizens Movement; neighborhood movement; Neighborhood Planning Assemblies; Parisian sectional assemblies; Spanish anarchism; soviets; town meeting; U.S. Greens
democracy, bourgeois, 6, 31, 38, 40, 58, 191
democracy, intramovement, 13, 14, 30, 120, 303; in *CI*, 52, 53, 76; in Clamshell, 180–186; in German Greens, 235, 236, 247, 272, 276, 27
democracy, participatory, 91, 138, 166, 272; in Students for a Democratic Society (SDS), 85, 105, 113–114, 123
democracy, social, 16, 21, 285, 306. *See also* Social Democratic Party, German (SPD)
Democracy of Content, 52, 60, 76, 85
Democratic Alternative (Norway), 296, 311–312n43
Devall, Bill, 263, 269
Dewey, John, 18, 29, 157
dialectical naturalism, 211–212, 229n21, 246–247, 260–261, 273, 279n18. *See also* dialectical philosophy; dialectical reason; natural evolution
dialectical philosophy, 60, 79n9, 294; Aristotle and, 207; *CI* and, 58, 63; ecology and, 90–91, 96; in *Ecology of Freedom*, 209, 288; Frankfurt School and, 56, 76, 143–144, 289; urban history and, 73, 74. *See also* dialectical naturalism
dialectical reason, 57–58, 260, 289
Diderot, Denis, 36, 53, 74, 78
direct action, 134, 206, 238, 263, 273, 295; in antinuclear movement, 179, 181, 184, 186–187
Ditfurth, Jutta, 205, 242, 243, 285; Hessian Greens and, 234–236, 244, 245; as BuVo speaker, 246, 247–250. *See also* Ecological Left; German Greens
Dohrn, Bernardine, 120, 124
Dolgoff, Sam, 97, 98, 148, 240
domination, 35, 56, 94, 95; of nature, 143–144, 149, 166, 169, 190, 195, 209, 215; opposition to, 114, 115, 131, 138, 141, 144, 161, 165, 169, 186, 188–189, 209–210, 215, 239, 251, 286, 288; of women, 166, 195, 196, 290. *See also* hierarchy
Doyle, Bill, 220
Drapeau, Jean, 137, 163–164, 187–188
Drucker, Peter F., 291
Dubos, René, 86, 101n26, 152n20, 173n21

Earth Day, 135–136, 148
Earth First! (group and periodical), 263–264, 268–269, 271, 288
East Tremont, 3–8, 9–10, 12, 15, 17, 21, 22, 24; Cross Bronx Expressway and, 69, 82, 139
East Village Other, 97
Eastman, Max, 18
Ebermann, Thomas, 245, 248–249
Eccli, Eugene, 158, 160
eco-anarchism, 115, 134, 140, 144–145, 149–150, 192, 196, 199, 239, 276. *See also* anarchism; eco-decentralism
eco-communities, 94, 115, 188, 240. *See also* eco-decentralism
eco-decentralism, 72–73, 77, 83, 88–89, 140, 149, 159, 207, 276, 278, 315. *See also* agriculture, organic; decentralization; eco-anarchism; eco-communities; eco-technics; renewable energy; scale; self-sufficiency; town and country
ecofeminism, 195–196, 289, 291
Ecological Left (Germany), 285, 309n4
ecology (as radical political concept), ix, 109, 115, 123, 134, 150, 157, 159; anarchism and, 89–91, 92, 97, 98, 138, 145, 146, 147, 239, 241; cooptation of, 188–189; environmentalism and, 148–149, 196, 249; feminism and, 166, 195–196, 215, 222; in Germany, 206, 234–236, 263, 285; Marxism and, 190–191; NEAC and, 196. *See also* eco-anarchism; eco-decentralism; ecology crisis; ecology movement; social ecology

Ecology Action East (EAE), 134, 135, 141, 151n17, 184
ecology crisis, x, 145, 189, 216, 224, 263, 265, 270; deforestation and, ix, 109, 205, 217, 244, 269, 270; pollution and, ix, 217, 234, 236, 269, 270, 274; social nature and origin of, 138, 144, 149, 159, 269, 316; topsoil depletion and, 61, 62, 90, 109, 169; urban crisis and, 71, 85, 91, 162, 215–216. *See also* agriculture; cities; global warming; town and country
Ecology Montreal, 285, 309n4
ecology movement, ix, 149, 150–151, 199, 207, 209, 251, 263, 265, 269, 288, 289, 316. *See also* ecofeminism; ecology crisis; population, human; U.S. Greens; wilderness
economy, 51; eco-decentralism and, 70, 95, 160–161; money-free, 53, 98, 158–159, 163; moral, 209–210, 261, 279; municipalized, 227, 240; neighborhood, 70, 160, 163, 225, 233n83; postscarcity and, 95, 97; World Plan for, 63–65. *See also* capitalism; cooperative; irreducible minimum; Marxism; socialism; usufruct
eco-socialism, 235, 245, 248
eco-spirituality, 189, 258, 265, 266, 269, 272, 289, 291, 292, 297, 298
eco-technics, 133, 140, 158, 159, 165, 166, 168, 170, 182, 182, 187–188, 189. *See also* agriculture, organic; renewable energy; self-sufficiency
Edelstein, Mike, 168
Edo, Luis, 241, 242
Ehrlich, Howard, 139
Ehrlich, Paul, 149–150
Eichhorn, Emil, 304
Eiglad, Eirik, 296–297, 297, 311n43
Eisen, Dave, 29, 32, 40, 43, 45; in *CI*, 54, 74, 78, 87, 90–91
Eleventh Street Movement, 169–170
Elton, Charles, 89, 101n41
Ely, John, 212
Emergency Committee for Arms to Hungary, 75
energy. *See* eco-technics; fossil fuels; nuclear power; renewable energy; self-sufficiency

Engels, Friedrich, 14, 17, 45, 50n61, 56, 59, 70, 114, 190; *Communist Manifesto* by, 45, 50n61. *See also* Marxism
Enlightenment, European, 56, 95, 143, 144, 209; defense of, 262, 288–289, 290, 299, 300, 306
environmentalism, ix, 148, 149, 189, 191, 199, 256
Erikson, Erik (pseudonym for Weber), 55
ethical revolt, 57, 58, 65, 84, 85, 86, 106, 180, 190, 259, 275
ethics, xi, 145, 295; anarchism and, 84, 235, 293; ancient Greek philosophy and, 190, 207; counterculture and, 85, 98, 115, 126; instrumentalism contrasted with, 52, 56, 57, 143–144, 246; moral standpoint in, 199, 278, 295; nature philosophy and, 210–211, 213, 256; objective basis for, 210–212, 246, 261; politics and, 85, 214, 226, 228, 237, 262, 276–277; relativism in, 210; socialism and, 293, 295; Weber on, 57, 65. *See also* dialectical reason; ethical revolt

face-to-face democracy. *See* democracy, assembly
Fanon, Frantz, 112, 117
Farmer, James, 92
Fekete, John, 192, 213
Felt, W. Mark, 192–194
feminism, 131, 137, 163, 165–166, 179, 190, 195, 236, 251, 252, 273, 277, 289. *See also* ecofeminism; hierarchy
Fichte, Johann, 212
Fifth Estate (periodical), 288
Finzi, Paolo, 242
Fischer, Joschka, 243–244, 246, 247–248, 249–250, 257
Foreman, David, 263–264, 268–269, 270–271. *See also* deep ecology; Earth First!
fossil fuels, ix, 88, 315. *See also* automobile; cities
Foster, William Z., 10
Fotopoulos, Takis, 291
Fourth World Review (periodical), 270
Franco, Francisco, 12, 16, 19–20, 22, 23, 109, 110, 293, 302
Frankfurt School, 145, 160, 215; *CI* and, 56–57, 60, 76; dialectical philosophy and, 56–57; domination of nature and,

143–144, 210, 215; enlightenment and, 56, 143–144, 289; ethics and, 67, 210, 246, 248; pessimism of, 95, 107, 144, 210, 211. *See also* Adorno, Theodor; Horkheimer, Max; Marcuse, Herbert
French Anarchist Federation, 108
French Communist Party, 15, 119, 120
French Revolution (1789), 14, 94, 145, 146, 227, 293, 294, 295. *See also* Parisian sectional assemblies
French Revolution (1848), 198, 295
Fresh Ground Coffee House, 141, 142, 145, 157, 308
Friend, Gil, 159
Fuller, Buckminster, 161; geodesic domes of, 161, 163, 165, 185, 187, 198

Garb, Yaakov, 86
Garcia, Chino, 161, 193, 246. *See also* CHARAS
Gardner, Joyce, 94, 96, 105–106, 167
Geddes, Patrick, 71
General Motors, 40, 43, 44
general strike: in Berlin (1919), 304; in France (1968), 118–120; in Hungary (1956), 74–75; in Minneapolis (1934), 12; in San Francisco (1934), 12
German-American Bund, 22, 23
German Communist Party (KPD), 10, 12, 13, 34, 35, 42. *See also* Luxemburg, Rosa
German Greens, 242, 247, 251–252, 256, 258, 262; democracy in, 206, 235–236, 237–238, 244, 247–249, 250; *fundi-realo* conflict in, 242–250. *See also* Ditfurth, Jutta; Fischer, Joschka
Ginsberg, Allen, 96
Gitlin, Todd, 139, 285
global warming, ix, 88–89, 195, 269, 285, 306
Goddard College, 157–160, 164, 166–167, 180, 196–198, 225; Social Ecology Studies Program at, 159–160, 164–167, 173n23, 174n47. *See also* Institute for Social Ecology
Goldman, Al, 29, 32, 38, 40
Goldman, Emma, 3, 97
Goodman, Paul, 60, 84, 92, 120, 130n84, 213
Gorz, André, 140
Green Mountain Alliance, 180, 258

Green Perspectives (periodical), 267, 269; as *Left Green Perspectives*, 287, 297
Greens. *See* Burlington Greens; German Greens; Hessian Greens; New England Greens; Northern Vermont Greens; U.S. Greens
Grossman, Jack, 54, 55, 59, 75–76, 78
Gruhl, Herbert, 235
Guevara, Che, 112, 117, 139
Guma, Greg, 219, 224, 257
Gutkind, Erwin Anton, 83–84, 100n11, 159, 173n21, 291

Hague, Frank (Jersey City boss), 19, 22, 26n46
Hawkins, Howie, 181, 223, 256,, 258, 267, 270, 272, 273, 277, 278, 291
Hayden, Tom, 117, 134, 188
Hayes, Wayne, 100n8, 160–161, 167, 168, 183, 207, 215
Hayley, Ron, 239
health (as issue), 62, 64, 71, 72, 83, 87, 89, 90
Hegel, G.W.F., 54, 90, 140, 144, 212, 261, 271, 287, 303. *See also* dialectical philosophy
Heider, Ulrike, 289–290, 293, 298, 299
Heller, Chaia, 308
Henderson, David, 96
Herber, Lewis (pseudonym), 55, 114, 138
Herzen, Alexander, 1, 4
Hess, Karl, 158, 162, 166, 171, 176n75, 182, 187, 192, 202n43, 213
Hessian Greens, 235, 236, 243–244, 246, 248, 249–250. *See also* German Greens
hierarchy, 115–116, 122, 149, 164, 186–187, 189–191, 209, 263, 288; anarchism and, 121, 184, 206; in education, 157; emergence of, 142–144, 196, 209, 222, 288; ideological justifications for, 211; opposition to, 109, 115–116, 119, 124–125, 131, 133, 139, 141, 145, 205, 239, 251–252, 273; as root of ecology crisis, 142, 149, 187, 263. *See also* domination
history, 55; dialectical processes of, 43, 74, 75, 89, 94, 133, 142, 144–145, 168, 182, 183, 190–191, 209–211, 284–285; in Marxism, 12–13, 14, 17, 30, 43, 56, 60; natural, 211–212; progress in, 160, 165, 289, 299; revolutionary, 4, 21, 54, 95, 99, 108–110, 117, 168, 196, 226, 251, 258, 261, 278, 293–295, 304; urban, 70–71, 74, 159

Ho Chi Minh, 113, 114, 118, 123, 124
Hoffman, Abbie, 97
Hoffman, Allan, 91–92, 93, 94–96, 107, 112, 125–126, 132, 225; "Eighteen Rounds of Total Revolution" by, 111; totality and, 99, 105, 111, 115–116, 132
Hoffman, Harold (N.J. governor), 19
Hook, Sidney, 29, 79n9
Horkheimer, Max, 56, 60, 76, 210, 215, 246; *Dialectic of Enlightenment* by, 143–144, 208, 289; *Eclipse of Reason* by, 56, 208. *See also* Frankfurt School
Howard, Ebenezer, 71
Howard, Sir Albert, 72
humanism, 149, 263–264, 265, 266, 269, 270; ecological, 101n26. *See also* social ecology

Incontro anarchico (Venice, 1984), 241–242, 250, 254n23
Institute for Social Ecology (ISE), 166, 167; 1970s sessions of, 168–170, 178, 182–183, 187, 194–195, 197–198, 289; 1986 session of, 259–261, 279n19; 1990s sessions of, 271, 286, 290, 298–299; 2000s sessions, 305, 313n65; post-Goddard, 225, 246. *See also* agriculture, organic; eco-technics; renewable energy
Institute for Social Research. *See* Frankfurt School
instrumentalism, 52, 56, 57, 59, 62, 64, 84, 144, 148, 210, 215, 225, 288, 289, 299
International Communists of Germany (IKD), 35–37, 55
"Internationale, The" (song), 7, 8, 19, 119, 284
irreducible minimum, 142, 145, 154n54, 299

Jacobson, Annette, 54, 59
Jacques le Fataliste (Diderot), 36, 53, 74, 78
Jezer, Marty, 224
Jonas, Hans, 212

Kabouters, 146–147, 154n71
Kaluskaya (or Kalusky), Zeitel Carlat (grandmother), 1–6, 7, 207
Kalusky, Dan (uncle), 2, 4
Kalusky, Moishe (grandfather), 1–2
Keller, Robert (pseudonym), 55, 114, 138
Kelly, Petra, 236, 247, 313n68
Khayati, Mustapha, 109

Kiefer, Joseph, 182, 183, 225
King, Ynestra, 165–166, 195–196, 213
Kitto, H.D.F., 73
Koehnlein, Bill, 139
Kotler, Milton, 162, 163, 188
Kovel, Joel, 223, 300
Kowaluk, Lucia, 137–138, 152n29
KPD. *See* German Communist Party
Kronstadt rebellion, 15, 30, 138, 223
Kropotkin, Peter, 71, 146, 239, 272, 292, 293, 300, 301; *Fields, Factories and Workshops* by, 80n10, 83, 140
Kunin, Madeleine, 219
Kurdish freedom movement, 316–317, 317n1
Kurtz, Alan, 208, 217, 231n52

labor theory of value (Marx), 77–78
labor unions, 4, 8, 17, 18–19, 23, 31, 33, 37, 38–39, 40, 44, 97, 110, 120, 121, 242, 257, 261, 265, 302. *See also* Confederación Nacional del Trabajo (CNT); United Auto Workers (UAW); United Electrical Workers (UE)
labor strikes, 2, 3, 9, 12, 13, 15, 18, 33–34, 38–39, 41, 43, 44, 74, 116–120, 242, 257, 261, 304. *See also* general strike; labor unions
LaFrance, Ron, 169
Lasky, Burton, 83, 85, 86
Le Guin, Ursula K., 156–157
Lee, Dorothy, 142–143
Left Green Network, 272–273, 277
Legard, Sveinung, 311n43
Lenin, 5, 7, 10, 12–14, 17, 30, 31, 114, 120, 242, 284, 286, 293, 303. *See also* Russian Revolution (1917)
Lens, Sidney, 30
Lepper, John, 186
Leval, Gaston, 109
Levellers, 145, 213
Levine, David, 271
Lewis, John L., x, 18
Libertarian League, 84, 97–99
libertarian municipalism, 227, 237, 238–241, 242, 259, 272, 277–278, 285, 291–292, 295–297, 301–302, 306. *See also* communalism; Left Green Network; *Politics of Social Ecology, The* (Biehl); *Rise of Urbanization and the Decline of Citizenship, The*

libertarianism, right-wing (anarcho-capitalism), 158, 171, 187–188, 213, 219
Lichtheim, George, 29, 287
Lieberman, Archie, 31, 34
Liebknecht, Karl, 5, 9, 10, 13, 74
Light, Andrew, 300
Lilienthal, David, 87
Limits to Growth, The (Club of Rome), 149–150
Living Theatre, 84, 91, 99
London, Jack, 32, 308
Lothstein, Arthur, 132
Love, Sam, 158, 167
Lovins, Amory, 188
Lower East Side, 3; bohemians in, 87, 96–97, 99; CHARAS in, 161, 165, 169–170, 187, 192–193, 225; counterculture in, 105, 107–108, 116, 125, 141
Ludd, Harry (pseudonym), 55
Lukacs, Georg, 190
Lunen, William (pseudonym for Weber), 55
Luxemburg, Rosa, 5, 7, 34, 41, 46, 74, 242; *Junius Pamphlet* by, 37, 50n61; murder of, 9, 10, 13, 304, 307; "socialism or barbarism" and, 37, 42, 90, 135

Macdonald, Dwight, 29–30, 45
Macdougal, Phil, 61–62
Mahnke, Erhard, 275–277
Malina, Judith, 84, 91, 96, 108, 112, 147–148
Manes, Chet, 59, 76, 77
Manes, Christopher, 264
Mao Zedong, 298; and New Left, 92, 112–114, 118, 123, 124, 139, 198
Maoism, 92, 105, 106, 119, 120, 193, 194. *See also* Progressive Labor (PL)
Marat/Sade (film), 107–108
Marcuse, Herbert, 107, 113, 134, 140; *Counterrevolution and Revolt* by, 157, 172n5; *Eros and Civilization* by, 94, 106; *Essay on Liberation* by, 126n11; *One-Dimensional Man* by, 95; *Reason and Revolution* by, 56, 94, 106
Margulis, Lynn, 211, 260
Marshall, Peter, 139, 301
Marx, Karl, xi, 7, 17, 56, 59, 95, 114, 168, 171, 190, 285, 287, 300; *Capital* by, 21, 74, 77–78, 142, 183; *Communist Manifesto* by, 45, 50n61; *Eighteenth Brumaire of Louis Napoleon* by, 199, 286; on town and country, 70. *See also* Marxism
Marxism, 2, 3, 10–12, 13, 21, 31, 38, 40, 147, 165, 190, 223–224, 286; academic, 190–191, 215–216, 291, 300; anarchism and, 84, 96, 115, 120–121, 139–141, 167, 191, 269, 292; authoritarianism of, 14, 40, 75, 84, 92, 96, 113–114, 146, 190; dialectical philosophy in, 56–67, 207; domination of nature and, 190; in France, 119–120, 123; in Germany, 35, 56–57, 59, 304–305; labor theory of value in, 77–78; New Left and, 92, 106–107, 113–114, 120–122, 139–140, 186, 193–194; rejections of, 44–46, 56, 139–141, 145–146, 180, 198, 245, 278; in Spain, 97; technology and, 60–61, 142; Weber and, 35, 42, 54, 56, 59–60. *See also* Communism; Engels, Friedrich; Marx, Karl; Maoism; proletariat; Trotskyism; vanguard
Marxism-Leninism. *See* Marxism
May-June events (Paris, 1968), 118–120
McClaughry, John, 219
McLarney, Bill, 160
McQuinn, Jason, 293
Mead, Margaret, 178
Meltzer, Albert, 241, 242
Mera, Cipriano, 109–110
Merchant, Carolyn, 291
Merrill, Richard, 132, 134, 167
Miller, Edward S., 192–194
Mills, Stephanie, 86
misanthropy, 263, 264, 269, 289. *See also* deep ecology
Mohawk, John, 169
Montreal, 137–138, 148, 163–164, 171–172, 187–188, 237, 267, 276, 285. *See also* Ecology Montreal; Montreal Citizens Movement
Montreal Citizens Movement (MCM), 163–164, 171–172, 187–188, 237, 276
Morea, Ben, 98–99, 107, 111–112, 116–117
Morley, Jim, 167–168, 183
Morris, David, 159, 162–163
Morris, William, 302–303
Morrow, Felix, 29, 32, 38, 40, 43, 45, 59; *Revolution and Counterrevolution Spain* by, 19
Moses, Robert, 69–70
Mumford, Lewis, 83, 84, 100n8, 101n26, 140; *Culture of Cities* by, 70–71
Muste, A. J., 64

Native Americans, 142–143, 145, 169, 179, 252, 270

nature: domination of, 143–144, 149, 166, 169, 190, 195, 209, 215; in eco-decentralism, 70, 72, 84, 88, 90, 95; evolution in, 89–90, 210–213, 261; as ground for ethics, 211–213, 246; harmony with, x, 90, 143, 209–210, 266, 270; humanity and, 72, 83, 95, 240, 264, 316; mysticism, 189, 258, 269, 271, 299; as natural resources, 143–144, 148–150, 235; philosophy of, 89–90, 190–192, 207–208, 210–212, 248, 259, 260–261, 273; wilderness as, 263, 265, 269, 270, 288, 292; women and, 195–196, 289. *See also* capitalism; ecology crisis; dialectical philosophy; social ecology

neighborhood movement, 147, 158, 187, 226–227, 239–240; in Adams-Morgan, 158–160, 162–163, 166, 192; in Amsterdam, 146; in Burlington, 208–209, 216–219; decline of, 188–189; on Lower East Side, 161–162, 169–170, 192, 231; in Montreal, 137–138, 163–164, 171–172, 187–188; radical, 162–163, 170. *See also* decentralization; town meeting

Neumann, Osha, 116, 125

New Alchemy Institute, 158, 159, 165, 173n13, 187

New England Anarchist Conference (NEAC), 196, 206

New England Greens, 256, 258

New Left, 105–106, 112, 114, 122, 125, 137, 157, 158; anarchism and, 98, 99, 120–121, 122, 180; Columbia University and, 117–118; *New Left Notes* of, 97, 120, 121; *Post-Scarcity Anarchism* and, 132, 138–139. *See also* democracy, participatory; Marxism; Students for a Democratic Society (SDS); Vietnam War

New York Federation of Anarchists, 95–99, 105–107, 111

New York World's Fair (1964), 92

Newton, Huey, 121

Nields, John, 193–194

nonviolent resistance, 85, 99; antiwar movement and, 112, 113, 117; antinuclear movement and, 179–181, 184–186, 247; CORE and, 92–93; ecofeminism and, 196, 213; peace movement and, 91, 206, 213–214, 222, 235, 236–238

Northern Vermont Greens, 258

nuclear freeze movement, 214, 216, 230n36, 247

nuclear power, 64, 72, 83, 85; at Chernobyl, 258; in Germany, 178, 205, 236, 244, 248–250; and megacities, 88; at Ravenswood, 87, 136; at Seabrook, 180, 181, 185–186; at Three Mile Island, 191. *See also* antinuclear movement

nuclear weapons, 140, 196; CI opposes, 64, 65, 84; nonviolent movement against, 85, 91, 137; Euromissiles as, 206, 213, 235, 236, 242. *See also* nuclear freeze movement; peace movement

O'Connor, James, 300
O'Neill, Calley, 182
Öcalan, Abdullah, 316–317
Ohly, Götz, 74
organic society, 143–145, 209, 288
Orwell, George, 19, 21
Osborn, Fairfield, 61–62

Paley, Grace, 166, 222
Palmer, Robert, 136, 154n71
Paris Commune (1871), 94, 99, 209, 227, 239, 261, 272, 295
Parisian sectional assemblies (1793), 115, 170, 227, 261, 272
peace movement, 85; against Euromissiles, 206, 213, 222, 236–237. *See also* nonviolent resistance; nuclear freeze movement
Peirats, José, 110
Perlman, Fredy, 146
Piccone, Paul, 215
Polanyi, Karl, 183, 256
Pomerleau, Antonio, 208, 217, 218, 220
Popular Front, 16–18, 20–21, 110, 180
population, human, 149, 263–265, 268–270, 288, 316. *See also* Ehrlich, Paul
postscarcity, 93–94, 95, 97, 98, 99, 107, 140. *See also* technology; utopia
potentiality (philosophical concept), xi, 57, 60, 74, 94, 114–115, 207–208, 211–213, 235, 246, 259, 261, 287, 288, 295, 306, 315. *See also* dialectical philosophy

Progressive Labor (PL), 92, 93, 106, 120, 122–124. *See also* Maoism

proletariat, 3, 18, 38, 40, 43, 92, 108, 120–121, 141, 190, 222–223, 285; Communist movement and, 5–7, 8–9, 12, 14, 16, 19, 45, 198; culture of, 70, 167, 168, 183, 189; in Germany, 10, 12, 36, 41, 74, 303–304; Marx and, 190; New Left and, 106–107, 113, 114, 122, 140, 190; nonrevolutionary nature of, 40–46, 56, 77, 83, 198; Spanish, 16–18, 19–21, 109–110, 297; students as, 124, 143, 16; Trotskyism and, 28, 30, 33–35, 37, 399. *See also* Marxism

Proudhon, Pierre-Joseph, 83, 239, 272

Provo, 109, 112, 146

Quakers, 91, 179–180, 184, 185

Radin, Paul, 142

Ramapo College of New Jersey, 160–161, 167–168, 183–184, 192, 206, 207, 214, 215, 226, 263

Ravenswood reactor, 87, 91, 180. *See also* antinuclear movement

RCA Institute, 60, 160

Read, Sir Herbert, 90

reason, 168, 189, 207, 208, 212, 295; defense of, 189, 259, 262, 269, 289, 298, 306, 316; dialectical, 57, 223, 259, 289; instrumental, 57, 144, 210, 289; irrationalism, 189, 195, 270, 271, 288–298, 306; rational society and, 57, 58, 60, 94, 98, 140, 199, 212, 278, 293, 295, 296, 306

Reed, Ishmael, 96

Reich, Wilhelm, 94, 103n75, 125

renewable energy, ix, xi, 156, 185, 188, 196, 227, 258, 288, 315; in Adams-Morgan, 163, 170; in Burlington, 209, 217, 274; in eco-decentralism, 72, 73, 88, 94, 95, 133, 140, 145, 188; at Goddard, 158–160; at Institute for Social Ecology, 161, 165, 166, 167, 168–169, 178, 182, 184, 187, 194–195; on Lower East Side, 169–170; at Ramapo, 168. *See also* eco-technics

retrogression, 41–42, 45, 46, 52, 55–59, 61, 63, 198

Revolutionary Shop Stewards (Germany), 303–304

Revolutionary Youth Movement (RYM), 122–124, 125, 194

Rice, Suki, 179

Richards, Vernon, 294

Riordan, Michael, 209

Roselle, Mike, 269

Roszak, Theodore, 86, 213

Rothbard, Murray, 158

Roussopoulos, Dimitrios, 137–138, 148, 163–164, 171–172, 188, 223–224, 238, 238–242, 285, 301. *See also* Black Rose Books; Ecology Montreal; Incontro Anarchico; Montreal Citizens Movement

Rudd, Mark, 117, 194

Ruiz, Pablo, 109–110

Russian Revolution (1905), 2, 284, 295

Russian Revolution (1917), 3, 13, 16–18, 22, 28–29, 32, 94, 145, 198, 227, 295, 297, 303; democratic soviets in, 75, 123, 284. *See also* Bolsheviks; Lenin; soviets; Stalin, Joseph; Trotsky

Rustin, Bayard, 92, 93

Sale, Kirkpatrick, 270

Salzman, Lorna, 272

Sanders, Bernard, 208–224, 237, 257, 268, 273–274; as legislator, 277. *See also* Burlington, Vermont

scale, x, 142, 171, 199; in agriculture, 62, 72, 83; in cities and communities, 70–71, 73, 84, 89, 145, 189, 216; eco-technics and, 133, 156, 165, 170, 219, 227; in manufacturing, 72, 95, 227; in Vermont, 141, 142, 206–207

Schecter, Stephen, 164, 174n46

Schibel, Karl-Ludwig, 145, 205, 224, 244, 245, 247, 248

Schumacher, E. F., 156

Schwartz, Jack, 77–78

Schweickart, Rusty, 132, 149

Schwerner, Mickey, 91, 92

Seeger, Pete, 185, 191

self-sufficiency, 70, 84, 107, 159, 160, 163, 169–171, 187, 217

Serge, Victor, 37

Service, Elman, 142

sexual liberation, 94, 96, 99, 105–106, 107, 108–109, 115

Shachtman, Max, 31, 32, 40, 41

Sharp, Rick, 221, 257

Shiloh, M. S. (pseudonym), 55

Shuttleworth, John, 134, 160

Sisco, Gary, 275, 278, 282n60
Situationism, 108, 109, 111–112, 119, 120, 139, 190
Social Democratic Party, German (SPD), 9, 10, 74, 243–244, 249, 291, 303, 304
social ecology, 171, 199, 278, 297–298, 301–302, 305, 306; basic tenets of, 144, 195, 263, 315; in Burlington, 208, 216, 217, 224; deep ecology and, 263–264, 269–271; Democratic Alternative and, 296, 312n43; *Ecology of Freedom* and, 209–210; German Greens and, 206, 235, 260–262; at Goddard College, 159–160, 164–166, 173n23; humanism and, 149, 266, 269, 270, 281n47; at Institute for Social Ecology, 166–167, 225; Kurdish movement and, 316–317; Left Green Network and, 272; in Montreal, 163; as political label, 84, 100n11, 101n26, 173n21, 237, 288, 290, 291–292, 312n47; at Ramapo, 168, 230n42; U.S. Greens and, 258, 266. *See also* agriculture, organic; eco-decentralism; ethics; humanism; Institute for Social Ecology; libertarian municipalism; nature, philosophy; renewable energy; Social Ecology Studies Program
Social Ecology Studies Program, 159–160, 164–166
socialism, 3–4, 46, 59, 168, 225, 278; anarchism and, 292–293, 300; Bolsheviks and, 6–7, 284; Burlington and, 208–209, 216–217, 219–220, 222, 257, 267; CI and, 52–54, 59–60; end of ideal of, 284–286, 306, 316; as ethical standard, 287, 293, 295; Frankfurt School and, 56; German Greens and, 235, 245, 248; Luxemburg and, 37, 41–42, 90, 135; Marxian, 12, 13, 38, 45, 57, 60, 84, 304; Montreal Citizens Movement and, 163–164, 171, 187–188; municipalism and, 227; New Left and, 97, 99, 113–114, 123, 126; proletarian, 12, 43–45, 198; Russian populism and, 1, 3; Stalin and, 9, 13, 15, 16, 23, 29, 31–32, 33, 55, 116; technology and, 60, 93–94; Vermont and, 141. *See also* cooperation; Marxism; irreducible minimum; postscarcity; usufruct; utopia
Socialist Revolutionary Party (Russian), 2, 145

sociobiology, 149
Soviet Union, 6–7, 9–10, 13–18, 21, 24, 31, 32, 42–43, 55–56, 75, 115, 123, 206, 284, 286. *See also* Bolsheviks; Russian Revolution (1917)
soviets, 39, 75, 94, 123, 261, 284
Spanish anarchism, 29–31, 97–98, 121–122, 142, 146, 180–181, 186, 193, 272, 292, 301; affinity groups in, 99, 121–122, 146, 180, 184, 186; Bookchin's passion for, 21, 98, 111, 146, 293; historians of, 294, 305; trade unions in, 19–20, 121. *See also* Confederación Nacional del Trabajo (CNT); Spanish Revolution (1936–37)
Spanish Communist Party, 16–17, 19, 20, 98, 110, 302
Spanish Revolution (1936–37), 16–17, 19–21, 20, 84, 97, 98, 109–111, 146, 193, 209, 294, 227, 304, 305; Barcelona street battle (May 1937) in, 19–21, 20, 98, 111, 193; collectives in, 20, 97, 109, 121, 146, 272, 292, 302, 304; expatriates from, 111–12, 193; militias of, 17, 20, 97, 109, 110, 121. *See also* Confederación Nacional del Trabajo (CNT); Spanish anarchism
Spartacus League (Germany), 303, 304
Spretnak, Charlene, 247, 266
squatters (European movement), 145, 205, 206, 235, 241
Squatters' Park (New York), 136–137, 184
Stalin, Joseph, 8, 9, 36, 198, 294, 29; Hitler and, 10, 15, 23, 24, 28–30, 31–328; retrogression and, 42, 43, 55–56, 74; show trials of, 16–18, 21; Trotsky and, 5, 13, 34. *See also* Soviet Union; Stalinism
Stalinism, 15, 17–19, 21, 23, 31–34, 40, 45, 55–56, 74–75, 97–98, 139, 302, 306. *See also* Stalin, Joseph; *specific Communist parties*
state, 35, 110, 188, 272, 306; anarchist rejection of, 3, 74, 95, 115, 187, 301; confederal alternative to, 226–227, 228, 239, 240, 241, 249, 252; in history, 89, 110, 143–144, 209, 211, 259; municipality as brake on, 228, 240, 241, 277, 296; municipality as miniature, 173, 227, 239, 250–251, 301; oppressive nature of, 35, 57, 70, 79, 83, 95, 114, 115, 145, 191, 249, 269; revolution against, 58, 116, 117; society without, 84, 144, 146, 172, 207, 227, 239,

306; Stalinist, 31, 55, 114, 272, 284. *See also* domination; libertarian municipalism
state capitalism, 55
statecraft, 239, 240, 247, 249, 258, 272
Stritzler, Suzanne, 222
student movement (1960s), 106, 109, 113–114, 117, 118–120, 122, 123, 125, 135, 137, 140, 145, 147
Students for a Democratic Society (SDS), 85, 97, 105–106, 113–114, 117, 120, 122–124, 125, 139, 158, 186, 193–194, 196. *See also* New Left; Progressive Labor (PL); student movement (1960s)
Socialist Workers Party (SWP), 23–24, 29–34, 37–40, 43–44, 96; SWP Minority in, 40, 41, 43–44, 45, 54, 59

Taoism, 91, 189, 259, 271, 291; *Tao Te Ching*, 259
technology, 93, 136, 186, 266; alternative, 88, 98, 106, 156, 163, 164–165, 169, 188; automation and, 60, 89, 93; computerization and, 89, 288; cybernation and, 95; miniaturization and, 72, 89, 95; mistrust of, 86, 93, 133, 139–140, 144, 265, 269, 293, 298, 316; as precondition for freedom, 60–61, 72, 93, 94, 108, 116, 133, 149, 159, 165, 210; toil minimized, 96, 109. *See also* eco-technics; postscarcity; renewable energy; scale
Todd, John, 158, 167. *See also* New Alchemy Institute
Tokar, Brian, 139, 308, 313n65; in Burlington, 196, 208, 212, 216–217, 219, 231n52; in Clamshell, 181, 186–187; Greens and, 236, 271
Torch Bookstore, 97
town and country, 70–73, 83, 88–89, 140, 148, 190. *See also* eco-decentralism
town meeting, 227, 290; in American tradition, 170–171, 206, 246, 247; in Burlington, 216, 218, 274; as citizens' assembly, xi, 162, 163, 166, 171; in Vermont, 207, 213–214, 219, 252, 272. *See also* democracy, assembly
Trotsky, Leon, 5, 16, 29, 60; on Boss Hague, 19, 22; dogmatism of, 31–32; New York SWP and, 29–30; on rethinking, 46, 56, 73; Stalin and, 5, 17–18, 19, 21–22, 34; Transitional Program (1938) by, 40, 42; on Weber, 36, 37. *See also* Russian Revolution (1917); Trotskyism
Trotskyism, 21–22, 28–50, 93, 96, 97, 119, 223, 285; in East Tremont, 10, 13–14, 15, 16. *See also* Socialist Workers Party (SWP)

U.S. Greens, 246, 247, 272, 273, 277. *See also* Burlington Greens; Left Green Network; New England Greens; Northern Vermont Greens
United Auto Workers (UAW), 18, 39, 40, 41, 43–44
United Electrical Workers (UE), 18, 19, 31, 33, 34, 39, 40
United States v. W. Mark Felt and Edward S. Miller, 192–194
Up Against the Wall Motherfucker (UAWMF), 116, 117, 125
usufruct, 142, 143, 145, 209, 299
utopia, x, xi, xii, 69, 115, 167, 168, 192, 199, 246, 276, 295, 316; anarchism and, 53, 97, 145, 156–157, 213, 224; counterculture and, 98, 99, 125–126, 136, 145; at Goddard, 160, 165; potentialities for, 60–61, 64–65, 73–74, 76, 88, 94, 132, 235, 256; socialism and, 14, 19, 284; Vermont and, 141, 261, 262. *See also* eco-decentralism; postscarcity

Van Duijn, Roel, 109, 146
Van Heijenoort, Jean, 29, 32, 38, 39, 40, 45
Vaneigem, Raoul, 109, 112; "Totality for Kids" by, 111
vanguard, 121, 140; Black Panthers as, 122, 123; Communist Parties as, 8, 12, 14, 92; Marxism and, 119, 122, 141; New Left and, 92, 106, 113, 115, 121
Verlaan, Tony, 108
Vermont, xi, 141, 145, 161, 170, 180, 196, 245, 258; constitution of, 256, 259; political culture of, 206–207, 219–220, 262; solar power in, 165, 168. *See also* Burlington; Burlington Greens; Burlington, waterfront; Clamshell Alliance; Goddard College; Institute for Social Ecology; nuclear freeze movement; town meeting; Vermont Council for Democracy
Vermont Council for Democracy, 219–220

Viennese uprising (1934), 12
Vietnam War, 92, 105–106, 108, 112–114, 117, 122, 137, 147, 198. *See* New Left; nonviolent resistance; SDS
Vogt, William, 61–62, 86
Vrba, Elisabeth, 211

Watson, David, 288, 290, 300; *Beyond Bookchin* by, 298, 299
Weather Underground, 125, 192, 193, 194
Weber, Josef, 34–38, 39–40, 41–43, 45–46; "Capitalist Barbarism or Socialism" by, 41, 46, 52; in *CI*, 52–63, 65, 69, 71, 73–79; "Great Utopia" by, 60, 63; legacy of, 83–85, 96, 132, 198, 224, 267; "Three Theses" by, 37–38
Widmer, Kingsley, 300

Witlin, Frances, 93
Women for Life on Earth, 196, 213
Woodcock, George, 84, 184, 301
World War II, 28–43, 92, 93, 198

Young Communist League (YCL), 8, 12–1; impact of, on Bookchin, 69, 96, 186, 198, 224–225, 2565; in Popular Front period, 16–18, 21–23
Young People's Socialist League (YPSL), 13, 23
Young Pioneers of America, 6–9, 13

Zander, Ernest (pseudonym for Weber), 55
Zerzan, John, 293
Zieran, Manfred, 235, 255n54
Zimmerman, Michael, 290, 298